Praise for
The Bush Agenda

"Juhasz has pulled down the financial pants of the con artists, tinker-toy imperialists, cash-crazed 'consultants' and other war-mongers-for-profit to expose exactly how they've made Iraqis and Americans bleed billions in cash while real blood flowed. Bravo for Juhasz!"

—Greg Palast
bestselling author of *Armed Madhouse*

"*The Bush Agenda* is a devastating indictment of the collusion between government and big business that has turned the United States—once respected as Savior of Democracy—into a feared and hated empire. Packed with facts and insider stories, it is a resounding call to action."

—John Perkins
author of *Confessions of an Economic Hit Man*

"*The Bush Agenda* is essential reading for anyone bewildered by the president's reckless pursuit of foreign policies so harmful to the values, the economic interest, and the security of the American people. Juhasz lucidly lays out the reasons, combining solid analysis and data with vivid, if infuriating, illustrations of where the Bush Agenda's adherents want to take our country."

—Congressman John Conyers Jr.

"*The Bush Agenda* lays out the 'noble cause' for which George Bush asked our sons and daughters to give their lives: to open Iraq to U.S. corporate control. All potential military recruits should read this book and then decide if Halliburton and Chevron are worth fighting for."

—Cindy Sheehan
peace mom

"Antonia Juhasz has captured the disturbing truth behind the Bush revolution in this riveting new book. *The Bush Agenda* sets out in disturbing detail the real motivation behind the invasion of Iraq and the unprecedented power grab now taking place in the Middle East."

—Maude Barlow
national chairperson of the Council of Canadians and coauthor of *Blue Gold*

"*The Bush Agenda* is a gut-wrenching account of precisely who sold us this war in Iraq and who has cashed in from the ensuing tragedy. Refreshingly, it doesn't leave us merely cursing the corporate profiteers—it points the way to a people-oriented agenda that would leave the neocons and corporate looters in the dustbin of history. Read it and take action!"

—Medea Benjamin
cofounder of Global Exchange and CODEPINK: Women for Peace

"Antonia Juhasz took the reporter's classic maxim to heart and followed the money to deliver one of the crispest, most insightful books yet to expose the Bush regime. . . . In these difficult times, it's encouraging that *The Bush Agenda* concludes with stubborn optimism about prospects for political progress in the United States."

—*The Georgia Straight* (Canada)

"A meticulous expose of corporate America's intentions in the Gulf."

—*The Organizer* (India)

"An exhilarating and disturbing read . . ."

—*Red Pepper Magazine* (United Kingdom)

"An infuriating book. It infuriates the way truth does when you get a cold splash of it and you learn how you've been taken. This is also a hopeful book. How the author manages this alchemy of outrage and uplift requires close attention and willingness to learn . . ."

—Gary Corseri
dissidentvoice.org

"Culls the essentials of highly technical economic analysis and renders them accessible to a more general readership, precisely what is needed today. . . ."

—Russell Branca
Book/Mark Quarterly Review, New York

THE BU$H AGENDA

Invading the World, One Economy at a Time

ANTONIA JUHASZ

HC

An Imprint of HarperCollins*Publishers*

A hardcover edition of this book was published in 2006 by HarperCollins Publishers.

First paperback edition published 2007.

Designed by: Publications Development Company of Texas

Printed on acid-free paper

The Library of Congress has catalogued the hardcover edition as follows:

Juhasz, Antonia.
The Bush agenda : invading the world, one economy at a time / Antonia Juhasz.—1st ed.
p. cm.
Includes bibliographical references and index.
ISBN-13: 978-0-06-084687-9 (acid-free paper)
ISBN-10: 0-06-084687-9 (acid-free paper)
1. United States—Economic policy—2001– 2. Free trade—United States. 3. United States—Foreign economic relations. 4. Globalization. 5. Bush, George W. (George Walker), 1946– I. Title.
HC106.83.J84 2006
337.73—dc22

2006043769

ISBN: 978-0-06-087878-8 (pbk.)
ISBN-10: 0-06-087878-9 (pbk.)

07 08 09 10 11 RRD 9 8 7 6 5 4 3 2 1

To my family, for your unending support: Joseph, Suzanne, Alex, Jenny, Christina, Linda, Paul, Branny, Emma, Eliza, Simone, and Gabriel.

To Sunny, for your love.

To my community, for the hope you give me.

To the movement, because change is afoot.

The current flows fast and furious. It issues in a spate of words from the loudspeakers and the politicians. Every day they tell us that we are a free people fighting to defend freedom. That is the current that has whirled the young airman up into the sky and keeps him circulating there among the clouds. Down here, with a roof to cover us and a gas mask handy, it is our business to puncture gas bags and discover the seeds of truth.

—*Virginia Woolf,*
"Thoughts on Peace in an Air Raid," August 1940

CONTENTS

ONE

THE BUSH AGENDA

An uncharacteristically somber George Walker Bush approached the podium of the Great Hall of the United Nations on September 14, 2005. As the president stood in midtown Manhattan to address the gathered members of the General Assembly, much of the U.S. Gulf Coast lay buried beneath a sea of water, mud, waste, sand, and debris. Two days before, the bodies of forty-five people had been discovered in a flooded New Orleans hospital, adding to a death toll that already exceeded a thousand. Over one million people were without homes, including tens of thousands just recently released from the New Orleans Convention Center and Superdome, where they were forced to stay for almost a week without food, water, or electricity while outdoor temperatures exceeded a sweltering 100 degrees.

The president selected the sixtieth anniversary of the founding of the United Nations (UN) for his first speech before the international community in the wake of the storm. It was a fitting choice given that the 2005 UN Summit was dedicated to the global eradication of poverty. The storm had forced the world's wealthiest nation to take notice of the destitution in its own midst when Katrina struck an area where more than one million people, or nearly one-fifth of the population, lived in poverty. Katrina's $200 billion price tag was rising, earning the storm the dubious distinction as the most expensive natural disaster in U.S. history. In response, 115 nations (including Rwanda and Ethiopia, two of the poorest countries in the world), all of whom were represented at the UN Summit, donated money or other forms of assistance to the United States in its hour of need.

This would be President George W. Bush's fifth address before the UN General Assembly. Two months after September 11, 2001, he established an annual tradition of addressing the Assembly within days of the anniversary of the terrorist attacks and just miles from ground zero. The president has used each speech to put forward his international agenda squarely within the context of 9/11. It was with these speeches that Bush made the case for war beyond Afghanistan, into Iraq, and against all states that harbor terrorists; he laid out the criteria for those who are "with" versus those who are "against us" as he built a "coalition of the willing"; and he affirmed his commitment to expanded international trade policies in the name of fighting terrorism and spreading freedom.

To those who watched the president's previous UN addresses, it was clear that in September 2005, recent events were weighing heavily on him. On the same day that bodies were found in the flooded New Orleans hospital, the president's leading federal official for emergency management, Michael D. Brown, was forced to resign amid widespread criticism of the administration's failure to prepare for the highly anticipated arrival of Hurricane Katrina and to adequately respond to its aftermath. Though the president spoke of directing federal funds to the local communities affected by Katrina, it had only recently been revealed that companies such as Halliburton and Bechtel, located in Texas and California respectively, with intimate connections to his administration were receiving multimillion-dollar reconstruction contracts while local companies were shut out. The president personally faced growing charges of political and corporate cronyism, mismanagement, and even racism in his response to the storm, contributing to the lowest job approval ratings (41 percent) of his presidency at the time, and the feeling expressed by a majority of Americans polled that the president was not to be trusted in a time of crisis. Potentially even more distressing to Bush were the nearly two-thirds of Americans who no longer approved of the way he was handling the central pillar of his presidency—the Iraq war—and the majority who wanted U.S. troops immediately withdrawn.[1]

The president, visibly tired, spent much of the speech looking down at his notes. His familiar easy swagger, comfortable grin, and animated gestures were all but missing. True to form, however, he made no alteration to his message. Bush spent a mere ninety-five seconds of the twenty-five-minute speech discussing the hurricane. He noted the devastation, thanked the gathered nations for their support, and moved on. Then, as he had done every year for the previous four years, the president devoted the bulk of his address to just two topics. The first, not surprisingly, was the war on terror, including the war in Iraq. The second was the expansion of free trade. Once again, Bush offered these two policies, war and free trade, as twin solutions to virtually all of the world's problems—from global poverty to international health crises, including AIDS, malaria, and the Avian flu—and as the means to achieving a better world.

The president described the benefits of war and his administration's commitment to it by assuring his listeners that "all of us will live in a safer world" if we stay the course in Iraq and complete the war effort. The United States and all "civilized nations" would "continue to take the fight to the terrorists" and "defeat the terrorists on the battlefield." As for free trade, Bush explained that the United States would also defeat the terrorists by fighting poverty and "the surest path to greater wealth is greater trade. . . . By expanding trade, we spread hope and opportunity to the corners of the world, and we strike a blow against the terrorists. . . . Our agenda for freer trade is part of our agenda for a freer world."

The agenda has been refined by President Bush and leading members and allies of his administration over decades, dating back most notably to the administration of his father, George Herbert Walker Bush. Its leading framers include men who served in the administrations of both father and son, such as Dick Cheney, Donald Rumsfeld, Paul Wolfowitz, Zalmay Khalilzad, Robert Zoellick, and Scooter Libby. Decades of joint writing, refining, and advocating for a set of clear economic and military principles reached its fullest articulation and most aggressive implementation under the administration of George W. Bush—what I call "The Bush Agenda." This agenda predates the current president, however, and its advocates certainly hope it will outlast him.

Within the Bush Agenda, "freer trade for a freer world" refers to specific economic policies designed especially to support key U.S. multinational corporations that are used as veritable weapons of war, both in the war on terror and in the administration's broader struggle to spread its vision of a freer and safer world. Often, these economic policies are applied in tandem with America's military forces, as was the case in the March 2003 invasion and ongoing occupation of Iraq. To date, the Iraq War represents the fullest and most relentless application of the Bush Agenda. The "freer and safer world" envisioned by Bush and his administration is ultimately one of an ever-expanding American empire driven forward by the growing powers of the nation's largest multinational corporations and unrivaled military.

Free trade is shorthand for a number of economic policies that expand the rights of multinational corporations and investors to operate in more locations, under fewer regulations, with less commitment to any specific location. Advocates contend that these companies and individuals, freed of burdensome government regulations, will amass great wealth and become engines of economic growth. Their wealth, in turn, will filter down through the economy, enriching even the very poorest members. One common image offered to depict the benefits of free trade is of a rising tide of wealth lifting all boats in its wake.

Critics, including myself, refer to the same policies as *corporate globalization,* pointing out that while they do generate vast wealth for certain multinational corporations and investors, those benefits rarely spread throughout a society. Instead, governments are restricted from using policies proven to benefit small local business, workers, consumers, or the environment, while being required to expand policies that benefit multinational corporations. The result is increased economic inequality both within and between nations, and greater economic and political insecurity, including job loss, poverty, and even disease. While the policies *free* multinational corporations from government regulation, they *cost* the rest of society a vast amount of economic and social security.

For example, in September 2005, the World Health Organization warned that a new free trade agreement being negotiated by the United States and the Andean nations of Bolivia, Colombia, El Salvador, and Venezuela could increase the price of medicines by 200 percent in these Andean countries.[2] The U.S.–Andean Free Trade Agreement would require the countries to implement intellectual property rules already in effect in the United States and favored by global pharmaceutical companies that restrict government regulations supporting the local manufacture of generic medicines—the same medicines used to cure or alleviate the diseases cited by President Bush in his UN speech, including HIV, AIDS, and malaria. The report warned that the price increase could either make the medicines inaccessible to those in need or force those who pay into poverty.

One year prior to the September 11 attacks, the U.S. Central Intelligence Agency predicted increased religious extremism and violence as a result of increasing global inequality, warning that "the rising tide of the global economy will create many economic winners, but it will not lift all boats. . . . [It will] spawn conflicts at home and abroad, ensuring an even wider gap between regional winners and losers than exists today. . . . Regions, countries, and groups feeling left behind will face deepening economic stagnation, political instability, and cultural alienation. They will foster political, ethnic, ideological, and religious extremism, *along with the violence that often accompanies it*" (emphasis added).[3]

The specific free trade policies advanced by the Bush administration are not new. Their modern roots trace to the end of World War II and the founding of the now dominant global financial institutions, the International Monetary Fund and the World Bank. They have been the preferred international economic tools of U.S. presidents for decades, especially over the last thirty-five years and in response to the growth of the Organization of Petroleum Exporting Countries (OPEC) and its attempts to control the global oil economy. The key difference between Bush and his recent predecessors is that Bush has directly aligned economic might with military force and has applied both with a more radical, unilateral, and audacious approach. As a result, the Bush Agenda has generated the greatest level of violent opposition to the United States and its allies in recent history and made the world a far more dangerous place. If the Bush Agenda is allowed to stay its course, the poverty, inequality, hostility, and violence it generates will intensify and grow.

Of course, the Bush Agenda does have supporters, especially corporate allies that have both shaped and benefited from the administration's economic and military policies. Many of those allies are found in the energy sector, while many from the energy sector are found in the Bush administration. In the 2000 election cycle, the oil and gas industry donated over thirteen times more money to the Bush/Cheney campaign than to its challenger—nearly $2 million versus just over $140,000. In 2004, the industry gave more than nine times more to Bush/Cheney.[4] The Bush administration itself represents the first time in history that the president, vice president, and secretary of state are all former energy company officials. In fact, the only other U.S. president to come from the oil and gas industry was Bush's father.

The Bush years have been a record-breaking bonanza for the oil industry. The twenty-nine major oil and gas firms in the United States earned $43 billion in profits in 2003 and $68 billion in 2004. Oil profits were so high in 2005, that the top three companies alone (ExxonMobil, Chevron, and ConocoPhillips) earned nearly $64 billion between them, more than half of which went to Texas-based ExxonMobil, which recorded the single most profitable year of any corporation in world history in both 2004 and 2005.[5]

Companies such as Halliburton and Chevron, which respectively count the vice president and secretary of state as former officials, are key allies to the Bush Agenda. The Bechtel Corporation, the largest engineering company in the world, with extensive work in the oil and gas field, has exercised influence over the Bush Agenda through its current and past executives, including current board member and former company president, George Shultz, Ronald Reagan's secretary of state. Lockheed Martin, the country's largest military contractor and the world's largest arms exporter, has also played a lead role, with no fewer than sixteen current and past company officials having held positions within the Bush administration.

The George W. Bush years have been remarkably rewarding for each of these companies, particularly in the post-Iraq invasion period. Indeed, each company has a long history in Iraq, played a lead through company executives past and present in advocating for war against Iraq in 2003, and has since profited greatly from that war. Chevron had its most profitable year in its 125-year history in 2004, earning $13.3 billion—nearly double its profits from the year before. The record did not last long, however, as 2005 brought more than $14 billion in profits. Bechtel's revenue increased from $11.6 billion in 2002 to $16.3 billion in 2003, to $17.4 billion in 2004. Halliburton's stock price has nearly quadrupled in value from March 2003 to January 2006, while Lockheed's stocks more than tripled from early 2000 to January 2006. Vice President Cheney is a stockholder in both Halliburton and Lockheed.

The past twenty-five years of U.S. economic engagement with Iraq, culminating in the 2003 invasion and the ongoing occupation, provide the most glaring example of the Bush Agenda and its imperial ambitions. President Ronald Reagan, and to an even greater extent President George H.W. Bush, focused U.S. economic policy toward Iraq on an increasingly intimate and profitable economic *engagement.* President George W. Bush has gone farther, and in so doing, revealed his imperial aims, by seeking and largely achieving economic control *over and within* Iraq's economy. The Bush administration used the military invasion of Iraq to oust its leader, replace its government, implement new economic, political, and oil laws, and write a new constitution. Through the ongoing U.S. military occupation, the Bush administration seeks to ensure that both the new government and the new economic structure stay firmly in place.

The new economic laws have fundamentally transformed Iraq's economy, applying some of the most radical, sought-after corporate globalization policies in the world and overturning existing laws on trade, public services, banking, taxes, agriculture, investment, foreign ownership, media, and oil, among others. The new laws lock in sweeping advantages to U.S. corporations, including greater U.S. access to, and corporate control of, Iraq's oil. And the benefits have already begun to flow. Between 2003 and 2004 alone, the value of U.S. imports of Iraqi oil increased by 86 percent and then increased again in the first three quarters of 2005.[6]

Thus, advocates of the Bush Agenda have succeeded in spreading corporate globalization policy to Iraq—securing both short- and long-term profits for U.S. corporations—and establishing an Iraqi government that is more favorable than the last ten years of Saddam Hussein's regime to the continued advancement of the agenda. But Iraq is only the beginning.

With the encouragement of Bechtel, Chevron, Halliburton, Lockheed Martin, and others, the Bush administration has begun to expand its agenda to countries across the Middle East with the U.S.–Middle East Free Trade Area. Insulated by oil revenues, the countries of the Middle East have been largely immune from the need to sign free trade agreements. With the invasion and occupation of Iraq, however, the Bush administration has demonstrated the lengths to which it will go to fulfill its interests. Worried about "regime change" spreading to their countries, an unprecedented number of Middle Eastern governments are participating in these free trade negotiations, which are progressing rapidly.

President Bush reaffirmed his commitment to expanding free trade policy in his 2005 UN address. Much of the speech was in fact devoted to the World Trade Organization (WTO) ministerial meeting in Hong Kong three months later. The president argued "the lives and futures of millions of the world's poorest citizens hang in the balance—and so we must bring the [WTO] trade talks to a successful conclusion." Founded in 1995, the WTO has 148 member governments and is the most powerful global institution writing and enforcing the rules of corporate globalization. Headquartered in Geneva, Switzerland, the WTO administers agreements on issues as broad and far-reaching as agriculture, telecommunications, government procurement, and services on behalf of its members. It provides a forum for expanding these agreements and negotiating new ones. The WTO monitors the internal laws of its members, arbitrates disputes between governments over its rules, and enforces its rulings through the imposition of sanctions. Every two years, the WTO holds ministerial level meetings at which high-ranking government officials finalize negotiations on existing and newly proposed WTO rules.

Before the WTO, multination trade rules dealt largely with the movement of goods between countries, primarily tariffs, which are taxes applied to goods as they enter or exit a country, and quotas, which dictate the number of a specific product that can enter or leave a country. While the WTO continued to regulate these aspects of trade, it went further, moving inside of countries and regulating their internal laws. Every law or government policy that has the potential, whether intended or not, to impact foreign companies or investors is open to WTO regulation.

Because of its unprecedented reach, the WTO has generated some of the most vigorous opposition of any global institution. Millions of people the world-over claim to be victims of "economic violence" brought on their lives by WTO policies in the form of inequality, insecurity, dislocation, and poverty. As described by the CIA earlier, economic violence can breed physical violence just as readily as military force, but the cause and effect are often more difficult to discern. Sometimes, however, the causal link is made painfully clear. Such was the case when Lee Kyung Hae committed suicide in protest of the harm inflicted on him, his family, and his community by the WTO. Lee's suicide offers a graphic warning of the potential costs associated with the implementation of corporate globalization policies in the United States, Iraq, the Middle East, and across the world as the Bush Agenda seeks to expand the American empire.

"THE WTO KILLS FARMERS"—THE SUICIDE OF LEE KYUNG HAE

On September 10, 2003, the opening day of the fifth ministerial meeting of the WTO in Cancun, Mexico, the sun was shining, the temperature was hot, and the sky was clear and blue. Ten thousand people, mostly farmers and campesiños from across Mexico and Central America, joined by several hundred gringos from the North and others from around the world, gathered to march in protest of the WTO. A Mexican marching band played traditional mariachi music, while elderly women carrying children walked alongside Guatemalan trade unionists marching in single file next to university students bearing giant puppets emblazoned with denunciations of the WTO and celebrating their home communities. All sang protest songs in tune with the band. The mood was joyful but also full of purpose.

The march wound through the streets of central Cancun, where those who service the hotel industry live, until it approached Kilometer Zero—the beginning of the hotel zone. "The Zone" is where tourists on spring break drink frothy piña coladas and dance topless on the bar at Señor Frogs. It is also where government trade ministers come together for WTO ministerial meetings. The pace slowed and the mood began to darken as the marchers reached Kilometer Zero. Waiting to meet them there were heavily armed Mexican riot police, cement barricades, and a ten-foot-high metal fence. The cold stares of the military-clad police reflected their determination to keep the marchers away from the tourists and trade ministers. The standoff that ensued took place directly beneath a brightly colored billboard with a man's broadly smiling face and a caption that read "Bienvenidos a Cancun!"

Many of the marchers held back, trying to keep a safe distance between themselves and the police. Yet, the sixty or so farmers, teachers, and trade unionists who had traveled all the way from South Korea were not deterred. Decked out in matching hats, T-shirts, and vests imprinted with bright anti-WTO slogans and carrying a giant colorful handmade paper dragon, they were impossible to miss. Their average age looked to be about fifty; most had closely cropped hair and wore conservative clothing beneath their anti-WTO apparel. They were the first group to approach the fence shouting "No No WTO! No No WTO!" Hundreds of the marchers soon followed. When two men from their group climbed up on top of the fence and raised their fists, the crowd cheered and took up the chant.

None of us knew at that moment what we were witnessing. We looked on, still cheering, as one of the South Korean men wearing a sign across his chest that read "The WTO Kills Farmers" removed a Swiss Army knife from his pocket, deftly stabbed it into his chest, crumpled into a ball and fell from the top of the fence to the ground. I shouted with others for "Los medicos!" but to no avail. Lee Kyung Hae had dealt himself a lethal wound that punctured his lung and heart. He died at the local hospital a few hours later while hundreds mourned outside.

Lee's suicide took place on *Chusok,* the Korean Day of the Dead. It was an act in protest of WTO agricultural policies that had bankrupted his farm, impoverished his family, and devoured his community. He left a note that read, "It is better that a single person sacrifices [his] life for ten people, than ten people sacrifice their lives for just one."[7]

A farmer from Taesong-Ri, a small rural village in South Korea, Lee owned and worked an eighty-acre farm with his wife and three daughters. There they raised cows and grew rice, corn, and other vegetables. Lee's ability to make a living off his farm fell virtually in proportion to the steady increase in his government's implementation of the WTO Agriculture Agreement. In 1999, his farm was foreclosed. Lee was not alone. As farmer after farmer was forced from the land, the population of Taesong-Ri dropped from fifteen hundred in the mid-1980s to just a few hundred in 2003.

In a statement published in *Korea AgraFood,* the country's leading agriculture magazine, in April 2003, Lee explained the devastation that the WTO had brought to his life (translated from Korean):

> I am crying out my words to you that have boiled for such a long time in my body. . . . Exclude the Agriculture Agreement from the WTO system. . . . It is true that Korean agricultural reform programs increased the productivity of individual farms. However it is also a fact that increased productivity simply added more volume to an oversupplied market in which imported goods occupied the lowest price portion. Since then, we have never been paid over our production costs. Sometimes, there are sudden price drops of four-times below the normal trend. How would you feel if your salary suddenly dropped by a half without clearly knowing the reason why?[8]

Lee refers to WTO rules that permit nations to subsidize agricultural producers and exporters, while denying governments the ability to provide price supports, market protections, or subsidized inputs—such as fertilizer, seeds, or tools—to their farmers. The all too common result is that small farmers, the world over, are undercut by cheaper subsidized products and then pushed off their land into poverty, while control of the world's food supply concentrates into the hands of those who are able to export their crops on the world market—an ever-shrinking number of giant multinational agriculture corporations. If farmers like Lee go to their local and national governments to complain, they are told that the WTO sets the rules and the government must comply.

Not only is the WTO's destructive impact on small farmers and communities tragically widespread, but so too is the desperate response, commonly referred to as an "epidemic of suicide among small farmers." Over 20,000 farmers committed suicide in India in response to identical policies between 1997 and 2003. In just one instance, 1,600 farmers in the Andhra Pradesh district killed themselves by drinking their own pesticides.[9]

Four days after Lee's suicide, on September 14, 2003, the fifth ministerial meeting of the WTO collapsed in failure. While Lee's act brought the plight of millions of farmers to global attention, it was already well known to their governments. Thus, developing country delegates were prepared to refuse to sign any WTO ministerial agreement that did not provide significant protections for their countries from the agricultural agreement. Not only were such protections denied them, but the European Union and the United States refused to negotiate any deals without an agreement on investment protections for multinational corporations. Two days earlier, seventy developing countries had submitted a letter to the chair of the WTO negotiations, demanding that the foreign investment agreement be withdrawn altogether because of the damage it would do to their small businesses, local producers, and domestic economic development policies. The demand was rejected.

Ronald Sanders, the trade minister for Antigua and Barbuda in the Caribbean, told a reporter, "My government has a duty to care for its people. Were we to accept this document [the September 13, 2003,

WTO draft ministerial text] we would deserve our people's condemnation. For we would not only have gained no relief for them, we would have condemned them to a life of perpetual underdevelopment. And that does not enjoy the support of my government."[10]

At a press conference one hour after the WTO ministerial collapsed, I watched as U.S. Trade Representative Robert Zoellick, the highest-ranking American official representing the United States in trade negotiations and now the undersecretary of state, was barely able to contain his contempt and condescension as he explained that the talks had failed because developing countries were "posturing" on agriculture and "did not know how to negotiate." He made no mention of Lee.

The day after Lee's suicide, I stood on a rain-soaked patch of grass on Kilometer Zero. The shock still weighed heavily on me as I stared at a makeshift memorial that had emerged over the course of the previous night and early morning. There were white candles, flowers, and handwritten signs in Korean and English. I was looking at a framed black and white photograph of Lee when a South Korean man joined me. For a few minutes, we stood there side by side, looking on in silence. Finally, he turned to me and pointed at a small white button on my bag, which had been there for well over a year. It read "No Blood for Oil" and had an image of a black oil spout with red blood spurting out of the top. He reached to his vest and removed a yellow button on which a black machine gun broken in two sprouted white daisies. When he offered it to me, I removed my button and we exchanged our symbols of war and peace.

It had been six months since President Bush launched a preemptive war against Iraq and four months since he had declared, "Mission Accomplished." The occupation of Iraq had begun—and so, too, the reconstruction: Not the reconstruction of vital public services such as water, electricity, or public security, but the radical reconstruction of Iraq's entire economy. The economic invasion was at hand. President Bush used the tragedies of September 11 and then the war against Iraq as justifications for implementing radical corporate globalization policies in Iraq that were surely making Robert Zoellick green with envy.

Apparently, what the U.S. Trade Representative failed to achieve through international negotiation, the U.S. Administrator of the Coalition Provisional Authority succeeded in achieving through military invasion.

Standing in the rain in Cancun, all I could think was, *God, this is only the beginning.*

I had come to Cancun as part of the International Forum on Globalization (IFG). The IFG, formed in 1994, has been described as "One of the most serious and respected groups of experts dedicated to analyzing and generating alternative proposals to the prevailing economic model promoted by international financial agencies," by the *La Jornada* newspaper of Mexico. Naomi Klein, author of *No Logo,* has called the IFG "the brain trust of the [anticorporate globalization] movement."

My personal opposition to corporate globalization policies, such as those imposed by the WTO, dated back seven years earlier to my work as a legislative assistant to Congressman John Conyers Jr., the ranking member of the House Judiciary Committee, the dean of the Congressional Black Caucus, a civil rights leader, and a representative of Detroit, Michigan, for more than forty years. How to describe Congressman Conyers? Well, the fact that we referred to him as "JC" might be illustrative. You might remember him from Michael Moore's *Fahrenheit 9/11* in which he is shown placing his hand on Moore's shoulder, wearing a slight smirk, as he begins to lecture Moore on the realities of the U.S. Congress with, "Well my son . . ." This is quintessential JC—calm, soft-spoken, looking you straight in the eye, and always ready to tell you just how things are in the real world. I received many such lectures in my day.

My policy expertise when I was hired by Congressman Conyers was derived from more than five years of work in the field, which included aiding in the development of an alternative poverty measure called the Self-Sufficiency Standard that is currently in use by several state governments; a Masters degree in public policy from Georgetown University; and an undergraduate degree in public policy from Brown

University. I was assigned virtually all of Congressman Conyers' domestic economic policy work, including Social Security, Medicaid, Medicare, welfare, job training, housing, and child care.

My duties were quickly overwhelmed with the fight to defeat President Clinton's welfare reform bill. The bill dramatically reduced access to funding and services for women and children, particularly our constituents, who comprised the largest welfare recipient population in the United States. When we lost, and the bill became law, it was my responsibility to explain to our constituents why we had cut them off and what we were going to do about the fact that they could no longer afford to feed or shelter their children. Although I worked in the congressman's Washington DC office, because I was the only person on staff familiar with the details of the legislation, I took every individual phone call from desperate women across Detroit. Needless to say, it did not take long for me to start searching for policy alternatives. I looked at community block grants and other funding sources that could be directed specifically to low-income communities, particularly low-income women of color.

I discussed some of these ideas with a colleague who told me that they were all fine and good, but that the Clinton administration was negotiating an agreement called the Multilateral Agreement on Investment (MAI) at the Organization for Economic Cooperation and Development (OECD) that would make them moot. The OECD is most often described as a "club" in which the world's twenty-seven wealthiest nations meet to discuss economic policy and occasionally write new policies, which are then either accepted or rejected by the nations' legislators. I immediately contacted the United States Trade Representatives (USTR) office, the U.S. government office responsible for all international trade negotiations, to get more information about the MAI. To my surprise and confusion, the USTR responded that there was no such agreement. So I decided to do some research on my own.

I discovered that not only was there a MAI, but I could read the draft text of the agreement on the website of the consumer rights organization, Public Citizen. I would later learn that the text had been "liberated" by a French government official who opposed the MAI and believed that the public should at least be able to read it. The United States was one of the lead countries pursuing this proposed foreign investment agreement. Under its rules, legislation could not be used to direct financial benefits to women or minority-owned businesses in Detroit unless the city provided the same benefits to any foreign multinational corporation that asked. Obviously, Detroit would not be able to support both its low-income residents and every multinational corporation that came knocking. However, the opposite was permitted: Legislation could provide benefits, tax breaks, special funding, and the like to foreign corporations while denying such benefits to Detroit's locally owned businesses. Furthermore, if a foreign-owned corporation did come to Detroit, we could not, for example, require that it hire local people, use local products, or invest any of its money locally.

I called USTR back. This time the response I got led to my future career choice. I was told that, yes, there is an MAI, and yes, the United States is negotiating it, but no, I could not learn more because Congressman Conyers was not on the appropriate committees, and I was not—and I quote—"to worry [my] pretty little head about it." That was all I needed to hear. I realized that the democratic institutions in which I believed so firmly were being leapfrogged by corporations that cared little about democratic process, local economic development, or the need to provide the low-income women of Detroit access to financial resources. Worse still, the government I worked for was complicit in the process.

The impact of this realization on me personally and on my career was quick and profound. In response, I left the Hill to work for the Preamble Center for Public Policy, a research, education, and advocacy organization, where I coordinated their project to defeat the MAI. In 1998, I was part of the first successful global movement of people and governments to defeat an international investment agreement. At the urging of my colleagues, I returned briefly to Capital Hill to serve as a legislative assistant, this time working specifically on international trade, finance, and military spending issues for Congressman Elijah Cummings (D-MD). Meanwhile, the advocates of the MAI transferred the agreement to the WTO (where it originated) for consideration at the Seattle 1999 WTO ministerial meeting. This transfer brought all those who had worked to defeat the MAI, and the rest of the anticorporate globalization movement, to Seattle, contributing to the historic collapse of the ministerial that has since been dubbed the "Battle of Seattle." The following year, I joined the IFG. In 2005, I left the IFG to become a Visiting Scholar with the Institute for Policy Studies based in Washington DC.

I have given testimony before the U.S. Congress. I have met with members of the Clinton and Bush administrations, the president of Lebanon, elected officials from Canada, South Africa, the European Union, and Bolivia, among other countries and regions. I have attended meetings of the OECD, WTO, World Bank, and IMF. I have learned from and worked with people around the nation and the world who are struggling against the impacts of corporate globalization on their lives. I have helped write and implement alternatives.

It was the Bush administration that forced me to turn my attention so fully to war. Just days after the September 11, 2001, terrorist attacks, the administration publicly announced that it would advance its preexisting corporate globalization agenda under the guise of fighting the war on terror and that free trade was in fact a weapon in that war.

COUNTERING TERROR WITH TRADE

On September 20, 2001, U.S. Trade Representative Robert Zoellick announced that the Bush administration would be "countering terror with trade." In a *Washington Post* Op-Ed, Zoellick argued that "free trade" and "freedom" are inextricably linked and that trade "promotes the values at the heart of this protected struggle." In the name of fighting terror, he called for the passage of a series of corporate globalization agreements—including negotiations to expand the WTO and Fast Track authority—which had already been the topic of serious Congressional debate and conflict.

"Fast Track" refers to legislation that allows the president to move trade bills through Congress quickly by overriding core aspects of the democratic process such as committee deliberations, full congressional debate, and the ability to offer amendments. The administration had tried unsuccessfully to pass such legislation for over a year. Now, however, a new opportunity presented itself—9/11. Literally wrapping the administration in the flag, Zoellick declared that "Congress, working with the Bush administration, has an opportunity to shape history by raising the flag of American economic leadership. The terrorists deliberately chose the World Trade towers as their target. While their blow toppled the towers, it cannot and will not shake the foundation of world trade and freedom."[11]

Congress and the public decried Zoellick's opportunism. One memorable condemnation came from New York Congressman Charlie Rangel, a senior member of the Democratic Party, a leader of the 1960s civil rights movement, and the ranking member on the Committee on Ways and Means. Rangel has an imposing physique and a raspy voice that gives the impression that he has spent the last twenty-four hours yelling at someone. He also has a quiet charm that is at least as disarming as his physical presence. He raised a powerful voice against Zoellick when he said that "to appeal to patriotism in an effort to force Congress to move on Fast Track by claiming it is needed to fight terrorism would be laughable it if weren't so serious."[12] Unfortunately, Zoellick was not alone in his position: It was administration policy.

Four months later, President Bush delivered what was arguably one of the most important State of the Union addresses in fifty years—the first after 9/11. In the speech, the president repeated Zoellick's characterization of 9/11 as an opportunity to expand free trade and free markets. He, too, called on Congress to pass his corporate globalization agenda in the spirit of recovery from 9/11. In the closing moments of his speech, the president explained that "in this moment of opportunity, a common danger is erasing old rivalries. . . . In every region, free markets and free trade and free societies are proving their power to lift lives. Together with friends and allies from Europe to Asia and Africa to Latin America, we will demonstrate that the forces of terror cannot stop the momentum of freedom."[13]

The mantra, soon to be repeated in speech after speech by President Bush and his subordinates in the buildup to war, was that his administration would be "trading in freedom." "Free trade" and "free markets" were synonymous with "freedom," and the United States was willing to implement this theory with military force. It was pure imperial ambition, which the advocates of the Bush Agenda had been waiting for decades to implement.

TWO

AMBITIONS OF EMPIRE

> We will actively work to bring the hope of democracy, development, free markets, and free trade to every corner of the world.
>
> —*President George W. Bush, National Security Strategy of the United States of America, September 17, 2002*

> The failure to prepare for tomorrow's challenges will ensure that the current *Pax Americana* comes to an early end.
>
> —*The Project for the New American Century, September 2000*[1]

Historically, Americans have considered themselves vehemently anti-Empire. Remember how the worst name that Ronald Reagan could give the Soviet Union was "The Evil Empire"? We have never liked to think of ourselves as either living in—or in pursuit of—an Empire. Clearly, this attitude has quite a bit to do with our roots as a nation born from revolution against an Empire and as a people whose heritage links them to every continent on the planet. Historian Arthur Schlesinger recently wrote: "The imperial dream has encountered consistent indifference and recurrent resistance through American history. The record hardly sustains the thesis of a people red hot for Empire."[2] It seems, however, that "the people" have been left out of the discussion in which advocates of the Bush Agenda have decided not only that America *is* an Empire, but that this is a position that should be embraced, expanded on, and defended using any and all means available.

Typical of the pro-Empire crowd is a piece written in 2003 by Robert D. Kaplan, prominent conservative and correspondent for the *Atlantic Monthly.* "It is a cliché these days to observe that the United States now possesses a global empire," he says. "It is time to move beyond a statement of the obvious. . . . So how should we operate on a tactical level to manage an unruly world? What are the rules and what are the tools?"[3] Kaplan then provides a list of ten rules for governing the world. Kaplan's query demonstrates his support for a United States that is not just a mere lone superpower, but rather a lone superpower with imperial designs. The latter is a nation that intends to use its unique status to influence, control, and rule over other nations and people, even over the entire world—in other words, an *Empire.*

For his part, President Bush was so concerned about what "the people" were thinking, as a rising tide of supporters and critics alike referred to his government as an Empire, that he was forced to proclaim in his 2004 State of the Union Address that the United States did not even have "ambitions of Empire," much less an existing one. He repeated the very same claim one year later in the 2005 State of the Union, explaining that "the United States has no right, no desire, and no intention to impose our form of government on anyone else. That is one of the main differences between us and our enemies. They seek to impose and expand an Empire of oppression. . . ."

Someone must have failed to share the president's purported anti-imperialist leanings with the White House human resources office, as the building was teeming with people who specifically defined their twenty-first century agenda as the creation of a *Pax Americana.* The term is a reference to the *Pax Romana* and the belief that the Roman Empire brought peace to the world by establishing such a militarily and economically dominant Empire that no nation in the world sought to challenge its hegemony. Sadly, for both the United States and the world, a *Pax Americana* is exactly what the Bush administration is after. Bush is pursuing "an Empire of oppression"—one economy at a time.

PAX AMERICANA

> The assemblies of the people were for ever abolished, and the emperors were delivered from a dangerous multitude, who without restoring liberty, might have disturbed, and perhaps endangered, the established government.
>
> —*Edward Gibbon,* The Decline and Fall of the Roman Empire, *1788*[4]

The leading ranks of the Bush administration are littered with men who have written and spoken frankly about their pursuit of a *Pax Americana.* They appear to have reserved use of the actual term for their period of exile: those long dark years from 1993 through 2000 when Bill Clinton was president and they were pushed out of office and sequestered in groups such as the Project for the New American Century. The ideas of *Pax Americana* provide the theoretical justification for what would eventually emerge as the Bush Agenda—simply put: overwhelming and expansive U.S. military and economic global dominance to ensure world peace. In reality, this agenda is far less noble than it may initially appear, as was the *Pax Romana.* The Bush administration is certainly interested in military and economic dominance and in expanding its field of domination, but its goal is more accurately described as ensuring that it gets what it wants and has its interests met. Those interests, in turn, align more with key U.S. corporate players than with either the American public or world peace.

While world peace is the broad selling point, the war on terror provides the immediate justification for the Bush Agenda. However, the roots of the agenda were planted well before the twin towers were even a sparkle in Osama bin Laden's eye. There are three documents, which, building on one another, culminate in the Bush Agenda. They are the 1992 "Defense Planning Guidance," the 2000 "Rebuilding America's Defenses: Strategy, Forces and Resources for a New Century," and the 2002 "National Security Strategy of the United States of America."

Each follows the model of *Pax Americana* to justify ever-increasing military budgets and more exacting economic demands. Each is ostensibly a military planning document, but the administration of President George W. Bush adds a new twist—making corporate globalization policy a specific tool of national security strategy, on par with military defense as a weapon of war. While the military aspects of these documents have been widely reviewed, the economic policies have received scant public attention.

While the documents span more than twenty-five years and four administrations, the authors have remained remarkably consistent. New adherents have been brought on board and leadership roles have been swapped, but overall, it is clear that the formulation of the Bush Agenda is the culmination of decades of work, ideas, and planning by a relatively small group of people. While they may not agree on everything, these individuals agree on enough to pool their resources and set out their own rules for governing the world.

PAX ROMANA

For anyone who believes that (1) the United States is or could be an Empire; (2) this is or would be a good thing; and (3) America should be the best Empire it can be, the pursuit of a *Pax Americana* is a logical goal. It is worth taking a moment, however, to see if the *Pax Romana* is truly worth emulating—particularly in the manner envisioned by the Bush Agenda adherents.

Pax Americana refers to *Pax Romana,* "the Roman Peace." This term is used to describe the 200-year period from 27 BCE, when Rome transitioned from a Republic to an Empire with the ascent of Augustus Caesar, until the death of Marcus Aurelius in 180 CE. It was a period of relative stability *within* the Empire compared to the century of civil wars that both preceded and followed it. The constant deadly civil wars over who would rule Rome and whom Rome would rule were briefly held at bay.

One aspect of the *Pax Romana* to which the adherents of the Bush Agenda seem particularly drawn is best described by Edward Gibbon in *The Decline and Fall of the Roman Empire.* More than two hundred years after its publication, it remains one of the best-known histories of the era and required reading for all political science majors. Gibbon writes, "The terror of the Roman arms added weight and dignity to the moderation of the emperors. They preserved peace by a constant preparation for war; and while justice regulated their conduct, they announced to the nations on their confines that they were as little disposed to endure as to offer an injury."[5]

This idea of a beneficent yet all-powerful Empire, supported by the most menacing military in the world, recoiled in peace but always on the ready to strike in war, resonates throughout the writings and statements of Bush Agenda adherents—as does the supposition that this form of Empire is the ultimate guarantor of peace in the world. The adoption of *Pax Romana* by the supporters of the Bush Agenda is troubling on many counts. For one, the transition from Republic to Empire was far from benign.

C. Octavian, Julius Caesar's adopted son and heir, became Augustus Caesar in 27 BCE and was in effect, if not in name, Rome's first emperor. *Pax Romana* is most often equated with the reign of Augustus. His power was absolute and covered every aspect of Roman life. However, Augustus was also a shrewd politician who understood that the facade of representative government would have to be maintained to keep the wealthy political class in check. Therefore, Augustus maintained the senate, but purged it of all opponents, retaining mainly those whom he had personally appointed. The senate then took an oath of allegiance to Augustus and granted him authority of tribune for life. Only death parted Augustus from the seat of power that he held for over forty years. Feigning modesty, he refused his father's deification and instead took the title "Son of a God," while permitting the people to worship "his genius."

Augustus "saved the Empire, but in the long run spelled the death of representative institutions," explains historian Steven Kreis. "Augustus never did away with these institutions; he merely united them under one person—himself. He was consul, tribune, chief priest of the civic religion and the public censor. He ruled by personal prestige: he was *princeps* (first citizen among equals) and *pater patriae* (father of the country). He was the supreme ruler, the king, the emperor and his authority (*auctoritas*) was absolute."[6] Thus, the "republican" government went on much as before, only, as Gibbon explains, rather than represent the people, the government answered solely to Caesar. Not unexpectedly, subsequent emperors, lacking Augustus's skill and restraint, blatantly and brutally abused their unqualified power, bringing about the demise of the Empire.

Democracy was not the only ideal that was sacrificed to maintain the Empire; freedom took quite a beating as well. Rome ran on slave labor and limited citizenship. In fact, the number of slaves increased dramatically during the reign of Augustus and continued to increase for almost two centuries thereafter. At the time of Claudius (41–54 CE), it is estimated that the number of slaves was roughly equal to the total number of free inhabitants of the Empire.[7] Fathers could—and did—sell their children into slavery. Toward the end of the period, as the number of slaves decreased, laws were written to distinguish more clearly between classes of "freedmen" and the wealthy "exempt from the humiliating procedures and punishments to which the [lower income] were subjected."[8]

Advocates of the *Pax Romana* conveniently ignore the brutality that was required to establish the Empire in the first place. They also look past the violence that continued during Augustus's reign against those who did not wish to be under his yoke. Resistance, although less frequent than during other periods of the Empire, did occur and was brutally quelled. Furthermore, the maintenance of peace with Rome's neighbors was a two-way street. After a crushing defeat and the loss of three legions in Germania, Augustus demanded a halt to further military annexations. He adopted a strict policy of border security—rather than imperial expansion—which remained essentially unaltered for the remainder of Roman history (the one notable exception being the invasion of the province of Britain). Thus, the *Pax Romana* was dependent on an Empire that was no longer imperial. Few nations had reason to challenge an Empire that did not pose a threat.

Advocates of the Bush Agenda appear to have dispensed with the non-imperial nature of "peace" in favor of using the military to secure access to resources around the world and to spread their particular models of democracy and freedom. Just as the *Pax Romana* was a period of Romanization, so they envision a period of Americanization. And while they may not speak of literally annexing new nations, in the age of globalization, political and economic acquiescence no longer requires full imperial overthrow—as Michael Hardt and Antonio Negri so aptly explain in their 2000 book, *Empire.*

The second element of the *Pax Romana* to which the Bush team is drawn is the emperor's ownership of the Empire's resources. All those living within the confines of the Empire were required to pay taxes to Rome and make their resources—human, natural, and physical—fully available. Much of the ongoing resistance to the Empire therefore took place in areas rich in mineral wealth and not inclined to part with it, such as Gallica in the northeast of Hispania with its gold mines.

There is at least one economic aspect of the *Pax Romana* that the Bush team has rejected: the tremendous commitment made by Augustus to public works. Rather than focus vast percentages of wealth on military expansion, as the Bush administration has done, Augustus invested in the public. This aspect was likely the key to the Empire's internal peace. Augustus built and expanded water aqueducts so that they reached all the regions of the city, including the poorest. He built theaters, temples, bridges, public housing, public baths, roads, and more. He also invested in intellectual endeavors; many of the theoretical and philosophical works that are considered the canon of most academic institutions in the world were written at that time. The Bush administration, however, has reduced federal spending on the modern-day American equivalents of vital public needs, such as affordable public housing, health care, child care, programs to address hunger and poverty, as well as funding for universities and the arts.

The adoption of *Pax Americana*—either in name or in theory—by the advocates of the Bush Agenda may ultimately amount to little more than an intellectual diversion: It allows the administration to justify policies best described as "what's good for America is good for the world" in a slightly more digestible framework. Bush Agenda supporters argue that an unrivaled American superpower is in the best interest of the entire world because only such a superpower can guarantee world peace. Thus, any nation that stands against the interests of the United States is an enemy of peace. By this logic, everything that the United States does to support its own wealth, power, and growth is inherently best for the world and therefore cannot be challenged. As President Bush said on November 4, 2004 upon his reelection, "Our military has brought justice to the enemy, and honor to America. Our nation has defended itself, and served the freedom of all mankind."

Pax Americana justifies unilateral leadership, unlimited defense spending, a perpetual state of war, a disregard for democracy and freedom in their truest forms, and global economic dominance. In other words, it offers a grand justification for the Bush Agenda.

INTRODUCING THE AUTHORS: CHENEY, WOLFOWITZ, KHALILZAD, LIBBY, EDELMAN, AND POWELL

The clearest early rendering of the ideas that would become the Bush Agenda took shape in the 1992 "Defense Planning Guidance" (DPG), which was one of the final products of the George H. W. Bush administration (1989–1993): It provides much of the Bush Agenda's military framework. The DPG is a classified, internal planning guide for the Pentagon prepared approximately every two years. It is not intended for public consumption. It describes America's overall military strategy and represents "guidance" from the president and the secretary of defense to the four military services on how to prepare their budgets and forces for the future.

The 1992 DPG was written by six men who served in the administrations of both Bush presidents: Dick Cheney, Paul Wolfowitz, Zalmay Khalilzad, Scooter Libby, Eric Edelman, and Colin Powell. These men, together with later authors of key Bush Agenda documents, have known and worked with each other, shared and sculpted ideas, and refined their positions for well over a quarter of a century.

Vice President Dick Cheney, Defense Secretary Donald Rumsfeld (a later author), and World Bank President and former Deputy Defense Secretary Paul Wolfowitz have worked together in the administrations of Presidents Richard Nixon, Gerald Ford, Ronald Reagan, George

H.W. Bush, and George W. Bush. Over the years, Cheney, Rumsfeld, and Wolfowitz have increasingly demonstrated a commitment to the combined forces of absolute U.S. military dominance and corporate expansion that have found their fulfillment in the Bush Agenda.

At age twenty-eight, Cheney began his government career as Rumsfeld's assistant in the Nixon and then the Ford White Houses, eventually replacing his mentor as Ford's chief of staff when Rumsfeld became defense secretary. Under Ford, Cheney established his cold-warrior credentials as a steadfast opponent to détente with the Soviet Union. In fact, the only administration official credited with being more opposed to détente at the time was Rumsfeld. When the Republicans lost the White House in 1977 to Jimmy Carter, Cheney moved to the House of Representatives, while Rumsfeld went to work for multinational pharmaceutical company, G.D. Searle, eventually becoming its CEO. Cheney maintained a consistent conservative voting record for eleven years as Wyoming's sole member of Congress.

Wolfowitz, who also began his government service in the Nixon administration at the Arms Control and Disarmament Agency, was increasingly focusing his attention on two highly interrelated issues: U.S. dominance over oil in general, and U.S. dominance over the Middle East in particular. In fact, Wolfowitz has shown a greater commitment to these twin modes of building the *Pax Americana* than to any particular party affiliation. Thus, unlike most of his fellow authors, Wolfowitz continued his government career working in President Carter's Pentagon.

In 1977, two years before Saddam Hussein even became Iraq's leader, Wolfowitz conducted the first extensive study of the need for the United States to defend the Persian Gulf and U.S. access to its oil against an Iraqi invasion of Kuwait and Saudi Arabia. Then, in 1981, as Ronald Reagan's director of policy planning at the State Department, Wolfowitz focused his department on three areas: East-West relations, security of non-Communist nations, and Western oil supplies in the Persian Gulf region.[9] In 1982, Wolfowitz told a reporter that the United States regards the whole Mediterranean region as part of its "strategic access" zone to the Middle East and that the Soviet invasion of Afghanistan proved that Soviet forces could reach the Persian Gulf "merely by driving there," while the United States is 10,000 miles away and needs to secure its access to the "world's major oil producing region."[10] In 1989, under the administration of George H.W. Bush, Wolfowitz ordered a review of U.S. foreign policy toward the Persian Gulf that focused on how the United States would defend the oil fields of Saudi Arabia against an attack from Iraq.[11]

Wolfowitz has worked with U.S. Ambassador to Iraq Zalmay Khalilzad for more then twenty years in pursuit of greater access to the region's oil. In 1984, Khalilzad began his government service in a one-year State Department fellowship with Wolfowitz that turned into a full-time position, extending through the Reagan administration. An Afghan-American, Khalilzad served as a State Department adviser on the Soviet War in Afghanistan, when administration policy was to supply arms and other forms of support to the Afghan mujahideen, including Osama bin Laden and other al-Qaeda members. Khalilzad was also Reagan's adviser on the Iran-Iraq war, when administration policy was direct economic engagement with Saddam Hussein, indirect arms sales, and direct military intelligence to aid Hussein in the war against Iran.

Wolfowitz also brought on board Scooter Libby, who was chief of staff to Vice President Cheney until he resigned in October 2005 after being indicated on two counts of perjury, two counts of making false statements to federal investigators, and one count of obstruction of justice for his alleged role in publicly disclosing the identity of covert CIA operative Valerie Plame. Wolfowitz had been Libby's political science professor at Yale, so when Wolfowitz called Libby in 1981 to invite him to work for the State Department, Libby agreed. Rounding out "Team Wolfowitz" is Deputy Defense Secretary Eric Edelman. Edelman also joined the Reagan State Department in 1981, eventually serving as special assistant to Secretary of State George Shultz. Edelman proceeded to serve in every successive administration, including as assistant deputy undersecretary of defense for Wolfowitz in the Bush Sr. administration, when he helped draft the Defense Planning Guidance (DPG).

Cheney returned to the White House in 1989 to serve as defense secretary, while Rumsfeld, who had originally sought to run against Bush for the Republican nomination, chose to continue his now highly profitable career as multinational corporate CEO. Khalilzad and Libby also served as deputies for Wolfowitz for the drafting of the 1992 DPG.

The final lead author of the 1992 DPG is former Secretary of State Colin Powell, whose theories supply its military muscle. Throughout his political career, Powell has steadfastly argued that a world peace that serves U.S. interests is wholly dependent on the creation and maintenance of absolute U.S. global military superiority. As Reagan's national security affairs adviser from 1986 to 1989, Powell advocated for a superior U.S. military to outflank the Soviet threat, declaring in 1989, "We have to put a shingle outside our door saying, 'Superpower Lives Here.'"[12] Powell's views did not alter after the collapse of the Soviet Union, when, as twelfth chairman of the Joint Chiefs of Staff under Bush Sr. (and for nine months under President Clinton), he argued that the United States must remain the preeminent military power in order to ensure the peace and shape the emerging new world order in accordance with American interests. In 1992, Powell told the House Armed Services Committee that the United States must "deter any challenger from ever dreaming of challenging us on the world stage." Everyone should know that "there is no future in trying to challenge the armed forces of the United States."

Unlike the rest of the Bush Agenda team, however, Powell has steadfastly opposed an imperial use of the U.S. military, making him more of a contributor to what would eventually emerge as the Bush Agenda than an adherent to its philosophy. Yet, his role in advocating for the second invasion of Iraq and seeing it to fruition—even if he was only "following orders"—makes Powell an active participant in the *application* of the Bush Agenda as well.

President George H. W. Bush is missing from the list of 1992 DPG authors, even though his approval allowed it to move from draft to final product. That the final document did not emerge until the final days of his administration demonstrates that it was not an absolute reflection of the president's priorities. There are areas of the 1992 DPG, particularly its focus on oil, which reflect his interests. President George H. W. Bush made his personal fortune from the oil business and devoted much of his career to expanding the rights of private industry in the energy sector. For example, as vice president, Bush chaired President Reagan's energy sector deregulation and regulatory relief task forces. Then, in 1992, he earned the title "The Energy President" from former Enron CEO Kenneth Lay, likely due to his 1992 Energy Act, which mandated the deregulation of electricity, obliged utilities to carry privately marketed electricity, and permitted states to deregulate retail electricity.[13] In the end, however, the senior George Bush was less imperial than his subordinates, causing many to regard the DPG as a failed attempt by Cheney, Wolfowitz, and the others to steer the administration in a direction that it was not otherwise going.

THE BIRTH OF THE BUSH AGENDA: THE 1992 DEFENSE PLANNING GUIDANCE

As defense secretary and the senior Pentagon policy analyst, respectively, Dick Cheney and Paul Wolfowitz are rightly considered the fathers of the 1992 DPG. In March 1992, a draft copy of the DPG was leaked to the press. The reason for the leak, according to Barton Gellman of the *Washington Post,* was that "inside the U.S. defense planning establishment, there were people who thought this thing was nuts. . . . That's why they talked to me, and that's why they talked with the *New York Times.*"[14] The full draft of the DPG was never made public. However, the portions reprinted on March 8, 1992, in the *New York Times* provide ample insight into the ultimate intent of the DPG's authors.

The 1992 DPG was the first document to lay out America's role in a post-Soviet world—a world in which America was suddenly the sole superpower. In response, the public and Congress, including Senators Robert Byrd and Edward Kennedy, demanded a "peace dividend." They argued that, because the country was no longer at war, it was no longer necessary to spend as much on military defense; more money should go toward paying for vital domestic needs that had been long neglected due to the demands of the Cold War. For example, in 1991,

Senator Kennedy proposed legislation that would take $210 billion from defense over seven years to pay for universal health insurance and education and job programs.

The authors of the draft DPG disagreed. They called for the continuation of the war economy, including maintenance of existing troop levels and expansion of U.S. security commitments abroad. They envisioned a world in which the peace dividend was translated into the creation of a superpower so militarily and economically dominant that no other nation would even strive to compete against it, now or in the future.

As written in the draft DPG, they envisioned a system where "the world order is ultimately backed by the U.S." The United States would "show the leadership necessary to establish and protect a new order that holds the promise of convincing potential competitors that they need not aspire to a greater role or pursue a more aggressive posture to protect their *legitimate interests.* Second, in the non-defense areas, we must . . . discourage them [the advanced industrial nations] from challenging our leadership or seeking to overturn the *established political and economic order.* Finally, we must maintain the mechanisms for deterring potential competitors from *even aspiring* to a larger regional or global role" (emphasis added).

At the time it was written, Senator Joseph R. Biden Jr. called the draft DPG "literally a *Pax Americana.*" Moreover, the senator argued, "It won't work. You can be the world superpower and still be unable to maintain peace throughout the world."[15]

With the military victory of the 1991 Gulf War under their belts but the frustration of leaving Saddam Hussein in power still haunting them, the drafters set out an aggressive unilateral preemptive military agenda for the future. The 1992 draft DPG states that the overall objective of the United States in the Middle East is "to remain the predominant outside power in the region and preserve U.S. and Western access to the region's oil."

The draft DPG rejects alliances with other countries in favor of coalitions or "ad hoc assemblies," marking, at minimum, a historic political shift toward unilateralism. An alliance is a close association formed to advance *common interests,* whereas a coalition is defined as a *temporary* grouping. While the alliance implies shared goals and a long-term commitment, the coalition is situational, pragmatic, and short-term. Since World War I, the United States has referred to "alliances" with other nations to describe an ongoing commitment to the achievement of shared interests. The 1992 draft DPG marks the first formal departure from such a stance, but it would not become actively applied to administration *policy* until the administration of George W. Bush used the term the *coalition of the willing* to describe those nations involved in the invasion of Iraq.

The 1992 draft DPG also adopts the use of preemptive war. Like "coalitions," "preemption" did not become administration policy until George W. Bush became president. Preemption also marks a historic shift in national policy. For sixty years, the stated national security policy of the United States was "containment." Containment was adopted initially as the strategy for defeating the Soviet Union. Rather than engage the USSR militarily through, say, an invasion, the United States sought to isolate it, starve it of political, economic, and even social support, and stop it from spreading its influence to other nations. After the fall of the USSR, the strategy continued and was applied to other nations believed to threaten the interests and/or security of the United States. Preemption, however, is a strategy by which the United States may act militarily and unilaterally if necessary, to ensure that a potential threat does not become an actual threat. In other words, the military is not only used to contain or even deter a threat, but, through active engagement, to actually preempt it from taking form.

There was a strong negative outcry when the draft 1992 DPG was leaked to the public. Some of the gravest concerns were raised by foreign country governments who had considered themselves allies of the United States but now feared they had somehow become rivals. In response, the draft was reworked—this time, the more forceful *Pax Americana* language was expunged—and intentionally "leaked" to the press. Just to be on the safe side, however, Cheney withheld the final version until after the 1992 presidential elections were decided. In January 1993, just prior to the inauguration of President Bill Clinton, the newly titled "Defense Strategy for the 1990s" was released. Then, as vividly described by David Armstrong in *Harper's*, "Cheney and company nailed [it] to the door on their way out" of the White House.[16] None of the underlying theses of the original draft 1992 DPG were changed.

Before the DPG drafters reunited in 1997 to establish a virtual Bush government in waiting with the Project for the New American Century, each man pursued the ideas enshrined in the DPG in his own way. Cheney considered a run for president against Bill Clinton. In 1993 and 1994, he formed a Political Action Committee, which included Thomas Cruikshank, CEO of the Halliburton Corporation, the Texas-based energy services giant, and Stephen Bechtel, president of California's Bechtel Corporation, the largest engineering company in the world. The "Cheney for President" buttons he distributed are still highly sought after; the same, however, was not true of his candidacy. Cheney did not despair; in 1995, he accepted his first job in the private sector in more than twenty years as president and CEO of Halliburton.

Paul Wolfowitz left the government as well. Apparently, President Clinton did not satisfy his increasingly aggressive agenda for Iraq, the Middle East, and their oil. Wolfowitz spent the next eight years as dean of the Johns Hopkins School of Advanced International Studies justifying the first Gulf War and advocating on behalf of a second. This time, Wolfowitz argued, the result would include the overthrow of Saddam Hussein and the implementation of a pro-U.S. government. Typical of his many Op-Eds, congressional testimony, speeches, and articles on the subject is his piece in the spring 1994 issue of the *National Interest,* in which he writes, "Saddam will remain a threat to all the governments that supported us and particularly to the Arab Gulf States. . . . The United States and the entire industrialized world have an enormous stake in the security of the Persian Gulf, not primarily in order to save a few dollars per gallon of gasoline but rather because a hostile regime in control of those resources could wreak untold damage on the world's economy, and could apply that wealth to purposes that would endanger peace globally."

Zalmay Khalilzad founded the Center for Middle Eastern Studies at the Rand Corporation, a leading Washington DC conservative think tank, a position that allowed him great intellectual and political freedom. Not content merely to talk about expanding U.S. control over foreign oil, Khalilzad served as a paid adviser for the California-based Unocal Oil Corporation (purchased by Chevron in 2005). In the mid 1990s, he conducted risk analyses for Unocal on a proposed 890-mile, $2-billion, 1.9-billion-cubic-feet-per-day natural gas pipeline project, which would have extended through Turkmenistan, Afghanistan, and Pakistan. Khalilzad's work brought him into direct discussions with Afghanistan's Taliban and even led him to advocate publicly in support of the regime. In 1996, he penned a *Washington Post* Op-Ed titled "Afghanistan: Time to Reengage," in which he argued "The Taliban does not practice the anti-U.S. style of fundamentalism practiced by Iran. . . . We should . . . be willing to offer recognition and humanitarian assistance and to promote international economic reconstruction. . . . It is time for the United States to reengage [the Taliban]."[17] In 2001, the *Washington Post* reported, "Four years ago [December 1997] at a luxury Houston hotel, oil company adviser Zalmay Khalilzad was chatting pleasantly over dinner with leaders of Afghanistan's Taliban regime about their shared enthusiasm for a proposed multibillion-dollar pipeline deal."[18]

Of course, what they really wanted was to be back in the White House. In 1996, they tried and failed with the presidential campaign of Republican Senator Bob Dole. Donald Rumsfeld served as Dole's national campaign chairman, and Paul Wolfowitz was his principal deputy for foreign policy. Dole's 1996 defeat marked the beginning of a tide change: It was necessary to circle the wagons and develop a solidly Republican vision to carry the party in 2000.

THE BUSH WHITE HOUSE IN WAITING: THE PROJECT FOR THE NEW AMERICAN CENTURY

The Project for the New American Century (PNAC) was established in 1997 as an advocacy group dedicated to the proposition that "American leadership is good both for America and for the world." The signatories to the Project's original statement of purpose include six people who served in both Bush administrations: Dick Cheney, Paul Wolfowitz, Scooter Libby, Zalmay Khalilzad, Peter Rodman, and Paula Dobriansky. Five others served both Bush presidents in either formal or advisery positions: Richard Perle, Eliot Cohen, Francis Fukuyama, Dan Quayle, and Henry S. Rowen. In addition, Donald Rumsfeld, Robert Zoellick, Elliott Abrams, and Richard Armitage have all signed key Project letters. Former and current directors of PNAC include, respectively, U.S. Ambassador to the UN John Bolton and former vice president of Lockheed Martin and author of the 2000 Republican Party foreign policy platform, Bruce P. Jackson.

Through PNAC, Cheney et al. put forward their foreign policy objectives without the normal restraint placed on those who are actually public servants. The results are letters, statements, and reports that do away with the usual niceties of public discourse and go straight to the jugular. The Project's seminal report is the September 2000 "Rebuilding America's Defenses: Strategy, Forces and Resources for a New Century"—the second on our list of three key Bush Agenda documents.

"Rebuilding America's Defenses" declares that the primary strategic goal of the United States in the twenty-first century is to "preserve *Pax Americana.*" In fact, the phrases "Pax Americana" and the "American peace" are used nineteen times in all, more if you count the variations to the theme, such as "America's global leadership, and its role as the guarantor of the current great-power peace," or my personal favorite, "the benevolent order," to be secured by American leadership.

The 2000 "Rebuilding America's Defenses" specifically builds on the 1992 DPG, which, the authors argue, "provided a blueprint for maintaining U.S. preeminence, precluding the rise of a great power rival, and shaping the international security order in line with American principles and interests. . . . The basic tenets of the DPG, in our judgment, remain sound."[19]

While the bulk of the eighty-plus-page report is focused on expanding America's armed forces, military might is not the only consideration. The authors explain that "American containment strategy did not proceed from the assumption that the Cold War would be a purely military struggle . . . ; rather, the United States would seek to deter the

Soviets militarily while defeating them economically and ideologically over time." The authors indicate that America's superpower status comes from the combination of its military might, technological know-how, and its possession of the "world's largest economy." They also state that America's "political and economic principles are almost universally embraced" and that the "challenge for the coming century is to preserve and enhance this 'American peace.' " U.S. economic dominance is therefore to be neither trifled with nor ignored.

Military and economic dominance do not come cheap. The authors propose a $15 billion to $20 billion *increase* in total defense spending *annually.* After all, the 1990s were nothing more than a "decade of defense neglect," a wrong that PNAC was prepared to correct if only it could find an American president to carry its mantel. PNAC succeeded with George W. Bush. As we will see, many of the core aspects of what became the Bush Agenda are found in this paper—for example, the ouster of Saddam Hussein in Iraq and the maintenance of a permanent military presence in the Middle East, the definition of the Axis of Evil, the critical importance of a missile defense system and the strengthening of U.S. nuclear weapon capacity, and the willingness to act preemptively to defend America's interests, particularly in the Middle East.

The authors explain that "the United States has for decades sought to play a more permanent role in Gulf regional security. While the unresolved conflict with Iraq provides the immediate justification, the need for a substantial American force presence in the Gulf transcends the issue of the regime of Saddam Hussein." The problem of Iraq may have "transcended" Hussein, but the first step toward resolving the conflict was still Hussein's removal. Thus, on December 1, 1997, the cover of the conservative *Weekly Standard* magazine carried the headline SADDAM MUST GO—A HOW-TO GUIDE in bold, black, capital letters, covering a full three-fourths of the otherwise yellow and red page. The lead article, "Overthrow Him," was coauthored by Khalilzad and Wolfowitz.

In January 1998, three years before the "War on Terror," PNAC released an open letter to President Clinton, calling for the removal of

Saddam Hussein's regime from power.[20] The letter argued that diplomacy had failed and military engagement was required. In particular, it focused on the threat Hussein posed to "a significant portion of the world's supply of oil." Among the signatories were Donald Rumsfeld, Paul Wolfowitz, Richard Perle, Zalmay Khalilzad, Robert Zoellick, and John Bolton.

In 2000, PNAC's report established the Axis of Evil by explaining that "adversaries like Iran, Iraq and North Korea are rushing to develop ballistic missiles and nuclear weapons as a deterrent to American intervention in regions they seek to dominate." The report places Iran next in the shooting order after Iraq: "Over the long-term, Iran may well prove as large a threat to U.S. interests in the Gulf as Iraq has. And even should U.S.-Iranian relations improve, retaining forward-based forces in the region would still be an essential element in U.S. security strategy given the longstanding American interests in the region."

Clearly, these men of power and influence were not satisfied with sitting on the sidelines, writing letters, drafting reports, and writing opinion editorials. They wanted to be back in the White House. While Robert Dole did not pan out, their second try did.

Prior to his 2000 Presidential victory, George W. Bush had spent six years in government service as governor of Texas, nine years as owner of the Texas Rangers Baseball Team, and eleven years as an oil company executive. His affiliation with the DPG/PNAC crowd includes his father, of course, and his commitment to and deep connections with corporate America—particularly the oil and energy sectors. Bush began his professional career following in the footsteps of his father—literally in the same building. As his father had done in the 1950s, Bush launched his oil career from an office in the aptly named Petroleum Building in Midland, Texas. Bush founded two oil companies: Arbusto ("bush" in Spanish) in 1977 and Bush Exploration a few years later. In between, he made an unsuccessful run for the U.S. Congress. One reason Bush lost was because his opponent painted him as an East Coast transplant rather than a true Texan. After all, Bush was born in Connecticut, and he returned to the East Coast at age fifteen to attend the prestigious Phillips Andover Academy in Massachusetts, followed by college and business school at Yale and Harvard, respectively. As a result of his loss, Bush vowed that he "would never be out-Texaned again" in his political career.[21]

In 1984, Spectrum 7 Energy Corporation bought out Bush's oil companies. Although both of Bush's companies had failed, they were purchased and Bush was retained as Spectrum 7's chairman and CEO likely due to the fact that his father had recently become vice president of the United States. Spectrum 7, which owed more than $3 million in debts by 1986, was saved through a merger with Harken Energy, which retained Bush as a board member and consultant.[22] Tapping into his many friendly and profitable connections to the Saudi royals, Bush raised $25 million for Harken by bringing Saudi real estate tycoon Sheikh Abdullah Taha Bakhsh onto the board. Harken subsequently received a gas and oil exploration contract with Bahrain in 1989.[23]

After working on his father's presidential campaign as liaison to the religious right, Bush began his second career as an owner of the Texas Rangers baseball team in 1989. When he sold his shares in 1998, he earned more than $14.5 million in profit. In the interim, he was elected governor of Texas in 1994 and again in 1998.

Bush's policies as governor deeply favored corporations and the wealthy. For example, in 1999 Houston overtook Los Angeles as the smoggiest U.S. city. This was due in part to Bush's decision to allow Texas's more than 800 oil refineries and chemical plants (which were contributing 900,000 tons of pollution annually) to choose whether or not they would comply with the state's clean air regulations, rather than requiring them to do so. Bush cut taxes on corporations and individuals and increased policies directing financial benefits to both. As a result, by 2000, Texas ranked forty-ninth among states in overall taxes collected and fiftieth in per capita state spending. Economic inequality overall in Texas rose dramatically during Bush's tenure as the incomes of those at the top soared and those in the middle and bottom either stagnated or dropped. Texas then joined Louisiana and New York as the three states with the worst income inequality in the nation.[24]

Bush was very good to the oil and energy sectors, and they, in turn, were very good to him. Kenneth Lay and the Houston-based energy services company, the Enron Corporation, were Bush's all-time career patrons, contributing more money than any individual and company to his political campaigns. That is, before Enron was found guilty of defrauding its investors, including the State of California, of billions of dollars, which led to the company's 2001 bankruptcy and Lay's 2004 indictment on eleven counts of securities fraud (for which he is currently standing trial).

In fact, Bush had an unprecedented ability to raise money for his campaigns. Charles Lewis of the Center on Public Integrity put it this way: "George W. Bush has altogether redefined the parameters of political fundraising," shattering all previous fundraising records in 1999 to 2000 and again in 2003 to 2004.[25]

George W. Bush certainly had the right connections and the right politics, but, if there were any doubts that he belonged in the *Pax Americana* brotherhood, he put them all to rest with his 2002 National Security Strategy.

THE BUSH AGENDA ARRIVES: THE 2002 NATIONAL SECURITY STRATEGY

> Free markets and free trade are key priorities of our national security strategy.
>
> —President George W. Bush, National Security Strategy of the United States of America, September 17, 2002

With the interlopers out of office, an administrative reunion took place when the Bush family returned to the Oval Office on January 20, 2001. But times had changed, and so had presidents. Young George brought back his father's court but lost his father's restraint. The radical policies that eventually emerged picked up where the 1992 DPG left off and gave the Project for the New American Century its presidential herald. George W. set himself apart in at least one key arena—by elevating economic policy into a tool of war on par with soldiers, missiles, and jets. He specifically united military and corporate globalization policy into one mighty weapon of Empire.

Some aspects of the Bush Agenda can be found in the first year of George W.'s administration—including, most notably, an aggressive unilateralism demonstrated by the withdrawal from the Anti-Ballistic Missile Treaty with Russia; opposition to the Comprehensive Test Ban Treaty, which bans nuclear test explosions; and rejection of both the International Criminal Court and the Biological and Toxin Weapons Convention protocols. In addition, the administration immediately sought to expand U.S. military dominance in space with the continuation of Reagan's expensive space-based missile defense program. However, it was the events of September 11, 2001, that led President Bush and his administration to consolidate their ideas into one document and release it for public consumption.

Instead of writing a new classified Defense Planning Guidance, the president released the very public September 2002 National Security Strategy of the United States of America. The document itself comes bearing the presidential seal and a three-page introductory letter personally signed by Bush. Officially, the National Security Strategy (NSS) has just one author, President George W. Bush, though Condoleezza Rice, as national security adviser, is also recognized as a lead contributor. Unofficially, the document's authors include Cheney, Wolfowitz, Rumsfeld, Khalilzad, and the rest of the DPG/ PNAC crowd. The NSS is less an internal policy-planning document than a statement from the president to the American people that lays out his national security agenda, which is, in fact, almost as much about corporate globalization as it is about war. The National Security Strategy is the third and definitive document that establishes the Bush Agenda.

The National Security Strategy begins by declaring, once and for all, that the Cold War is over and the United States has won. This victory is interpreted to mean that "our way" is not only the right way but also the only way for the entire world: "The great struggles of the twentieth century between liberty and totalitarianism ended with a decisive victory of the forces of freedom—and a single sustainable model for national success: freedom, democracy, and free enterprise." The National Security Strategy adds, "We will actively work to bring the hope of democracy, development, free markets, and free trade to every corner of the world." If, for the moment, we take at face value that America's way is "freedom, democracy, and free enterprise," it becomes clear that the goal of the NSS is Americanization of the world, using the American model of government and the American model of economics.

In keeping with the 1992 DPG and the 2000 Project for the New American Century report, the National Security Strategy maintains that superpower status is not a reason to sit on our laurels. Quite the opposite: It is the reason for increased military strength to dissuade others from challenging U.S. supremacy. Furthermore, U.S. military strength is good for all because it guarantees peace: "The unparalleled strength of the United States armed forces and their forward presence, have maintained the peace in some of the world's most strategically vital regions."

Thus, the benevolent American order is established and justified in the following statements: "It is time to reaffirm the essential role of American military strength. We must build and maintain our defenses beyond challenge. . . . Our forces will be strong enough to dissuade potential adversaries from pursuing a military build-up in hopes of surpassing, or equaling, the power of the U.S. . . . The great strength of this nation must be used to promote a balance of power that favors freedom. . . . [The aim] is to help make the world not just safe but better." The *Pax Americana* is at hand.

If this argument for *Pax Americana,* best characterized as perpetual war for perpetual peace, seemed slightly suspect one year after the fall of the Soviet Union, it is now far more dubious in light of the fact that the United States is and has been the world's sole superpower for well over a decade. The National Security Strategy alludes to the possibility that Russia could reemerge as a threat and places India and China in the "potential rival" category, even as it lists all three countries as current allies. What, then, is the new justification for perpetual war? It is, of course, September 11 and the War on Terror. Unlike the Cold War, the War on Terror targets no single country or even group that threatens us. Instead, the United States must fight "shadowy networks of individuals," against whose threats we must be prepared to act "before they are fully formed" and for an "uncertain duration."

In order to fund a military that has the capacity to fight a phantom menace anywhere at any time, or everywhere all of the time, the annual U.S. defense budget under Bush (as reported by the U.S. Department of Defense comptroller) has steadily skyrocketed. Bush's first defense budget, at $317 billion in 2002 (set prior to September 11, 2001) was larger than those of the next twenty-five nations combined. The budgets steadily increased from there to $355 billion in 2003, $368 billion in 2004, $416 billion in 2005, and $419 billion in 2006. The price tag for the War on Terror itself goes even higher, in addition to the wars in Afghanistan and Iraq, as all three wars have been funded separately from the defense budget through supplemental spending bills to the amount of some $300 billion by 2006.

The National Security Strategy describes the *Pax Americana* military rattlesnake, coiled but always ready to strike: "This nation is peaceful, but fierce when stirred to anger." Unilateral preemptive action is now our due. America "will not hesitate to act alone, if necessary, to exercise our right of self-defense by acting preemptively. . . . The greater the threat, the greater is the risk of inaction—and the more compelling the case for taking anticipatory action to defend ourselves, even if uncertainty remains as to the time and place of the enemy's attack." As discussed earlier, the adoption of preemption is a radical shift in formal U.S. policy. Senator Edward Kennedy, squaring off against yet another Bush presidency, said that Bush's preemptive doctrine amounted to "a call for 21st century American imperialism that no other nation can or should accept."[26]

What Bush has proposed and since put into practice in Iraq is more than preemption; it is prevention. The distinction marks the difference between a legal and an illegal act of war. A preemptive war, allowed by international law under specified circumstances, is a war against a threat that is unquestionably imminent—the proverbial "troops massing at the border." They have not yet fired their first bullet, but it is obvious that they are about to do so. The United States has always reserved the right to defend itself militarily against such a threat. What has since been dubbed the "Bush Doctrine" of preemption marks a massive departure from the accepted definition of preemption in two ways.

First, there is a difference between maintaining the right to engage the military and stating that it is the national security policy of the United States to act preemptively against possible threats. The latter indicates a significantly lowered threshold for the use of the military, as opposed to other modes of defense, such as diplomacy. Second, what the National Security Strategy describes and what Bush put into practice in Iraq is not preemption. It is a preventive war. Prevention is a response to a *possible* or *potential* threat that could emerge at some point in the future from somewhere. Again, the threshold is lowered even further and the requirement for objective evidence of an imminent attack (e.g., troop movements) is altogether discarded. International law does not authorize preventive war because virtually any state can argue that someone somewhere is possibly thinking of doing it harm.

Six months before the invasion of Iraq, President Bush laid out his case for war in the National Security Strategy, although he named the country only once. He first argued that the newest threat faced by the United States is from terrorists and rogue states that brutalize their own people, display no regard for international law, threaten their neighbors, callously violate international treaties, are determined to acquire weapons of mass destruction, sponsor terrorism around the globe, reject basic human values, and hate the United States and everything for which it stands. He then explained:

> At the time of the Gulf War, we acquired irrefutable proof that Iraq's designs were not limited to the chemical weapons it had used against Iran and its own people, but also extended to the acquisition of nuclear weapons and biological agenda. In the past decade North Korea has become the world's principal purveyor of ballistic missiles and has tested increasingly capable missiles while developing its own WMD arsenal. These states' pursuit of, and global trade in, such weapons has become a looming threat to all nations. We must be prepared to stop rogue states and their terrorist clients before they are able to threaten or use weapons of mass destruction against the US and our allies and friends.

The rest of the military agenda is now well known. It includes North Korea and Iran as looming threats, the need for an advanced missile defense program, and a focus on increasing American nuclear power and weaponry. What is new about the National Security Strategy is the position it accords to corporate globalization policy as a weapon of war.

Trading in Freedom

Marie von Clausewitz released *Vom Kriege* (*On War*) in 1832, the year following the death of her husband, the book's author. Born in Prussia in 1780, Major General Carl von Clausewitz joined the Prussian army at the age of twelve and spent his entire life in military service. The last sixteen years of his life were spent writing his magnum opus, *On War*. In the last one hundred and fifty years, one would be hard-pressed to find a text of military theory more highly read, quoted, and used as the basis of national military strategy.

In a 2002 article in *Foreign Affairs*, Defense Secretary Rumsfeld quotes Clausewitz to explain how the Bush administration places economic policy on par with military operations as tools of war: "Wars in the twenty-first century will increasingly require all elements of national power: economic, diplomatic, financial, law enforcement, intelligence, and both overt and covert military operations. Clausewitz said, 'War is the continuation of politics by other means.' In this century, more of those means may not be military." With this quote, Rumsfeld defines a central tenant of the Bush Agenda: Economics is a weapon of war. With Clausewitz as a guide, economics should be considered a very serious weapon indeed.

Much of what is frequently referred to as the "Powell Doctrine" is more accurately attributed to Clausewitz, particularly the idea that war should be fought using force that is overwhelming and disproportionate to that used by the enemy. Clausewitz called it "utmost use of force" in war. If you are going to fight, then fight big and always fight to win: "For in such dangerous things as War, the errors which proceed from a spirit of benevolence are the worst. . . . War is an act of violence to compel the enemy to fulfill our will . . . all depends on our overthrowing the enemy, disarming him, and on that alone."[27]

Thus, if the Bush administration is going to fight wars with economics and finance as its weapons, it must fight to overthrow the enemy's existing order and replace it with its own models of free trade, free markets, and free enterprise.

The uniting of corporate globalization policy and military warfare is crystallized in the 2002 National Security Strategy. Unlike the earlier documents in which economic policy is mentioned mainly as an afterthought to military strategy, President Bush devotes a full one-third of his national security agenda to defining his global economic strategy. In fact, freedom itself is defined in economic—that is, free trade—terms: "If you can make something that others value, you should be able to sell it to them. This is real freedom." According to the president, the United States is now working to build a world that "trades in freedom." Apparently, the price for freedom is adoption of a corporate globalization economic model.

Under the heading, "Ignite a New Era of Global Economic Growth Through Free Markets and Free Trade," the National Security Strategy provides a list of policies that the administration will advance with other countries to loosen government regulations on corporations. They include: legal and regulatory policies; policies that encourage business investment, innovation, and entrepreneurial activity; tax policies—"particularly lower marginal tax rates that improve incentives for work and investment"; "sound fiscal polices to support business activity"; and, of course, free trade.

Specific proposals are saved for the president's free trade agenda. A comprehensive corporate globalization agenda, in which policies that free American corporations abroad are used as tools for advancing the administration's national security interests, clearly emerges as a central component of Bush's National Security Strategy. The president calls for completion of the WTO negotiations, which began in Doha, Qatar, in November 2001, completion of regional trade initiatives such as the Free Trade Area of the Americas, completion of a list of specific bilateral trade agreements, and renewal of Fast Track.

There is a special focus on "enhanced energy security . . . by working with our allies, trading partners, and energy producers to expand the sources and types of global energy supplied, especially in the Western Hemisphere, Africa, Central Asia, and the Caspian region." Conspicuously absent from this list, and from President Bush's entire National Security Strategy document, is the focus on access to Middle Eastern oil that is so prominent in the first two Bush Agenda documents. This absence is explained by the president's use of the NSS to advance a move toward war against Iraq just six months in the future. Like his father, George W. Bush appears to have realized that the American public is not swayed by arguments for wars fought for oil. However, wars fought in the name of overthrowing dictators and terrorists wielding weapons of mass destruction, particularly those said to be pointing at the United States or its close allies, are more palatable.

The National Security Strategy's focus on corporate globalization policy is not to the exclusion of issues such as health care, hunger, education, HIV and AIDS, worker protections, and the environment (particularly climate change). To the contrary, all are mentioned, and free trade is offered as the best tool to address each vital issue. As President Bush regularly argues and which the National Security Strategy repeats several times, "Free trade and free markets have proven their ability to lift whole societies out of poverty." Particular attention is paid to promoting the connection between trade and development policy in order to "strengthen property rights, competition, the rule of law, investment, the spread of knowledge, open societies, the efficient allocation of resources, and regional integration—all leading to growth, opportunity, and confidence in developing countries."

The National Security Strategy criticizes existing international development policy, particularly that of the World Bank, for not advocating strenuously enough for free trade policies. It then points to the 2002 UN International Conference on Financing for Development held in Monterrey, Mexico, where Bush introduced the Millennium Challenge Account (MCA). The Monterrey conference was unique in that criticism of corporate globalization seemed to come from everywhere. For example, Joseph Kahn, the principal trade reporter at the *New York Times* wrote, "Rather than an unstoppable force for development, globalization now seems more like an economic temptress, promising riches but often not delivering."[28]

President Bush joined in the criticism at Monterrey, and his conclusion was that the World Bank and IMF were too weak in their application of free trade policy and that they coddled "corrupt regimes." President Bush announced that the United States would demonstrate a new way of providing aid. In addition to its continued financial support of the World Bank and IMF, the United States would begin to disperse development aid through the MCA under the authority of the State and Treasury Departments. The MCA, the president explained, would "tie greater aid to political and legal and economic reforms." The reforms offered were the very same free trade policies described earlier.

The president returns to the MCA in his National Security Strategy. After recommending that the World Bank provide grants of aid rather than loans (an excellent suggestion), the NSS reiterates the president's bottom line: "Trade and investment are the real engines of economic growth. Even if government aid increases, most money for development must come from trade, domestic capital, and foreign investment." The United States will therefore give aid to those governments that adhere to its political and economic policies (i.e., non-corrupt regimes) particularly policies that support free trade.

Just three years after the 2002 National Security Strategy was written, President Bush appointed Deputy Defense Secretary Paul Wolfowitz to head the World Bank and Undersecretary of State for Arms Control and International Security John Bolton as U.S. Ambassador to the United Nations. In addition, he made U.S. Trade Representative Robert Zoellick undersecretary of state. Wolfowitz and Bolton now apply military strategy directly to international development aid and to the economic, social, and political decisions of the United Nations, while Zoellick more readily applies corporate globalization policy to the State Department's activities. The Bush Agenda's military-economic hybrid is in full force.

~

The three documents discussed here reveal the Bush Agenda as the culmination of a radical conservative agenda for a *Pax Americana,* which its proponents have been waiting to implement, some for well over a decade and others for more than thirty years. If Empires have one thing in common, it is their inevitable collapse. With the *Pax Romana* as a model, we in America can expect an increasingly undemocratic, brutal, and self-destructive government to take hold of our nation prior to its fall. George W. Bush has made a new mark on the *Pax Americana* legacy by adding corporate globalization policy to its arsenal—arguing that "free trade" will bring increased "freedom." Over the past fifty years, the record of such policies has in fact been the opposite: increased inequality, economic instability, and even violence.

THREE

A MODEL FOR FAILURE

CORPORATE GLOBALIZATION

The new rules of globalization—and the players writing them—focus on integrating global markets, neglecting the needs of people that markets cannot meet. The process is concentrating power and marginalizing the poor, both countries and people.

—*United Nations 1999 Human Development Report*

Next week's meeting of the International Monetary Fund will bring to Washington, DC, many of the same demonstrators who trashed the World Trade Organization in Seattle last fall. They'll say the IMF is arrogant. They'll say the IMF doesn't really listen to the developing countries it is supposed to help. They'll say the IMF is secretive and insulated from democratic accountability. They'll say the IMF's economic "remedies" often make things worse—turning slowdowns into recessions and recessions into depressions. And they'll have a point.

—*Joseph Stiglitz, Nobel Prize recipient, former World Bank senior vice president and chief economist, April 17, 2000*

At the center of the Bush Agenda is the argument that free trade will bring freedom, peace, and prosperity to the world. The facts, however, belie this assertion. The free trade policies of the Bush Agenda are torn from the pages of a corporate globalization playbook written fifty years ago. U.S. corporations, among others, have sought to implement these policies with varying degrees of success through international institutions, trade and investment agreements, and the like. The debilitating impact of these policies as they have been enforced around the world provides damning evidence as to what can be expected in Iraq, the Middle East, the United States, and globally, if the policies continue to be applied and expanded on by the Bush Agenda. The result will be increased inequality and devastation for people, communities, and nations, while extreme wealth and profit are concentrated into the hands of a shrinking pool of corporate and political elites.

As Carlos Andres Perez, the former president of Venezuela once said, the International Monetary Fund (IMF) practices "an economic totalitarianism which kills not with bullets but with famine."[1] Because George W. Bush has taken this agenda further than his predecessors, the potential devastation is even greater, going beyond even the instability, extremism, and violence described by the Central Intelligence Agency in chapter 1.

While born from genuine interest in creating a sustainable world economy after World War II, the formation of the World Bank and the IMF was dominated by a solid U.S. government and corporate agenda, an agenda that the institutions have increasingly come to serve. Developing countries sought alternatives from the outset at the United Nations but were ultimately unsuccessful. The twin 1970s oil crises led successive U.S. governments to use the World Bank and the IMF to secure new sources of oil abroad and to expand U.S. corporate access to that oil. As developing countries sank into debt, the policies demanded of them in exchange for World Bank and IMF loans became increasingly stringent—opening the door for the North American Free Trade Agreement (NAFTA) and the World Trade Organization (WTO). Indeed, anyone looking for the roots of the Bush administration's approach to reshaping countries from Iraq to the United States need only review the actions and impacts of the same economic policies as imposed by the World Bank, IMF, NAFTA, and WTO on nations such as Zambia, Russia, Argentina, Mexico, China, and South Africa—case studies that offer a clear template for the Bush Agenda.

THE INSTITUTIONS: IMF, WORLD BANK, AND WTO

The Strange Case of Harry Dexter White

> The IMF and World Bank resemble "much too closely the operation of power politics rather than of international cooperation, except that the power employed is financial instead of military and political."
>
> —Harry Dexter White, 1946[2]

Harry Dexter White, the mastermind behind the IMF and World Bank has been all but forgotten by history. But at the height of the 1950s Red Scare, he was front-page news. On November 23, 1953, a black-and-white photograph of White graced the cover of *Time* magazine under a yellow and red banner headline, "The Strange Case of Harry Dexter White."

This former hardware clerk rose to become one of the most influential economists in modern history, only to die under the public glare of the House Un-American Activities Committee. White died trying to bring the Soviet Union into the global economic fold that he himself had created. Maybe White was lucky. Forty years later, his dream was achieved, but much had changed in the intervening years. The Soviet Union was now Russia, and the IMF of 1998 bore only structural similarities to the IMF of White's day. Russia not only joined the IMF but became one of the institution's poster children: first for its potential, then for the rate and extremity of its economic collapse.

White's life was a classic American success story, at least until he was accused of being a Communist spy. The son of Russian immigrants, White was born in 1892 in Boston, Massachusetts. He worked in the family hardware store before joining the army and fighting in France in World War I. He returned to attend some of the finest academic institutions in the country: Columbia, Stanford, and Harvard, where he received a Ph.D. in economics in 1930. In the 1930s, as a professor at the University of Wisconsin specializing in international economics, White's work focused on centralized government control over foreign exchange rates and trade. His intelligence and hard work paid off. In 1934, he began a twelve-year career at the United States Treasury Department, eventually becoming assistant secretary. When the United States entered World War II in 1941, White was named chief international economist at the Treasury, and his immediate focus was planning for post-War economic recovery.

The most authoritative White biography is R. Bruce Craig's *Treasonable Doubt: The Harry Dexter White Spy Case.* According to Craig, White believed that the salvation of American capitalism and the avoidance of a postwar recession or even depression lay in the ability of the United States to exploit new international markets. In so doing, the entire global economic system would be sustained. White saw two viable methods to ensure the continued health of the American economy after World War II: continue federal government spending at near wartime levels or stimulate the expansion of international trade to create new markets for American goods.[3] He advocated for the latter. White argued that large amounts of American dollars placed in the hands of foreign countries through loans and aid would allow those nations to buy large quantities of American products. Foreign nations would have to eliminate trade barriers to American products, such as tariffs—taxes placed on goods when they enter a country. To be fair, and to help stimulate economic growth in other economies, the United States would do the same. However, White believed that strong domestic government policies were also vital to protect workers, non-export-oriented businesses, and consumers to ensure that the benefits were equally distributed. The proposal that he developed to put his ideas into practice, dubbed the "White Plan," became the blueprint for the IMF and the World Bank only a few years later.

With the end of World War II came the search for answers. First, the question was: "How could the world have allowed such things to happen?" and then it was: "How can the world keep such things from ever happening again?" There were many arenas for change. One was economic. It was commonly accepted that the global Depression of the 1920s bore significant responsibility for the catastrophic social and political upheavals that followed. There were, of course, competing theories as to which economic forces led to the Depression and how to counter them in the future. However, there were a few points of agreement: An economic downturn in one country should not be allowed to become a global economic disaster; global economic inequality must be addressed; and, at a minimum, the economies that were crushed during the war itself had to be speedily rebuilt so as not to incur lingering misfortune and subsequent anger.

In July 1944, President Franklin Roosevelt brought together representatives from forty-four governments to Bretton Woods, New Hampshire, for the UN Monetary and Financial Conference to come up with solutions. He also brought the White Plan. The negotiators met for three weeks. According to the World Bank's official archives, microphones were passed from speaker to speaker by the dutiful service of local Boy Scouts. Coca-Cola was available free of charge on the hotel veranda with greater reliability than clean water from the faucets. And at one point, the hotel's Arthur Murray dance instructor participated intently in the negotiations for several hours until delegates learned his true identity.

What ultimately emerged were institutions that followed White's Plan and met the interests of first the U.S. government and then its British allies. They would be known as the Bretton Woods Institutions: the International Monetary Fund (IMF) and the International Bank for Reconstruction and Development (IBRD)—now called the World Bank Group. Although not established at Bretton Woods, the World Trade Organization (WTO) is commonly considered a "Bretton Woods Institution" because its roots began at the meeting with the creation of its predecessor, the short-lived International Trade Organization (ITO).

The IMF was the focus of the 1944 Bretton Woods meeting. The conference invitation described the objective as the "formulation of a definite proposal for an international monetary fund and possibly for a bank." Many of the decisions made for the initial structure of the IMF were simply transferred to the World Bank, such as the distribution of voting power, board representation, and each country's financial contribution.

The IMF was created with the highly laudable goal of ensuring that economic trouble in one country does not become a global economic crisis. In order to achieve this goal, the IMF would manage a system of fixed exchange rates in which the value of the world's currencies was based on gold and the U.S. dollar. The IMF would also be on the look-out for countries facing economic turmoil. When an economy was in trouble, the IMF would loan it enough money to stave off a collapse and contain any potential damage to other national economies. The World Bank was originally envisioned as the source of reconstruction funds for war-torn Europe. However, President Truman preferred that the rebuilding effort remain exclusively in American hands and instituted the Marshall Plan to achieve this end. The World Bank then shifted focus to developing economies elsewhere in the world.

Both the IMF and the World Bank are lending institutions. They make loans at more favorable rates than commercial banks. Each institution was created with a central focus on expanding international trade. The IMF would ensure that economies had stable exchange rates with which to trade; the World Bank would ensure that countries had the necessary infrastructure to facilitate trade; and the International Trade Organization would control the rules of trade.

U.S. multinational corporations were at the heart of the system. As White's theory dictated, countries the world over would have more dollars, stable economies, and reduced trade barriers, all of which would increase their capacity to purchase U.S. goods. U.S. corporations, in turn, would gain access to foreign countries to build infrastructure and spur production. A rising tide of wealth would then lift all boats and all would be better off. White also presupposed that domestic governments would be able to guide and regulate the investment and operations of foreign companies and use labor, consumer, production, and other areas of government law to distribute the benefits equitably throughout society. By the 1980s, the very opposite would occur. The World Bank and IMF would be used specifically to restrict governments from implementing such controls on corporate behavior.

The United States Takes Control

Although U. S. Treasury Secretary Henry Morgenthau chaired the U.S. delegation, White was the lead negotiator behind the scenes. At Bretton

Woods, White squared off against the first of two men whom history would credit for his achievements—his intellectual icon, the venerable Lord John Maynard Keynes. As the author of one of the central texts of twentieth-century economic theory, *The General Theory of Employment, Interest and Money* (1936), Britain's Keynes was already considered the father of modern macroeconomics in 1944. In fact, White was himself a "Keynesian": An economist who argued on behalf of the modern welfare state, holding that government intervention in the market is necessary for capitalism to function at a social level. Keynes and White, therefore, agreed on many aspects of the shape of the new international economic system. But disagreement arose over one fundamental issue—U.S. global economic control, which White and the U.S. government wanted, but Keynes and the British government firmly opposed.

Britain's Keynes argued for the IMF to be a countervailing balance to American economic hegemony, while White demanded an "adjunct to American economic power, an agency that could promote the balanced growth of international trade in a way that preserved the central role of the U.S. dollar in international finance."[4] Although the more well-known Keynes has been credited as the brains behind the IMF and World Bank, it was White who won the battle at Bretton Woods. White's victory is best attributed to the economic and political might of the United States vis-à-vis the rest of the world, including Britain. As White himself noted, Britain was a "going nation"—a country in economic decline.

The White Plan included placing the U.S. dollar and its ties to gold at the center of the international monetary system. Thus, the U.S. economy became the anchor on which the entire system was now based. In addition, the White Plan included U.S. control over decision making at both the IMF and at the World Bank. Rather than adopt the UN's democratic structure, in which each country has one vote of equal weight, votes at the IMF and World Bank are weighted according to the size of a nation's economy and its financial contribution to the institutions. Because the United States not only had the largest economy but also the most money to put into the institutions, it received—and has maintained to this day—the greatest level of influence over all decisions at the IMF and World Bank. Finally, White also won the debate over where the institutions would be located. Keynes wanted less overt U.S. government control and lobbied for New York City (the home of the United Nations). White lobbied for and won Washington DC. In fact, the World Bank was initially housed inside the U.S. State Department. Finally, U.S. government officials were placed at the head of both institutions—White at the IMF and Eugene Meyer at the World Bank.

On April 12, 1945, President Roosevelt died and Vice President Truman assumed the presidency. Just four months later, Congress passed legislation establishing the World Bank and IMF.

The Roots of the World Trade Organization (or, The Short and Unhappy History of the International Trade Organization)

You have likely never heard of the International Trade Organization (ITO), the third Bretton Woods Institution. It never actually came into being because the U.S. Congress refused to ratify it. The U.S. State Department drew up the proposal for the ITO, which was discussed at Bretton Woods. The basic idea was agreed to by all those present at Bretton Woods: an international institution to write and guide the rules of international trade. It was also agreed that a separate international meeting would be held to work out the details. This meeting, the UN Conference on Trade and Employment, did not take place until 1947 and was held in Havana, Cuba. It differed significantly from the Bretton Woods meeting, primarily because of those who attended—representatives of fifty-six nations, almost all of which were developing countries. To say that criticism was leveled against the U.S. ITO proposal by the gathered nations is an extreme understatement. The participants saw the ITO as a thoroughly lopsided institution, with the United States and its business interests on one side and the rest of the world on the other. As historian Daniel Drache aptly summarized, "What had begun as an 'American project' did not remain one once the developing countries became involved in designing the ITO. They changed the U.S. agenda. Whatever Washington's intentions were originally, the final product was very different from the script that State Department officials had carefully prepared. In response, the Republican-dominated Congress opposed its ratification."[5]

The main issue of contention was investment rules. U.S. corporations wanted protection from domestic laws when they invested in foreign countries. In their view, the Havana ITO Charter did not provide such protections. In the ITO, trade was treated as just one tool among many to achieve economic development. The ITO also included agreements on full employment, breaking up corporate monopolies, commodity trade agreements to ensure the products of developing countries received fair treatment on the world market, and other protections for domestic markets.

While the details of the ITO were being hashed out, a General Agreement on Tariffs and Trade (GATT) was established as an interim international negotiating body for trade until the ITO was done. However, when the U.S. Congress rejected the ITO, it died, leaving the "interim" GATT as the major arbiter of world trade for nearly fifty years. After decades of negotiation, an international trade body was finally agreed on: the 1995 World Trade Organization. Like the IMF and World Bank, the WTO of 1995 was a radical departure from the institution envisioned by its original creators.

Over the years, the three Bretton Woods Institutions—the IMF, World Bank, and WTO—have attained an aura of inevitability. Those who challenge them are frequently painted as Luddites fighting the natural course of progress. But in light of their true history, it is clear that there was nothing inevitable about these organizations. They were created by individuals who chose specific policies in order to meet the interests of certain governments and financial players—such as U.S. banks and multinational corporations. They were not, as their supporters like to claim, the result of some natural process of human or governmental evolution. Alternative policies were discussed from their inception and continue to be proposed and acted on to this day.

In fact, the IMF and World Bank that emerged under President Harry Truman did not even meet White's expectations. Rather, in the words of biographer Craig, had White lived beyond 1948, he "no doubt would have been horrified" at the "State Department's use of the Bretton Woods Institutions as mere extensions of U.S. foreign policy."

Truman Takes Over

White was a Rooseveltian internationalist. As such, he believed that global economic stability was dependent on cooperation among all nations, but particularly among the three dominant global economies: the United States, the Soviet Union, and to a lesser degree, Britain. As early as 1939, White wrote that the time was at hand "to clear the decks for future economic collaboration between the two most powerful countries in the world"—the United States and the USSR. White is credited with bringing the Soviets to Bretton Woods as observers. However, the Soviets did not join the institutions but rather became increasingly skeptical of them as the Truman Presidency took hold and the United States was engulfed in anti-Communist hysteria. White, like many members of the former FDR administration, was a victim of the ensuing political witch hunts.

White was as unlikely a secret agent as one can imagine. A small, balding, bespectacled economist with a round face and tiny eyes, White wore conservative three-piece suits and a Charlie Chaplin mustache. But when America's brief flirtation with the Communists in the USSR ended with the conclusion of World War II, the Cold War began in earnest. Bolsheviks were apparently breeding in our bathrooms while shadowy networks of spies lurked behind every corner. The hallmark institution of the period was the House Un-American Activities Committee (HUAC). But before Senator Joe McCarthy would get the whole era attributed to him, there was Republican Congressman John Parnell Thomas.

Congressman John Parnell Thomas of New Jersey served in the infantry in World War I. Every day of his HUAC chairmanship from 1947 to 1948, he behaved as though he was doing battle against the evils of Communism. He opened the proceedings into the alleged Communist activity in Hollywood. His astute media know-how—not McCarthy's (who was not involved until 1951)—brought the likes of Gary Cooper, Ronald Reagan, Robert Montgomery, and Robert Taylor to serve as friendly witnesses before the committee. The interrogation of "hostile witnesses," those who refused to answer questions or declared their right to produce their art as they saw fit under the First Amendment, brought even more stars, including those who came to protest the hearings, such as Humphrey Bogart, Lauren Bacall, and Gene Kelly. America's attention was rapt.

Given this lineup, it is not particularly surprising that White's name was largely forgotten. However, he was not trumped by the stars. Just as Keynes stole the limelight at Bretton Woods, Alger Hiss is the government official most often associated with the HUAC. Hiss was a U.S. State Department official who served as the secretary general of the conference at which the United Nations was created. He was found guilty of two counts of perjury before the HUAC by a federal grand jury. Before his death, however, Hiss stated that he believed White was the real target of the HUAC because White was the most senior government official implicated by the committee.

In 1948, self-admitted Communist agent and FBI informant Elizabeth Bentley shocked a federal grand jury when she charged White with passing documents from federal agencies to the Soviets. These and other charges were bolstered by testimony of another admitted Communist agent, Whittaker Chambers, then a senior editor of *Time*. Chambers accused White of being a "fellow traveler" involved in "the communist conspiracy" with Hiss. Days later, White appeared at the HUAC. White stood before the jury and the gathered observers, put on his wire-rimmed glasses, and pulled a crumpled piece of paper from his jacket pocket. In his own defense, he read a prepared statement, which he referred to as his "American Creed": "I believe in freedom of religion, freedom of speech, freedom of thought, freedom of the press, freedom of criticism, and freedom of movement. I believe in the goal of equality of opportunity, and the right of each individual to follow the calling of his or her own choice. . . . This is my creed." When he finished, the room erupted in thunderous and sustained applause. This was White's last public appearance. A few days later, he died of a heart attack, leading the HUAC to drop the case against him and turn all of its energies to Hiss.

For his part, Congressman Thomas was forced to resign from Congress in 1950 after being convicted of salary fraud. He was putting nonexistent workers on the government payroll and appropriating their salaries for himself. When brought before the grand jury, Thomas claimed his rights under Article 1 of the U.S. Constitution, which states that all senators and representatives, except for "treason, felony and breach of the peace" shall be "privileged from arrest . . ." He was found guilty and sentenced to eighteen months in prison and a $10,000 fine. In a twist of fate that even Thomas probably found ironic, he was placed in the federal correctional institution in Danbury, Connecticut, with two men, Lester Cole and Ring Lardner Jr., who were serving terms for refusing to testify before him and the HUAC.

I do not know if White was a spy. I do know that he wanted the Soviet Union to attend the Bretton Woods meeting (which it did), to become a member of the IMF (which it did forty years later), and to participate in the creation of the post-World War II global economic architecture led by the United States (which it did not). According to the various histories of the case, it seems highly likely that White shared information with the Soviets that advanced these goals. Fifty years later, the institutions created by an alleged Soviet spy would become the vessels through which U.S. economic policy in the post-Soviet era would be pursued.

In the interim, the Bretton Woods Institutions were let loose on an unsuspecting world.

THIRD WORLD RISING

From the onset, the loudest criticism against and ideas for alternatives to the World Bank and IMF emerged from those countries that would be forced to live under their policies—the developing world. In the 1950s and 1960s, peoples' revolutionary movements around the world fought for and won the end of colonial rule. These newly independent nations demanded compensation for their years of servitude and the resources stolen from them. Chief among these demands was global redistribution of political power and economic wealth. They did not want loans, such as those offered by the IMF and World Bank, but direct payments of aid. They also wanted an equal seat at the global decision-making table. Developing countries turned to the United Nations as the best international institution to facilitate new rules of trade, economic development, and aid. Developing countries also created their own new institutions—including the Organization of Petroleum Exporting Countries (OPEC), founded in 1961. OPEC was formed to allow oil-exporting nations to stand as a bloc against an international oil company cartel from wealthy or "developed" nations that had heretofore dominated the oil sector and left the oil-rich nations financially poor. The framework for a new global economic architecture that would allow the poor and powerless to achieve parity with the rich and powerful was taking shape.

The United States and Britain, among other wealthy nations, viewed this emerging architecture as a threat, fearing that developing countries, with OPEC as a guide, would come together to control the world's strategic resources and create a significant political challenge to Northern hegemony (a common way to distinguish between blocs of wealthy versus blocs of poor nations, or developed versus developing nations, is to refer to Northern versus Southern nations). In response, in November 1975, the wealthiest nations in the world formed their own bloc: the Group of Seven Industrialized Countries, or the G7. The United States, the United Kingdom, Canada, France, Germany, Italy, and Japan were eventually joined by Russia years later, becoming today's G8. The G7 (and later the G8) served as a forum in which the wealthy countries could devise their own economic and political agenda outside of the more democratic United Nations.

Rather than support developing country demands for a strengthened UN, the United States chose increasingly to turn its money, time, and political attention to the World Bank and IMF, institutions where it maintained dominant control. In order to appease developing countries seeking to make the UN the arbiter of development programs and assistance, however, the United States, among others, offered instead new benefits at the World Bank and the IMF, including the International Development Association, which made soft loans to developing countries—essentially, more money at better rates. A split then ensued between those developing country governments who believed "something is better than nothing" and were willing to accept the World Bank/IMF option, and those who were determined to hold out for more at the UN.[6] The split among the developing nations, combined with the drain of money and attention from the North, led to the eventual demise of some, and the weakening of many, of the programs envisioned at the United Nations. These programs would have been run by and for the developing nations in collaboration with the North. Instead, while the World Bank and IMF were recast as "development institutions," they continued to be run by and largely for the Northern governments and their corporations.

OIL AS A WEAPON

> Many Americans "deeply resented that the greatest nation on the earth was being jerked around by a few desert states."
>
> —President Jimmy Carter, from his memoirs[7]

The 1970s changed the Bretton Woods Institutions forever. In 1971, President Richard Nixon took the U.S. dollar off the gold standard. The core function of the IMF, as established by Harry White at Bretton Woods, was to manage a system of fixed exchange rates based on the U.S. dollar and the gold standard. The end of the gold standard meant that floating exchange rates replaced fixed rates and that the value of every nation's currency was now determined by the market or by its government—not by the dollar, gold, or the IMF. Thus, one of the IMF's core functions was eliminated. In response, the focus of the IMF shifted to monitor more aggressively the exchange rate policies of countries and to increase its lending to countries facing economic difficulties.

In 1970, U.S. domestic oil production hit its peak, with output declining from this point on. This was the first year that the United States had to depend on foreign sources of oil to meet its domestic needs. It was also a year in which oil-exporting countries began to demand sovereignty over their oil, leading some of them to nationalize their oil companies. The Arab-Israeli war of 1973 came next. The United States, under Nixon's leadership, threw its support behind Israel. In response, the Arab nations of OPEC used oil as a weapon by imposing a full oil embargo against the United States in October 1973. OPEC ministers set the price of oil at $11.65 per barrel in December ($51.44 in 2005 dollars), a *468 percent* increase over the price in 1970, when oil was $1.80 per barrel ($9.07 in 2005).[8] The embargo ended in March 1974, but its impact on U.S. economic and political policy has yet to fade.

Oil had officially become a weapon of war, and President Jimmy Carter was the first to state the U.S. government's intention of fighting back. In 1977, Defense Secretary Harold Brown testified before Congress that "There is no more serious threat to the long-term security of the United States and to its allies than that which stems from the growing deficiency of secure and assured energy resources." The influence of Paul Wolfowitz could be heard in President Carter's 1980 State of the Union Address. In what would be dubbed "the Carter Doctrine," the president explained that U.S. strategic interest in the Persian Gulf is based on "the overwhelming dependence of the Western democracies on oil supplies from the Middle East. . . . Any attempt by an outside force to gain control of the Persian Gulf will be regarded as an assault on the vital interests of the United States of America and . . . will be repelled by any means necessary including the use of force."

Carter turned to the World Bank to find more oil and directed the institution to invest in oil for the first time.[9] The World Bank's Projects Database reveals that the result of Carter's decision was both quick and profound. In 1977, the World Bank began with a handful of oil and gas exploration projects. In 1980, the number of such projects skyrocketed to twenty, with an additional eight the following year. Carter was content to allow the recipient countries to maintain control over their oil. In fact, most of the World Bank loans were made to nationalized oil companies. Carter's priority was getting more oil out of the ground so that it could be sold to the United States.

REAGAN, THE WORLD BANK, AND OIL

President Ronald Reagan's attitude toward the World Bank, like virtually all of his economic policies, differed significantly from that of his predecessor. Reagan used the World Bank to force countries to change their laws so that U.S. corporations would gain direct access to their oil. Reagan's economic policies, as vividly portrayed in the 1987 film *Wall Street,* followed the tenet that vast and unequal wealth is not only good, but virtuous. Wealth is to be envied, fought for, and celebrated. As the film's protagonist, Gordon Gekko, declares in one of the most famous lines of the decade, "Greed is good." The greed of the 1980s, particularly among the already wealthy, however, was not a coincidence. It was the direct outcome of government policy that is the domestic application of international corporate globalization policy.

While *Wall Street* is fictional, Gekko's words are based on an actual speech given by Ivan Boesky, a Wall Street arbitrageur, at the University of California Haas Business School's 1986 graduation. Boesky was fined $100 million for insider trading the same year that he told the California graduating seniors, "Greed is all right, by the way. I want you to know that. I think greed is healthy. You can be greedy and still feel good about yourself." Director Oliver Stone's *Wall Street* offered a critique of such greed. *Dallas* and *Dynasty,* however, were genuine in their ardor. As Kevin Phillips astutely notes in *American Dynasty,* these two television series encapsulated both the decade and its most important resource: oil. *Dallas* ran from 1978 to 1991 and *Dynasty* from 1981 to 1989. Both shows depicted the extravagant wealth and luxurious lifestyles of families who made and increased their fortunes as the heads of U.S. oil companies in the 1980s Texas oil boom.

The policies that led to this boom are known as "Reaganomics," even though British Prime Minister Margaret Thatcher shared a great deal of the credit (and blame) for the model. It boils down to a shift of economic resources from the "have-nots" to the "haves." It is, in fact, the same economic theory underpinning corporate globalization: The wealth passed to the rich would generate more wealth, which would in turn trickle down to the rest of society, making everyone better off in the end. Of course, the results were nothing of the sort, neither in the United States nor in the world.

In the thirteen years preceding Reaganomics, income inequality in the United States was shrinking. Social welfare programs, unions, labor laws, anti-discrimination laws, and the like were raising the wealth of the lower income population, and progressive tax structures were redistributing wealth from the upper income brackets down. According to the U.S. Census Bureau, from 1967 to 1980 the poorest U.S. households increased their share of the total income pie by 6.5 percent, while the wealthiest decreased their share by nearly 10 percent. Reagan aggressively reversed this trend. He gutted social welfare programs, shifted the tax burden from the wealthy to middle and lower income groups, poured enormous sums of money into the military-industrial complex, and reduced labor protections. Thus, Census Bureau data revealed a massive redistribution of income from the poor to the wealthy between 1980 and 1990. The poorest Americans lost more than 10 percent of the income pie, while the wealthiest gained almost 20 percent.

It turned out that the "haves" seemed more interested in holding onto the wealth that they accumulated than in passing it along to the wayward masses. Reaganomics ensured that they would not be required to do so. Reagan and Thatcher's ideas were carried to the World Bank and IMF, paving the way for the 1995 creation of the WTO and igniting a global trend of increasing inequality both within and between nations that continues to this day.

In addition to providing a source of drama on *Dallas* and *Dynasty,* American oil companies provided a driving force behind the Reagan administration's foreign economic policies. Reagan sought better relations with foreign oil-rich nations, including Iraq, and seized the opportunity presented by the poor nations who emerged from the 1970s highly dependent on the World Bank and the IMF.

U.S. private commercial banks were awash in cash from OPEC countries and willingly lent large sums of money to developing nations. But in December 1978, the second oil shock hit. An Iranian oil embargo reduced world oil supplies by almost 5 percent and increased prices by 150 percent.[10] In the United States, inflation skyrocketed with interest rates following suit. U.S. banks raised interest rates on their loans and developing countries the world over faced significantly larger payments on their debts. In order to avoid defaulting to the banks, the developing nations turned to the IMF and World Bank for loans, thereby becoming doubly indebted, first to the private commercial banks and then to the international lending institutions. The result was a downward spiral of economic debt known as "the 1980s debt crisis."

Reagan continued to expand World Bank investments in oil and gas exploration, but he also used the institutions to force nations to change their laws so that foreign companies could gain increased access to their resources.

The increase in World Bank oil exploration projects under Reagan was monumental. The World Bank's Projects Database reveals that, from 1982 to 1984, the World Bank funded more than fifty-five oil and gas projects in every corner of the globe. These included specific "Petroleum Exploration and Promotion Projects" in countries as diverse as Bangladesh, Ethiopia, Guinea-Bissau, Guyana, Morocco, Nepal, Pakistan, Papua New Guinea, the Philippines, Senegal, Yemen, and Zambia. The World Bank's hunt for oil was on.

When oil was found, the World Bank ushered in U.S. oil companies, who then laid their roots and stayed in place. But oil was not the only resource of interest. There were also agricultural products, copper and other ores, timber, labor, capital, land, and dozens of other resources to be had. More than anything else, both the commercial banks and the treasury departments of wealthy nations wanted the money they had lent during the oil shocks back—with interest. The U.S. government, its corporations, and its banks, among other players, wanted access to these resources, and World Bank and IMF Structural Adjustment Programs became some of the most useful tools available to them.

Before the 1980s, IMF and World Bank funds had been lent for projects with relatively few strings attached. No more. The 1980s brought a new phase for these institutions: the age of the Structural Adjustment Program (SAP). Developing countries were in debt to both foreign commercial banks and the lending institutions. The banks wanted their money back, foreign companies wanted access, and the developing countries were not in a position to say no. In order to receive loans, countries now had to adhere to a series of strict conditions that would reduce domestic spending while increasing capital available to pay back loans. The conditions were always the same, regardless of the country in question. They all followed the same corporate globalization model: privatize government industries, eliminate restrictions on foreign ownership and investment, eliminate barriers to trade, eliminate government restrictions on foreign corporations, cut government spending, devalue the nation's currency, and focus development on exporting key resources such as oil, minerals, trees, agriculture products, luxury goods such as coffee and flowers, and the like.

Reagan focused on SAPs particularly as a tool that could force countries to open their oil sectors to foreign companies. Six months into the Reagan administration, the U.S. Treasury Department released a report on how the U.S. government could best take advantage of "the neutral stance of the World Bank" to expand U.S. corporate access to oil and reduce OPEC's power. The report argues that the World Bank is supposed to lend to oil and gas projects in order to "catalyze private investment flows. However, an examination of the Bank's oil and gas loans to date shows little catalytic effect."[11]

The paper explains that the World Bank could be used to eliminate political, financial, and other restrictions to private company oil development in the world's poorest countries. The World Bank could get countries to adopt laws that "establish the necessary climate to foster private sector investment in energy and other sectors." This would "enhance security of supplies and reduce OPEC market power over oil prices." And, finally, structural adjustment programs would help countries suffering financially due to the oil crisis pay off their debts to private commercial banks.

The World Bank began to implement the U.S. Treasury Department's plan immediately. In fact, the World Bank's principal counsel for energy and mining, William Onorato, explained that the Bank was financing oil projects and writing legal reforms for recipient governments with the specific purpose of increasing the presence of foreign oil companies shortly thereafter.[12]

A LEGACY OF DESTRUCTION

> Consider the facts and it soon becomes evident that the $1,000 billion spent on aid since the 1960's, with the efforts of advisors, foreign aid givers, the International Monetary Fund and the World Bank, have failed to attain the desired results. . . . It is little wonder that

> protestors have demonstrated so vehemently against the international organizations.
>
> —William Easterly, World Bank senior adviser, 2001[13]

From the 1980s forward, the policies of the World Bank and IMF have had serious, often tragic effects on the nations of the developing world—effects that have often led to popular resistance and protest. This was certainly the case in nations like Zambia, Russia, Argentina, and South Africa. Because the institutions have such a profound impact on the most basic areas of peoples' lives, from the cost of bread to the availability of electricity and water, people in loan-recipient nations become World Bank and IMF experts from an early age. They learn that to bring change, they must challenge not only their governments but also the international financial institutions behind them as well. Rarely is there much doubt over which party is more to blame, and rarely is there any debate over which country is ultimately pushing the financial institutions themselves.

Zambia Forced Backward

> Globalization appears to increase poverty and inequality. . . . The costs of adjusting to greater openness are borne exclusively by the poor, regardless of how long the adjustment takes.
>
> —The World Bank, 1999[14]

On February 18, 2004, the people of Zambia went on nationwide strike. Ninety percent of all of the people who work for the government left their posts, took to the streets, and demanded reform. Their immediate focus was opposition to a wage freeze and simultaneous increase in their taxes. The ultimate focus, however, was on the IMF and the World Bank and what the workers were calling the "recolonization of Zambia." Joyce Nonde, president of the Federation of Free Trade Unions of Zambia, expressed widespread sentiment when she said, "We feel that Government, through privatization of national assets, has surrendered our country into foreign hands on a silver plate and, as a union, we are prepared to stop this unfair economic arrangement. . . . [Privatization] is a tool being used by the IMF to recolonize us. It is something being forced down our throats and therefore it is highly undemocratic."[15]

Forty years earlier, the Zambian people brought more than half a century of British colonization to an end and gained their independence. They were not about to let the IMF and World Bank take over where the British had been stopped. After independence, Zambia became the second wealthiest nation in sub-Saharan Africa. It had a highly urbanized population, a strong manufacturing sector, thriving copper exports, and a government that played a large role in guiding the economy. But the 1973 oil shock brought with it a triple financial burden. First, it increased the cost of imported oil. Second, it increased the cost of all imported goods, as countries around the world responded to the increasing cost of oil. Finally, it led to a lower demand for Zambia's key export, copper. As a result, Zambia was forced to borrow from foreign donors. Its external debt rose from US $814 million in 1970 to US $3.2 billion in 1980. To help pay off these debts, Zambia turned to the IMF and World Bank.

The World Bank provided $6.6 million in loans for a Petroleum Exploration Promotion Project in Zambia in 1982. For better or for worse, no oil was found. Between 1983 and 1987, the World Bank and IMF applied a Structural Adjustment Program to Zambia's loans. The impacts of these and subsequent World Bank and IMF conditions are brilliantly described in a 2004 report by Lishala Situmbeko, the former finance minister of Zambia and current economist at the Bank of Zambia, and Jack Jones Zulu, an economist at the University of Zambia.[16]

The SAPs required several changes in the Zambian economy, all of which were designed to force the government to free as much of its money as possible from domestic spending in order to pay back its loans. The Zambian government was required to eliminate price supports for goods such as corn and fertilizer (which had been used to keep the cost of such items affordable and reliable); devalue the currency in order to make exports more attractive on the global market; eliminate its barriers to imports, such as tariffs (which had been used to protect the domestic industries from foreign competition); reduce government spending by freezing wages for public sector workers; and loosen government control over interest rates so that investors would be more apt to put money into the economy.

The results were grim. The Zambian economy (as measured by the gross domestic product) did not grow at all from 1983 to 1986. The main reason was that local production came to a near standstill because it could not compete with the newly introduced foreign competition. Reduced price controls sent prices skyrocketing and then took inflation (newly loosened from government control) with it. Simultaneous increased prices and wage freezes meant that people could no longer afford food, which brought the now globally infamous "IMF food riots" to Zambia. The need for more money to meet basic necessities led to a ninefold increase in the number of twelve- to fourteen-year-old children working to support their families between 1980 and 1986. Finally, the federal budget deficit and the trade deficit both increased.

Under intense public pressure, the Zambian government told the IMF and the World Bank that it would not accept this punishment and abandoned the SAPs in 1987. But this liberation experiment lasted just one year before the donors rebelled. Not only were IMF and World Bank funds suspended—so, too, were funds from Zambia's other international lenders, who withdrew their money until Zambia followed, at a minimum, its IMF conditions. Zambia could not face the elimination of all of its foreign loans and was subsequently forced back into line. When people ask: *Why do countries accept these conditions?*—this is the answer.

Perhaps as retribution for Zambia's indiscretion, the IMF and World Bank saddled the country with new and significantly more imposing conditions in 1991. This was also, coincidentally, the same year that one-party rule officially ended and a new multiparty democratic government came to power. But the new government's hands were tied by the SAPs. The three key conditions placed on Zambia by the SAPs were identical to those imposed by the Bush administration on Iraq fifteen years later: privatization, trade liberalization, and agricultural liberalization.

The World Bank and IMF have rarely met a government-run enterprise that they did not prefer to see run by the private sector. A central requirement made by the institutions, therefore, is that governments must "privatize" or turn public entities—such as drinking and sewage water systems that are owned, operated and/or managed (usually all three) by a government—over to private entities such as corporations. Under its SAP, Zambia privatized 257 former government enterprises in just five years and was hailed by the World Bank for its efforts. Unfortunately, while privatization did benefit a few enterprises, far more state-run companies and the services they provided were simply eliminated when the private sector proved incapable of or simply uninterested in taking them over. At the same time, increased "trade liberalization" meant that Zambia had to reduce tariffs—in this case taxes applied to goods as they entered Zambia—which forced Zambian companies to face global competition suddenly for the first time. The combined impact of privatization and tariff reductions led some companies simply to pack up, move to neighboring countries, and sell their products back to Zambia as imports—depriving Zambia of employment and revenue.

The Zambian textile industry was decimated by the tariff reductions. It simply was not prepared to compete with the flood of cheap imports, mainly from Northern developed nations. Zambia had 140 textile manufacturing firms in 1991. After the tariff reductions, only eight firms remained in 2002. Overall, manufacturing employment fell by 43 percent from 1991 to 1998. This is called "deindustrialization," the all-too-common impact of corporate globalization policies the world over.

Tariffs on foreign products not only protected local Zambian industries, but were a crucial source of government income. During the 1990s, total income from tariffs fell by more than 50 percent. The IMF made the government fire ten thousand public sector workers and freeze the wages of those who held onto their jobs. The reason? To reduce government spending. However, with so many people out of work, fewer people had income to tax. In addition, the money that the government once earned from its companies was quickly disappearing. Thus, overall government revenue fell by more than 30 percent between 1990 and 1998. The World Bank and IMF simultaneously required cuts in domestic government spending. Overall, government spending decreased by almost half through the 1990s.

The World Bank has increasingly turned to the imposition of "user fees" on public services—requiring people who want access to health care, water, electricity, education, or the like, to pay a fee first. In wealthy nations where more people have expendable income, user fees for those who can afford them make sense. However, such fees are thoroughly incompatible with the poverty experienced by 73 percent of the people in Zambia. With incomes falling and unemployment rising, fees were simply beyond the reach of most families. In 1994, the World Bank itself was forced to admit that, following the introduction of user fees, outpatient attendance fell by almost 60 percent and delivery services by over 20 percent in urban Lusaka, and that "vulnerable groups" were simply denied access to health services. Despite this finding, in 2004, the World Bank told the Zambian Ministry of Health to "pursue improvement in cost recovery through user fees."[17] After attending a United Nations AIDS conference in Lusaka, one British doctor commented, "It is no coincidence that the HIV crisis has gone hand in hand with the debt crisis" in Zambia.[18]

Then there is the issue of food, or lack thereof. IMF food riots happen because the IMF puts conditions on its loans that force governments to abandon the needs of its people—particularly its most needy populations—in favor of its foreign creditors. By restricting the ability of governments to control prices and interest rates, for example, prices rise. Imagine the price of a loaf of bread increasing from $2.00 to $60.00 in less than three years. What would it do to your ability to feed your family? This is what happened in Zambia between 1990 and 1993.[19] As the World Bank itself reported in 2000, the removal of subsidies on maize and fertilizer under the World Bank/IMF SAP led to "stagnation and regression instead of helping Zambia's agriculture sector."

The end result was to be expected. In 1970, when IMF and World Bank loans first began, life expectancy in Zambia was 49.7 years. In 2001, life expectancy was 33.4 years—the lowest of any country in the world.

Why did this happen? Foreign creditors wanted their money back. In this context, many of the "reforms" make sense. They are all geared toward freeing as much money as possible from the national budget to go to debt servicing rather than internal expenses. Zambia was far from alone. From 1985 to 1992, poor nations in the Southern hemisphere paid some $280 billion *more* in debt service to creditors in wealthy Northern countries (such as the World Bank and IMF) than they received in new loans or aid. As a result, gross national product (GNP) rose an average 1 percent in Southern nations in the 1980s. It fell in sub-Saharan Africa by 1.2 percent while it rose by 2.3 percent in Northern nations.[20] Anyone who believed that the populace would somehow patiently endure the dislocation and pain caused by the reforms was mistaken. The people of these Southern countries were not willing to let themselves or their family members starve, die of AIDS, lose health care, or lose twenty years from their lives so that wealthy foreign creditors could get their cash back. Fewer people and countries are.

The Collapse of the Russian Economy

On November 9, 1989, the Berlin Wall crumbled. On Christmas Day 1991, the red flag was lowered from the Kremlin, and by the end of December the Union of Soviet Socialist Republics was no more. The United States declared victory and the Cold War officially came to an end. It took just three months for the U.S. House of Representatives to introduce the "Freedom for Russia and Emerging Eurasian Democracies and Open Markets Support Act of 1992."

As in Iraq almost fifteen years later, the oil sector was the most gleaming prize on the Russian horizon. Russia sits on approximately 5 percent of the world's known oil reserves. However, some estimates put its potential reserves as high as 14 percent.[21] All of this oil was controlled by the state, and U.S. oil companies wanted in. The economists, for their part, saw an opportunity to transform the world's most renowned state-controlled Communist economy into a market-driven capitalist one. As economist David Kotz wrote, "American economists do not often get a chance to conduct an experiment, at least not on a large scale. In 1992, they were suddenly handed Russia. At conferences on Russia that year the glee was evident, as U.S. economists discussed how to remake Russia's economy and society. They had found an ideal opportunity to demonstrate that once a country burdened by a state run economy was relieved of its burden, a prosperous free market system would spring to life."[22]

There were two competing views, however, on how best to accomplish the task of completely restructuring Russia's economy. On one side, there were those who wanted to proceed cautiously, to follow a more democratic path that included the Russian people, and to base decisions on Russian history and past economic experience. On the other side, there were those who followed the standard IMF approach, or the more Clausewitzian model: "Do it fast, do it now, and do it all the way." Joseph Stiglitz described the latter group as economists who "typically had little knowledge of the history or details of the Russian economy and didn't believe they needed any."[23] Guess who won?

George H.W. Bush put his full weight behind the "Freedom for Russia and Emerging Eurasian Democracies and Open Markets Support Act of 1992." On June 1, 1992, the White House issued a press release to express its support of the bill. The press release could just as easily have been written by Ari Fleischer and the White House of George W. Bush about the fall of Saddam Hussein's Iraq. It explained that the collapse of the Soviet Union "provides America with a once-in-a-century opportunity to help freedom take root and flourish." Democracy and open markets, in this case in Russia and Eurasia, "will directly enhance our national security. The growth of freedom there will create business and investment opportunities for Americans and multiply the opportunities for friendship between our peoples."

The Freedom Support Act became law in August 1992. It specifically called for an increase in the U.S. total contribution to the IMF so that the IMF could provide more resources to Russia. Ever since Harry Dexter White created the IMF and ran it through its first year, the U.S. Treasury Department has been the acknowledged power and decision-making body behind the institution. Thus, working through the IMF gave the United States a facade of internationalism that Reagan had sought from the World Bank a decade earlier.

On June 1, 1992, forty years after White's death and the same day the White House issued its press release, Russia was admitted into both the IMF and the World Bank. Shortly thereafter, the loans started to flow. In August, the Bank released a description of its loan of $760 million to support reforms for the transition to a market economy: "The reforms include privatization and restructuring of state-owned enterprises, promotion of foreign direct investment, pro-competition and anti-monopoly policies, reform of financial institutions and the commercial banking sector, and establishment of a social safety net to protect those who may be affected by the reforms. Subsequent World Bank operations will support efforts by the government to deepen the programs for privatization and social protection, as well as extend reforms to include agriculture and the energy sector." As we will see, the focus on the "social safety net" would not prove nearly as in-depth as that on the restructuring of the economy.

When it came to the actual restructuring, the World Bank took a backseat to the IMF. IMF loans to Russia began on August 5, 1992 with $719 million. This increased to over $1 billion a year in both 1993 and 1994, then tripled to just under $3.6 billion in 1995, followed by $2.5 billion in 1996, $1.5 billion in 1997, and then up to $4.6 billion in 1998—the year the Russian economy collapsed. The following year, Russia received less than half a million from the IMF and was then cut off.

The transformation of the Russian economy was fast and furious. At the IMF, they call it "shock therapy." Strobe Talbott, then in charge of the noneconomic aspects of Russian policy at the U.S. State Department, eventually admitted that Russia had experienced "too much shock and too little therapy."[24] Others put it a little more bluntly. In the words of economist Mark Weisbrot, "The IMF has presided over one of the worst economic declines in modern history."[25]

First the IMF required that the government eliminate all price supports. This sent prices skyrocketing. Russians quickly spent all of their savings, which in turn led to a 520 percent increase in inflation in the first three months alone. Millions of people saw their life savings and pensions eviscerated virtually overnight. To curb inflation, the IMF required the government to slam on the monetary and fiscal brakes, bringing about a massive depression. Within four years of reform, the average income fell by 50 percent.

Next was rapid mass privatization. Among the many problems with the forced privatization was the apparent ignorance about how vertically integrated the Russian economy was. Thus, when one firm was closed, ten others soon followed. Privatization led to the elimination of government revenue from its once profitable enterprises. It also led to mass lay-offs. Russian production was not ready to compete with a world market. From 1992 to 1998, Russian output declined by more than 40 percent. All of this amounted to deindustrialization.

Just as the IMF had done in Zambia, the IMF made the faulty assumption that the Russian government would get its income, once derived from government enterprises, from taxes. But with Russians out of work, there was no one to tax. The IMF also assumed that foreign private investors would rush in to buy up Russian businesses. But the companies waited to see what the impact of the restructuring would be, and while they waited, the government had no income base. Neither the privatized companies nor the remaining state-controlled industries were earning income. As a result, millions of workers were denied wages and pensions were delayed or nonexistent. The IMF also required mass cuts in government spending, which led to immediate cuts in services.

According to Stiglitz, "The IMF and [U.S.] Treasury had rejiggered Russia's economic incentives, all right—but the wrong way. . . . While only two percent of the population had lived in poverty even at the end of the dismal Soviet period, 'reform' saw poverty rates soar to almost fifty percent, with more than half of Russia's children living below the poverty line."[26] Male life expectancy subsequently declined from 65.5 years before "reform" to 57 years in 1998.

Boris Kagarlitsky, a senior research fellow at the Institute for Comparative Studies of the Russian Academy of Sciences, told the U.S. House Banking Committee on September 10, 1998, "It is quite possible that the chief concern of the IMF decision-makers was not the success of Russia but the prosperity of the Western financial community which made a lot of money out of our crisis." He concluded, "There is one thing we need from the West now—for it to leave us in peace. We need it to stop imposing economic policies that are ruinous for us, while using the pretext of giving us aid."

There were, of course, winners. In this case, Russia's well-placed oil and gas magnates, bankers, speculators, and real estate operatives—the new Russian "oligarchs"—stepped in to take advantage of the newly "reformed" economy. They made fortunes exporting oil, speculating in securities, and lending money to the government.[27] With the redistributive tools of the government eliminated by the IMF and the World Bank, wealth was generated—it just stayed at the top.

So, where was the foreign investment? Why didn't the West rush in? Well, Western companies were rushing in, but only to the oil sector. The IMF and World Bank rewrote Russia's laws to allow foreign companies access to the region's oil. Then, beginning in 1993, the World Bank began providing financing in the form of loans, credits, equity, or guarantees that benefited many U.S. companies, including Halliburton in Azerbaijan, Georgia, Kazakhstan, and Russia; Chevron in Kazakhstan and Russia; and Bechtel in Azerbaijan, Georgia, Kazakhstan, Russia, and Turkmenistan. Other companies aided by the World Bank in the former Soviet Union include Total, ExxonMobil, and Enron.

Once the World Bank started financing oil projects in the region, its support exploded. In 1993, the Bank provided loans or other financing for three oil projects in Russia. In 1994, it supported two projects in Russia and one in Kazakhstan. In 1995, financing went to Azerbaijan and Russia. In 1996, it financed two oil fields in Kazakhstan and one in Russia. In 1997, it supported two pipelines in Azerbaijan and one in Georgia. In 1998, it provided funding for oil field and pipeline development in Azerbaijan and for an oil field in Georgia. In 1999, it supported Russia's Kirtayel and Bitech–Silur oil fields, and the list continues to this day.

The Caspian region is believed to possess approximately 10 percent of the world's oil. Today, Paul Wolfowitz sits at the helm of the World Bank, and funding for oil and gas projects can be expected to increase, as can pressure on the region's governments to implement policies that grant greater access and ownership rights to foreign companies while reducing regulations on their operations.

The economic policies that the IMF, the World Bank, and the U.S. government implemented in post-Soviet Russia are responsible for Russia's 1998 financial collapse. These are the same policies that the Bush administration forced on Iraq in the wake of the 2003 invasion. The two countries provide stark parallels. Both were economies heavily controlled by the government. Both were forced to transition to market-controlled economies virtually overnight. Both have been described by American political and economic leaders as "experiments" to demonstrate that American economic policy can turn around whole regions. Both have a wealth of oil lying just beneath the surface of their soil—taunting Americans without allowing them access. Both have been forced to change their laws in order to grant U.S. corporations increased access to their resources. Hopefully for Iraq, the parallels between the two nations do not also include Russia's subsequent economic ruin.

Argentina Survives the IMF

What happens when the foreign corporations do show up to buy the formerly state-owned enterprises and foreign investment does pour in? Unfortunately, even when foreign companies do arrive, the economic benefits to the host countries are often severely limited. The World Bank itself found that, in 1992 alone, roughly half of the new foreign investment by global corporations in the South quickly left those countries as profits. Rather than invest profits locally, the foreign companies sent their profits home. In addition, the World Bank reported that more than 70 percent of investment flows in 1998 went to only ten developing countries: China, Brazil, Mexico, Singapore, Thailand, Poland, Argentina, South Korea, Malaysia, and Chile.[28] Argentina is one of the most often cited examples of what happens to the few "lucky" countries that attract foreign corporations and investors. Argentina was once *the* IMF success story, then it became *the* IMF failure, and now it has become *the* cause for hope among people the world over who are struggling against the impacts of IMF and World Bank policies.

In the early 1990s, Argentina followed IMF dictums to the letter. It privatized state-owned industries, liberalized trade and financial markets, eliminated capital controls, and cut government spending. The government even tied its currency to the U.S. dollar. Argentina was not a poor country when its "reforms" began. Thus, foreign corporations rushed to Argentina, buying companies, opening branches of their stores and banks, and purchasing social services such as hospitals. Foreign financial institutions lent Argentina billions of dollars. There were even streets named after U.S. banks.

This was a fully internationalized economy, if ever there was one. But when the value of the U.S. dollar began to rise in the mid-1990s, Argentine exports were no longer competitive and industry began to decline, causing unemployment to increase. The World Bank had led Argentina to privatize its social security program, causing government revenues to decline as contributions once made to social security were diverted to private pension funds. With its revenue falling, the government turned to the IMF for help. In return for its money, the IMF demanded deep cuts in public spending that further reduced domestic demand and stoked social unrest. Millions lost health coverage, as private international insurers pressured their local providers to cut costs. Argentine banks now owned by foreign firms cut back lending to small and medium-size enterprises. Stripped of protections, private employers were pressured to become lean and mean through mass layoffs.

The IMF rules succeeded in opening the Argentine market to foreign investors, but did nothing to require those investors to make commitments to the health and welfare of the Argentine economy in return. Thus, freed to come and go as they wished due to the elimination of capital controls, foreign companies and investors simply pulled their money out when the going got tough and moved on to the next hot developing country market—leaving the Argentineans to clean up the mess.

In December 2001, the government of Argentina closed its banks and froze assets in a last ditch effort to stave off total financial collapse. Before doing so, however, armored cars went to the banks in the dead of night, filled up with money, and sped off to deliver what remained to foreign creditors. Overnight, everyone in Argentina was cut off from the money in their bank accounts and the crisis was spread to the middle and upper classes. People rallied at the banks and tried to break them down to get inside. The government decreed a state of emergency that suspended democratic liberties, such as the right to gather and demonstrate. As Argentinean activist Patricio McCabe describes it, "On the day that the state issued its most restrictive orders, a huge, completely spontaneous demonstration broke out in Buenos Aires, and people poured in to the streets, chanting a slogan that no one had heard before: *Que se Vayan Todos!* (All of them must go!)"[29] They also banged on pans with spoons to demonstrate that their pots were empty of food.

The people of Argentina had had enough. They took to the streets and demanded that the government resign. On December 21, it did. The next president lasted just one week before the people rejected his continuation of IMF policy. A total of four presidents fell in succession. Then, on January 3, 2002, Alberto Duhalde won a good deal of favor when he formally defaulted on the government's foreign debt. As the economy and the government struggled to put back the pieces, the people of Argentina began organizing for themselves. The model that emerged is most often referred to as horizontalism.

The most important component of horizontalism is probably the neighborhood assembly, where decisions such as trash collection, repair of potholes or road signs, school boards, and even city budgets are made. The assembly is a form of "direct democracy" in which people participate directly in making political and economic decisions that affect their daily lives. For example, when the foreign owners of companies abandoned Argentina, many workers simply took over the factories and began running them themselves. Most are worker-run cooperatives in which decisions are made through assemblies in which all workers have equal decision-making authority about pay, production schedules, materials, distribution, health benefits, and the like.

In addition, when the foreign companies left and stopped selling their products or providing their services, major cities in Argentina adopted extensive barter systems in which people barter dentistry services for haircuts, aerobics classes for massages, food for clothing.

There are barter currencies and banks that lend these currencies without fees. The banks barter their services just like everyone else. There are also kitchens, day-care centers, health clinics, computer programming classes, schools, and community centers run by professionals, those who were once paid in pesos for such services, who now volunteer and barter their services instead.

Why Corporations Want Free Trade: The Case of Wal-Mart

Wal-Mart prides itself on its folksy small town persona, derived from its roots as a five-and-dime store run by founder Sam Walton in Rogers, Arkansas. While the name and the family remain, little else about today's Wal-Mart resembles that first small store in 1962. According to *Forbes,* in 2005, five Walton family members, Christy, Jim, S. Robson, Alice, and Helen, were among the top ten wealthiest people in America and the tenth through the thirteenth wealthiest people on earth (a few Walton family members are tied). Wal-Mart is the largest company in the world, the largest employer, and the nineteenth largest economy on the planet. Wal-Mart has more than 6,100 stores with 1.6 million employees. Its 2004 sales were more than $285 billion, while profits were the highest in Wal-Mart history, topping $10 billion, up more than 13 percent from the year before.

In the words of Wal-Mart's chief financial officer, Tom Schoewe, Wal-Mart's $256 billion in sales in 2003 were equal to "one IBM, one Hewlett Packard, one Dell computer, one Microsoft and one Cisco System—and oh, by the way, after that we got $2 billion left over."[30] In 2002, Wal-Mart's sales were larger than the individual GDP's of Sweden, Austria, Norway, Poland, Saudi Arabia, Turkey, Denmark, Indonesia, Hong Kong, Greece, Finland, and hundreds of other nations. And Wal-Mart just keeps growing, at a rate of more than 18 percent a year—more four times that of the United States (3.9 percent) and the world (4 percent).

Wal-Mart has achieved its number-one status by mastering the use of international trade agreements. This has enabled the company to enter and dominate markets with its stores and to use those suppliers most willing to pick up, close shop, and scour the planet for the cheapest places to make products—successfully pitting the poor against the poorer in its pursuit of ever-falling prices.

Wal-Mart did not open a single store outside of the United States until 1991 in Mexico. As late as 1995, the company reports that imports accounted for no more than 6 percent of the products sold in its U.S. stores. Although many dispute that figure as far too low, there is no debate over the dramatic changes in Wal-Mart's operations in the past decade. In 2003, consulting firm Retail Forward estimated that 50 to 60 percent of the merchandise sold in Wal-Mart's U.S. stores was made overseas.[31] In addition, Wal-Mart now owns more than 2,400 stores in fifteen countries outside of the United States.[32] Two trade agreements facilitated this growth: the North American Free Trade Agreement (NAFTA) signed in 1994 and the World Trade Organization (WTO) established in 1995.

WAL-MART AND THE NORTH AMERICAN FREE TRADE AGREEMENT. In 1994, the United States, Mexico, and Canada signed the most far-reaching multilateral trade and investment agreement of its time. NAFTA was signed under the promise that it would create jobs and economic benefits across all three NAFTA countries. As with all trade agreements, benefits did accrue, but certainly not to all. As Jorge Castaneda, Mexico's former foreign secretary, observed, NAFTA was "an accord among magnates and potentates: an agreement for the rich and powerful . . . effectively excluding ordinary people in all three societies."[33] One thing that NAFTA unquestionably accomplished was to allow U.S. businesses to find a cheaper location to produce their products.

People are often surprised to discover the extreme difficulty of determining where companies actually make their goods. There are no federal reporting requirements short of product labels. This means, for example, if I wanted to learn how many products sold at Wal-Mart were made in Mexico (given that the company has thus far refused to tell me), I would have to visit every one of its more than 6,000 stores, personally turn over each item, and look for the product label. When such production information is made publicly available, it is almost always acquired in spite of the company's policies. The sources tend to be nonprofit organizations in the United States collaborating directly with workers in the production facilities abroad.

I have always found this to be one of the most glaring examples of the failure of corporate globalization. Adam Smith, the eighteenth-century economist, philosopher, and founder of modern capitalist theory, asserted that the key tenet of the free market and the only way for the invisible hand to operate is if consumers have "perfect information," which means complete access to all information about the goods they may wish to buy. In the age of globalization, production is so far removed from the consumer and so hidden by the producer that it is virtually impossible for consumers to make informed decisions and hold companies accountable for destructive, harmful, unjust, unethical, or even illegal practices. Wal-Mart has perfected the art of removing production from consumption and both the NAFTA and the WTO have facilitated this sleight of hand.

NAFTA eliminated tariffs and other import controls on goods moving between the three countries. This meant that U.S. companies could send products to be assembled in Mexican factories, where labor is cheap, environmental protections weak, taxes low, and protections from further regulation and government oversight even greater than in the United States, and then send the finished products back home to sell at prices far cheaper than if the goods were produced domestically. These factories are called maquiladoras. Many maquiladoras emerged in the export processing zones right on the border between the United States and Mexico. The export processing zones are almost completely divorced from the local Mexican economy, with even fewer environmental and labor protections, virtually no taxes, and fewer requirements placed on producers than in the rest of the country. From 1990 to 2001, the number of maquiladora factories across Mexico more than doubled from 1,700 to 3,600 plants, with 2,700 of these located in the export processing zones.[34]

The result, according to the U.S. Congressional Research Service, is that U.S. imports from Mexico increased by 229 percent between 1993 and 2001. While U.S. exports to Mexico increased 144 percent, 60 percent of these were components being shipped to the maquiladora factories for processing, yielding little or no benefit to the Mexican economy or consumer. Laws that would have addressed this problem, such as requiring a certain amount of domestic content in production, a certain amount of local investment, or a transfer of new technologies, and the like, were stripped away by the NAFTA—exactly the same laws that have been stripped away in Iraq.

The shift of so much production to Mexico came at a huge cost to American workers. The U.S. Census Bureau calculates that the United States lost three million manufacturing jobs—one in six jobs in that sector—during the NAFTA years. Some 525,000 U.S. workers have been specifically certified as "NAFTA job-loss victims." These are workers determined by the U.S. government as eligible for NAFTA job-loss assistance. This new pool of unemployed workers then put downward pressure on all American jobs, contributing to the fast pass increase of income inequality in the United States. In 1997, *Business Week* reported that PaineWebber was advising investors to follow a Tiffany/Wal-Mart strategy and "avoid companies that serve the middle of the consumer market, and for good reason. The middle class who once seemed to include almost everyone is no longer growing in terms of numbers or purchasing power. Instead, it's the top and bottom ends that are swelling . . . the '90s have seen a greater polarization of income in the U.S. than at any point since the end of World War II."[35] A 2001 U.S. Census Bureau report confirmed this assessment, finding that in 1999, real hourly wages for workers with less than a college education had eroded to an all-time low since World War II, while real wages for those with college or advanced degrees had steadily increased to the greatest level since 1973. This trend of rapidly increasing income inequality is a global phenomenon.

Back in Mexico, the dramatic rise of the maquiladoras coincided with the near collapse of the farming sector because of the NAFTA. The elimination of agricultural tariffs and quotas on products such as corn, which had accounted for 60 percent of all Mexican farming, allowed cheap, heavily government-subsidized U.S. products to flood the Mexican market. Similar provisions are found in the WTO's Agriculture Agreement. The price paid to Mexican farmers for corn dropped by 70 percent.[36] In addition, in order to join the NAFTA, Mexico was forced to eliminate one of the most important victories of the Mexican Revolution, the Ejidos program, which granted guaranteed land-rights to indigenous people. The combined impact was that Mexican farmers could not compete and could not keep their land. One and a half million farmers and their families were forced from their land and subsequently found themselves in search of work. The more unemployed workers there were, the more U.S. companies could demand stiff sacrifices in return for jobs. Maquiladoras have since become synonymous with "sweat shop labor"—with human rights abuses rampant, unionization unheard of, and long hours, low pay, no benefits, and unsafe working conditions the norm. In addition, some 80 percent of maquiladora workers are women, most of them young women who frequently face sex and age discrimination in the workplace.

The result is that average real wages in Mexican manufacturing are lower today than they were before NAFTA; the minimum wage has declined by 20 percent and hovers at around $4/day, and half of the nation now lives in poverty. As unemployment, low wages, and poverty grow in Mexico, Wal-Mart's low prices look more and more attractive to consumers.

After the implementation of NAFTA, Wal-Mart became the largest retailer in all three NAFTA countries. NAFTA made this possible in several ways. For example, NAFTA investment rules eliminated many of the requirements on how and where Wal-Mart could operate in Mexico. The elimination of tariffs made the shipment of products in and out of Mexico tax-free. NAFTA also stopped the Mexican government from applying requirements that would force Wal-Mart to partner with a local business, sell products made by Mexican-owned companies, or invest a certain amount of its profits locally.

After visiting Mexico in 2003, Tim Weiner of the *New York Times* wrote, "Wal-Mart's power is changing Mexico in the same way it changed the economic landscape of the United States, and with the same formula: cut prices relentlessly, pump up productivity, pay low wages, ban unions, give suppliers the tightest possible profit margins and sell everything under the sun for less than the guy next door."[37] Today, Wal-Mart is the largest private employer in Mexico. It has 683 stores and does more business than the entire tourism industry. It sells $6 billion worth of food a year, more than anyone else in Mexico. As Weiner wrote, "It sells more of almost everything than almost anyone."

In the United States, Wal-Mart's Supercenter grocery store employees earn an average annual salary of $14,000, which is $1,000 dollars below the poverty line for a family of three. Wal-Mart does not allow its workers to form unions. If employees come too close to unionizing, Wal-Mart simply closes up shop and moves elsewhere. Unionized grocery workers in the United States earn an average of one-third more than Wal-Mart workers—about $18,000 a year. Because Wal-Mart is the world's largest employer, its labor practices put downward pressure on the rights of all workers. In 2004, when Wal-Mart announced the opening of forty Supercenters in Southern California, local rival supermarket chains were rightfully worried. They reacted by demanding a two-year freeze on the salaries of their own workers. The stores also lowered pay for newly hired workers and demanded that employees pay more of their own money towards their health insurance benefits.

In response, the unions went on strike. One of the companies, Safeway, has a giant store across the street from my apartment. Every weekend hundreds of people from across the city protested in front of the store in support of the striking workers. San Francisco Mayor Gavin Newsom even participated: He stood on a flatbed truck directly in front of the entrance to the store and voiced his opposition to the company's unfair labor practices. In the end, the union was unsuccessful. While Safeway and the other supermarkets are smaller than Wal-Mart, they are larger than the unions. The stores could afford to lose profits in California for weeks, even months on end, because they were making enough money in their stores in other states and in other countries to subsist. The union was forced to cave.

Wal-Mart and the WTO. Free trade allows corporations to be fickle in choosing their partners. If they no longer enjoy the benefits of one nation, they can pick up and move on to the next without any thought to commitment. Wal-Mart has mastered this skill, and the WTO, particularly China's membership, helped it to do so.

Every year I worked for Congress, one of the most difficult, intractable, and morally wrenching trade debates was whether to grant "preferential trade status" to China. Until 2000, the United States applied uniquely high tariffs on goods imported from China in opposition to the human rights abuses inflicted on the Chinese by their government. U.S. corporations looked to China and saw 1.2 billion potential new customers and workers in a country where unionization is illegal, workers are cheap and disciplined, unemployment is rampant, and environmental protections are nil. Human rights advocates saw a move that would increase the powerlessness of China's workers not only at the hands of the Chinese government but also at the hands of U.S. corporations as well. Under what one Capital Hill publication called a corporate "Blitz for Free Trade," an odd coalition of CEOs, President Clinton, and the Republican House that had voted to impeach him pulled together to "normalize trade" with China just two months before the 2000 presidential election. One year later, in December 2001, China became a member of the WTO.

The result has been nothing short of phenomenal, but not in the way that U.S. business advertised. Rather than open a new market of Chinese consumers, it opened a new market for the world's most aggrieved workers. The U.S. trade deficit with China more than doubled from $70 billion in 1999 to $162 billion today. This means that the United States is buying $162 billion more from China than we are selling to it. A large percentage of these purchases are made by U.S. companies that build products in China and then ship them to the United States, like Wal-Mart's suppliers.

Because NAFTA makes it illegal for the Mexican government to require any sort of commitment on the part of foreign companies, when things started looking good in China, U.S. producers picked up and moved out. A full third of the eight hundred thousand manufacturing jobs initially created under NAFTA have since disappeared. Due to the WTO's elimination of tariffs and quotas on products entering and leaving China, and removal of many of the restrictions on which companies can operate there, in the last five years Wal-Mart alone has doubled its imports from China. It even opened its global procurement center in Shenzhen, China. In 2002, it bought approximately $12 billion in merchandise from China, 20 percent more than in 2001, which represented nearly 10 percent of all Chinese exports to the United States. Wal-Mart is the single largest U.S. importer of Chinese consumer goods, surpassing the trade volume of entire countries, such as Germany and Russia.[38]

In 2003, the *Los Angeles Times* visited the Gladpeer Garment Factory in the southern Chinese city of Dongguan where the reporters met twenty-year-old Ping Quixia.[39] Quixia was one of 1,200 workers, mostly women, making women's underwear and other garments for Wal-Mart, earning about $55 a month and living eight to a room in cramped dormitories. "In southern China," the paper reports, "Wal-Mart has found all the ingredients it needs to keep its 'everyday low prices' among the lowest in the world. Although labor costs more here than it does in Bangladesh, China offers other advantages: low-cost raw materials; modern factories, highways and ports; and helpful government officials."

Wal-Mart has more than three thousand supplier factories in China and the number is expected to rise. But that doesn't mean workers in China enjoy job security. The managing director of the Gladpeer Garment Factory said that he is likely to reduce employment in Dongguan and open a new factory in Guangxi province, where labor, electricity, housing, and taxes are even cheaper: "Competition is intense, and our biggest single issue is cost. . . . That's why we're going to Guangxi."

The manager's complaint is just as readily heard across the United States, as Wal-Mart suppliers are forced by the company to cut costs annually. In order to do so, many have closed shop in the United States and moved abroad. Ken Eaton, head of Wal-Mart's global procurement division, told the *Los Angeles Times* that in purchasing fabrics such as denim and khaki, Wal-Mart would approach three to five mills around the world to get the cheapest production price. "We'll be putting our global muscle on them," Eaten explained.

The outcome of such competition is a global pool of workers and producers pitted against one another in a race to the bottom. It resembles a boxing match without protective rules, in an arena filled with literally millions of able and eager boxers waiting to replace any fighter who falls in the process.

Resistance in South Africa

On August 25, 2002, I sat alone in the lobby of Witwatersrand University in the heart of downtown Johannesburg. The organization I worked for, the International Forum on Globalization, had co-convened a public teach-in with the university and several South African organizations to discuss international trade, the World Bank, IMF, WTO, privatization, and alternatives.

More than a decade after the end of apartheid, Johannesburg remained effectively a fully segregated city. White people lived in the city and black people lived in the townships surrounding the city's borders. To counter the segregation and to ensure that all those who wanted to could attend our free event, we had arranged with the local organizations for buses to bring people from the townships to the university. The buses were running, but at 9:00 A.M. the lobby was still empty.

Then, out of nowhere, I heard a distant "humph."

At first, I thought it was the police who had been appearing every so often in the courtyard just outside the conference hall from the moment we arrived, but I didn't see anyone approaching. Then I heard it again—"Humph." The sound brought a few of my more quizzical colleagues into the lobby. We all turned toward the doors leading to the outer courtyard from where the noise appeared to be steadily approaching.

Suddenly all at once, hundreds of people burst through the doors in a blast of sound, movement, and color. The men, women and children—all African—entered the lobby moving in unison stomping one foot forward, freezing in place with knees bent, and then came the "humph" from deep within their throats. They moved through the lobby, the men brandishing body-length sticks above their heads, wearing matching bright red and yellow cotton T-shirts with "APF" emblazoned across the front. The Anti-Privatization Forum was an event cosponsor and the leading South African organization fighting World Bank/IMF water and electricity privatization policy. They paraded in this way through the lobby, through the doors to the theater, down the aisle and into the theater rows. There they stood, cheering and singing. Virginia Setshedi of the APF took to the stage and led the crowd in an increasingly celebratory series of chants and songs. The teach-in had begun!

We met for nine hours and heard from speakers such as Wangari Matthai of Kenya's Greenbelt Movement, who would be awarded the Nobel Peace Prize two years later; Oronto Douglas, a Nigerian lawyer who had defended human rights legend Ken Saro-Wiwa against Shell Oil and the Nigerian government; M.P. Giyose, renowned leader of South Africa's antiapartheid movement; Ela Gandhi of the South African National Congress; Vandana Shiva, one of India's most acclaimed authors and activists; and Naomi Klein, author of the bestselling book *No Logo.*

The central issue of discussion was the impact of World Bank and IMF policy on South Africa, particularly the privatization of water and electricity.

SOUTH AFRICAN STRUCTURAL ADJUSTMENT

> We did not fight for liberation only to sell it to the highest bidder!
>
> —*Sign carried during a two-day strike of one quarter of South Africa's national workforce, 2001*[40]

On February 11, 1990, Nelson Mandela ended twenty-seven years of imprisonment. Four years later, one of the most powerful movements in history, embraced by people around the world, achieved the abolition of South African apartheid. In 1994, Mandela and his African National Congress (ANC) party won the first free and democratic elections in South African history. People around the world were glued to their television sets as black South Africans willingly stood in lines several hundred people long for their first opportunity to cast a ballot.

Tragically, the economic transformation did not share the democratic freedom of the political process.

In 1993, the IMF lent South Africa $850 million, conditioned with a set of classic structural adjustment provisions: tariff reduction, cuts in government spending, and deep public sector wage reductions. Three years later, the World Bank followed suit, expanding the Structural Adjustment Program with money, two of its economists, and a name—"Growth, Employment, and Redistribution" (GEAR). GEAR called for commercialization and then privatization of all of South Africa's companies and services, cuts to corporate taxation, more cuts to government spending, more tariff cuts, including in newly emerging sectors such as textiles and food products, and liberalization of capital controls and foreign exchange rates.[41] As described by Patrick Bond, a leading economist at Witwatersrand University: "The reality is that South Africa has witnessed the replacement of racial apartheid with what is increasingly referred to as class apartheid—systemic underdevelopment and segregation of the oppressed majority through structured economic political, legal, and cultural practices."[42]

As with Argentina, the lifting of capital controls devastated the South African economy and may be one of the most significant obstacles to ending economic, not just racial, apartheid. Eliminating barriers to the outflow of capital meant that the owners of capital—white people—were permitted to remove all of their capital from the country (much of which went to London) while maintaining their residency in South Africa. Corporations were also allowed to pick up shop and move headquarters and production facilities out of the country. Individual and corporate wealth grew, but it benefited London's economy, not South Africa's. The result was a massive outflow of capital without any concurrent attraction of new capital. According to Patrick Bond, the "white capital flight," in combination with the massive job loss, led black African household income to fall 19 percent from 1995 to 2000, while white household income rose 15 percent. The poorest half of all South Africans earned just 9.7 percent of national income in 2000, down from 11.4 percent in 1995. The richest 20 percent of South Africans earned 65 percent of all income.[43] In 2003,

South Africa earned the dubious distinction of having the greatest income inequality of any nation other than Guatemala.

In the first year of GEAR's implementation, South Africa lost more than 100,000 jobs. Unemployment rose from 16 percent in 1995 to 30 percent in 2002. With the addition of "frustrated job-seekers," those who had given up searching for work, unemployment was at 43 percent.[44] Unemployment figures in most of the country's provinces have continued to hover near 50 percent since the late 1990s.

Much of the job loss was caused by privatization. For example, more than 50,000 workers lost their jobs through the partial privatizations of the state-owned telephone company and the electricity sector. The most devastating impact of privatization on South Africa, as on Zambia, was the dramatic reduction of vital services, including phone lines, water, electricity, and health care, to those most desperately in need. For example, of the thirteen million people given access to a fixed telephone line for the first time after 1994, ten million were disconnected because they could not pay their bills—due to dramatically higher prices after the phone company was partially privatized.

Beginning in the late 1990s, the World Bank made a total of seven loans to South Africa totaling approximately $130 million. The loans required that South Africa adopt 100 percent cost recovery for previously free services. Through user fees, the state would retrieve 100 percent of its cost. At the same time, many of the services were being privatized and the prices increased. The combination of 100 percent cost recovery and privatization meant that water, electricity, and health care suddenly became priced completely out of reach for millions of South Africans.

More than ten million of the nation's poor and primarily black population had their water cut off from 1994 to 2002—more households than the government had managed to connect to water since the end of apartheid in 1990. The same number had their electricity disconnected. Two million people were evicted from their homes due to their inability to pay utility bills.[45] In Soweto, I personally saw children playing on the ground outside of their homes just inches away from small streams of raw sewage. The sewage ran directly into rivers that had become the primary drinking source to millions who could no longer afford clean water. As a result, the country experienced its worst-ever cholera outbreak, which affected more than 140,000 people between the years 2000 and 2002. In the late 1990s, diarrhea killed 43,000 children a year, mainly due to the lack of clean water.

Finally, as in Zambia, 100 percent cost recovery and government cuts in health care have priced millions of South Africans out of access to care and have directly contributed to skyrocketing increases in preventable infectious diseases, including HIV, AIDS, tuberculosis, cholera, and malaria.

Trevor Manuel, South Africa's former finance minister, may have earned the award for the grandest understatement of the year when he admitted in 2002, "We have undertaken a policy of very substantial macroeconomic reform. But the rewards are few."[46]

Thus, many of South Africa's leading antiapartheid organizers who were ready to rejoice after 1994 have since been forced to fight "economic apartheid." Trevor Ngwane, chairperson of the Anti-Privatization Forum, was expelled from the African National Congress for opposing its privatization policies. He was a cohost of our Johannesburg teach-in and helped orchestrate a conclusion to the teach-in that was even more memorable than its beginning.

Protest March. At 6:00 P.M., Trevor Ngwane and Dennis Brutus, also a leading antiapartheid activist sentenced to eighteen months hard labor at Robben Island Prison alongside Nelson Mandela, organized an impromptu rally on the steps of the University Hall. Ngwane and Brutus reported that seventy-seven members of the South African Landless Peoples' Movement had been arrested the day before and were still being held in jail for a peaceful protest against World Bank and IMF policies. Several hundred people from the teach-in eventually joined the rally and we decided to walk together from the Hall to the jail as a sign of solidarity. The sun was setting as the march began, and candles were passed from hand to hand to light the way.

The march progressed slowly. After walking about fifty yards, we saw a line of South African riot police in the distance. Fully armed, with fingers poised on the triggers of their weapons, they were not going to let us exit the campus. It seemed that we were not allowed to rally outside of the University itself. We continued to walk toward the campus exit and the police, when suddenly we heard a loud bang. Almost 99.9 percent of the marchers were South African, and this was not their first experience with the blast sound of a gunshot. They *knew* to be afraid—so when they ran, I ran. But when no other shots were heard, the crowd slowed to a jog, then to a walk, and then began to turn around and cautiously head back to continue the march. Few people spoke. As we returned to the police line, we learned that the shot we'd heard earlier had been a concussion grenade—metal canisters that explode and make an ear-crushing "boom" to disperse crowds. The grenade had been mistakenly fired directly into the crowd, badly injuring a young woman from Canada.

At the exit to the campus, a standoff ensued. We insisted on our right to march to the jail, while the police firmly disagreed. Dozens of television, newspaper, and print reporters gathered to relay the news of the march, the Landless Peoples' Movement, and the costs of privatization policy in South Africa. When we decided that it was time to go, Trevor declared, "We will return to fight another day." And indeed, the next day we were joined by 20,000 people, including the newly released members of the Landless Peoples' Movement, in protest against privatization and corporate globalization in a march from the impoverished Alexandra township to the wealthy city of Standton, organized to coincide with the UN World Summit on Sustainable Development.

THE LIST GOES ON . . .

The prior examples of the devastating effects of corporate globalization policy come from just four countries among a countless number from which to choose. By the late 1980s, more than seventy countries had been submitted to World Bank and IMF Structural Adjustment Programs. In Nigeria in 1989, after dozens of people were killed and hundreds arrested in riots and strikes against the IMF, the Nigerian government was forced to offer a welfare program called an IMF "SAP Relief Package." In 1996 in Jordan, riots broke out after the IMF demanded the removal of food subsidies that had caused the price of a loaf of bread to triple. Such protests are so common around the world that they have been named "IMF food riots." In 1998, Indonesian President Habibie made a public plea for his entire nation to fast two times a week in response to IMF required cuts in the subsidies of basic goods. In 1999, more than five thousand villagers in Thailand occupied the World Bank–funded Pak Mun dam site and established a settlement in protest of the dam's destruction of their fisheries, livelihoods, and communities. Over the past thirty-five years, more than ten million people worldwide have been forcefully displaced from their homes by World Bank dam construction projects. In Paraguay in 2000, a forty-eight-hour general strike was called against government plans to privatize the telephone, water, and railroad companies. A presidential spokesperson responded by telling the strikers that the "policies are non-negotiable," because the government needed to meet IMF privatization requirements in order to get $400 million in World Bank loans. In India in 2000, more than one million electricity workers protested against a proposed bill that would implement a World Bank plan to privatize the power sector. In Thailand in 2001, hundreds of farmers and students protested outside a meeting of the WTO in Chiang Mai. They dumped potatoes, garlic, onions, and soybeans in the lobby to demonstrate how the Agriculture Agreement has harmed them. In Ecuador in 2003, over 200,000 teachers staged a mass hunger strike in protest of the government's signing of a new IMF loan agreement. In 2005 in Yemen, thousands of people protested against fuel price increases that came as part of a World Bank/IMF economic program. In 2005 in Hong Kong, tens of thousands of people from across Asia and around the world protested against the WTO's Sixth Ministerial Meeting.

The list goes on.

~

If the devastating effects of corporate globalization policy have been so heavily proven, reported on, and resisted in so many different places for such a long period of time, why do they still persist? The fact is, there are many for whom the policies work extremely well. The next chapter discusses one set of victors, the corporations, focusing on the histories and activities of four companies that are key players in the Bush Agenda, particularly as it has been applied in Iraq: Bechtel, Chevron, Halliburton, and Lockheed Martin.

FOUR

THE CORPORATIONS

BECHTEL, CHEVRON, HALLIBURTON, AND LOCKHEED MARTIN

As the CEO of Halliburton Energy Services, the board of directors doesn't sit me down and say "John, make this a better planet." They want us to make and create wealth for our shareholders and employees. So the only way we can adopt a sustainability agenda is it must create sustainable wealth for all our shareholders.

—*John Gibson, Halliburton president and CEO, February 2004*[1]

Corporate globalization continues as a model, despite the devastation it is known to cause, for one very simple reason: It works—only not in the way that its advocates promise. It has successfully restricted the ability of governments and communities all over the world to regulate the activities of multinational corporations. As a result, these companies have been freed to scour the globe in search of the cheapest locations to produce, the most abundant natural resources, and the most business-friendly governments. Under the theory of "what is good for corporations is good for everyone," laws the world over, including in the United States, have been rewritten specifically to benefit corporate growth and profit. The result is an unprecedented shift in global economic power from countries to corporations and a concurrent concentration of wealth in the hands of an ever-shrinking group of the ultra-wealthy and powerful.

In many ways, corporations have supplanted governments as the dominant economic force in the world. In 2002, corporations represented fifty-two of the hundred largest economies in the world.[2] The sales of the largest 200 corporations are growing faster than overall global economic activity. The largest corporations (by revenue) are heavily concentrated in the United States. According to *Forbes* magazine, in 2005, 75 of the 200 largest corporations were U.S.-based—more than the next three nations combined: Japan (28), France (21), and Germany (20). Therefore, if corporate globalization worked not only to make corporations rich but also to make *everyone* better off, as its advocates promise, we would expect its benefits to be most evident in the United States. The theory is that wealth generated by these companies should trickle down the American economic ladder, lift the poor out of poverty, and spread wealth across the nation. The reality is that the wealth being generated by the companies is locked firmly at the top. The rich are getting richer and the poor are not only getting poorer but are also increasing in number. In fact, since 1983, the only segment of the U.S. population that has experienced large wealth gains is the richest 20 percent.[3]

Just compare CEO pay to that of the average worker. Twenty years ago, U.S. corporate CEOs earned on average forty-two times more than production workers. Today, they earn a whopping *431 times more.*[4] Put another way, if the average production worker's pay had kept pace with that of CEOs, he or she would be taking home more than $110,000 a year instead of less than $28,000. Likewise, the average minimum wage earner would be taking home over $23 an hour instead of $5.15. Imagine what one could do with an extra $18 an hour—perhaps afford a better place to live, child care during work hours, or health insurance for the entire family.

Instead, in 2004, while seventy-six more Americans became billionaires, poverty increased (both the number of poor people and their percentage of the total population), real median earnings of full-time workers fell, median household income fell, fewer people had health insurance, and more people were living in poverty even though they had jobs.[5] How do low pay and no health insurance translate in real-life terms? The 2005 UN Human Development Report found that the infant mortality rate in the United States is comparable to that in Malaysia—a country with a quarter the income. The UN report also found that infant death rates are higher for black children in Washington, DC, than for children in Kerala, India.

Inequality as a result of corporate globalization policy is not restricted to the United States; it is a growing global phenomenon. Professor Robert Wade of the London School of Economics, the most respected economist following this trend, reported in the *Economist* magazine, "Global inequality is worsening rapidly. . . . Technological change and financial liberalization result in a disproportionately fast increase in the number of households at the extreme rich end, without shrinking the distribution at the poor end. . . . The richest 10 percent pulled away from the median, while the poorest 10 percent fell away from the median, falling absolutely and by a large amount."[6] Corporate globalization persists as a model, therefore, because it works for a specific group of people.

Of course, when it comes to the distribution of benefits between corporations and governments, or between CEOs and the rest of us, it can be very difficult to determine where one group ends and the other begins with the Bush administration. The president, vice president, and the secretaries of the U.S. Defense, Energy, Treasury, and Commerce Departments are all former corporate CEOs. The secretaries of the U.S. State, Labor, Housing and Urban Development, and Transportation Departments are all former corporate executives or directors.

The oil sector is prominently represented in both the Bush Agenda and the Bush administration. It is worth repeating that, for the first time in American history, the president, vice president, and secretary of state are all former energy company officials, and the only other U.S. president to come from the oil or energy business was George W. Bush's father. More specifically, both the president and vice president are former chief executives of Texas-based oil services companies with deep financial and political ties to the Middle East. Furthermore, both the president and secretary of state have more experience as oil executives than they do as public servants. Prior to his 2000 presidential election, George W. Bush had spent eleven years in the oil business and just six in government service. Similarly, before she became Bush's national security adviser, Condoleezza Rice had spent ten years on the board of directors of Chevron, but less than three working for the government. Of these three people, only Vice President Cheney can claim more experience in the government than as an oil company executive, with some thirty years as a federal employee versus five years as CEO and president of the world's largest oil services company, Halliburton.

The defense industry is also well represented in the Bush administration—particularly Lockheed Martin, which boasts no fewer then sixteen current and past executives and directors (including the vice president's wife) who have served in George W. Bush's administration. The construction industry also has a strong presence, most notably the Bechtel Corporation, with key Bechtel executives—including company CEO Riley Bechtel—situated in influential positions in the administration.

Chevron, Halliburton, Lockheed Martin, and Bechtel represent three key pillars of the Bush Agenda: oil, war, and building the infrastructure of corporate globalization. As detailed in the next chapters, not only have their past and present executives directly shaped the Bush Agenda, but the companies continue to profit from its implementation today.

For example, while Cheney was secretary of defense for George H.W. Bush, his Defense Department paid the Halliburton Corporation $9 million to study whether the military's logistics services should be privatized. After Halliburton determined that the services should be privatized, it was awarded the first privatization contract. Three years later, Cheney became CEO of Halliburton. From this position, he advocated for war against Iraq while his company simultaneously conducted at least $73 million of work for Hussein. As vice president, he has ushered the country into war against Iraq, while Halliburton has received U.S. government contracts in Iraq worth nearly $11 billion to perform services privatized under Cheney's watch. Halliburton's stocks have tripled since the Iraq invasion. Cheney is a stockholder of both Halliburton and Lockheed Martin shares.

Bruce Jackson, an executive at Lockheed Martin from 1992 to 2002, is described as "the nexus between the defense industry and the neoconservatives. He translates us to them, and them to us."[7] In 2000, while still at Lockheed Martin, Jackson authored the Republican Party foreign policy platform. In 2002, Jackson cofounded the Committee for the Liberation of Iraq which lobbied for the overthrow of Saddam Hussein—a policy that was ultimately followed by President George W. Bush. Lockheed Martin's stocks, like Halliburton's, have tripled since the invasion. The company has also seen an $11 billion increase in sales and contracts worth more than $5.6 million to perform work for the U.S. Air Force in Iraq.[8] Today, Jackson is a director of the Project for the New American Century.

George Shultz served as the president and director of the Bechtel Company from 1974 to 1982. After leaving Bechtel, he became President Ronald Reagan's secretary of state. At the State Department, he worked forcefully to increase economic ties between the United States and Iraq in general and specifically on behalf of an oil pipeline from Iraq to Jordan that Bechtel would build. After leaving the State Department, Shultz aggressively advocated for war against Iraq as the chairman of Jackson's Committee for the Liberation of Iraq. Today Shultz is back at Bechtel as a director, where he is enjoying the company's nearly $3 billion in Iraq reconstruction contracts. Shultz is also a former member of the board of the Chevron Corporation, and he played a large role in securing a seat on this board for Condoleezza Rice.

Condoleezza Rice was once George H.W. Bush's National Security Council director of Soviet and East European affairs, where she helped draft corporate globalization policy for the newly liberated Soviet Union. Next, she sat on the board of directors of Chevron, where she was the principal expert on Kazakhstan during a period in which the company moved aggressively into the Caspian region, thanks in large part to economic policies she helped to write. She also chaired Chevron's Committee on Public Policy when Chevron was one of the only U.S. oil companies with a contract to sell Iraqi oil under the United Nation's oil-for-food program with Hussein. Today, she is the secretary of state, as Chevron moves into position to take advantage of new radically opened Caspian, Afghani, and Iraqi oil sectors. Already, 2004 and 2005 marked the most profitable years in Chevron's 126-year history.

Oil is a common thread linking all four companies. Each owes its roots and a good deal (in some cases, all) of its longevity and fortunes to oil—from their mutual beginnings in the California oil boom, which was made possible by high tariffs on imported oil, to their more recent histories in Iraq. Each company traces its corporate history approximately one hundred years back to a pioneering founding father who discovered his calling at the dawn of the twentieth century in California. Each company has grown, thrived, and survived using a similar formula: a knack for mergers, acquisitions, and consolidations; a revolving door with the U.S. government, allowing the companies to both write and be rewarded by U.S. legislation; a mastery of international trade; an apparent tendency to ignore laws that do not serve the companies' interests; and a willingness, in the words of Dick Cheney, to "go where the oil is" regardless of the political, social, economic, or environmental impacts of the pursuit.

These histories reveal a pattern of destructive, even deadly, corporate behavior that strongly contradicts the claim that increased corporate globalization, particularly less government regulation of corporate activity, is in the public interest. In the previous chapter I posited the question, "What happens when foreign investment does come to a country?" The answer provided in this chapter is that without strong government regulation, and even in spite of it, the results can be disastrous.

IN THE BEGINNING, THERE WAS OIL

. . . that he may bring out of the earth oil to make a cheerful countenance.

—*Psalm 104:15*[9]

For thousands of years, oil has literally oozed out of the rocks of North America and of the Middle East: thus the name, petroleum—"rock oil." The many varied uses of petroleum are well documented in the Bible, *Stories from the Thousand and One Arabian Nights,* Native American histories, and from the inscriptions of the earliest civilizations. Industrial, medicinal, and household uses for petroleum—everything from paving roads, construction, illumination, and weapons of war to beautifying the skin—were discovered centuries ago. Native American tribes in California, such as the Yokuts, Chumash, Achomawi, and Maidu, used oil to glue and waterproof arrows, hatchets, knives, canoes, and baskets. They even used oil to create dice for gambling. As described by Edwin Black in his tome, *Banking on Baghdad: Inside Iraq's 7,000-Year History of War, Profit, and Conflict,* Babylon's ziggurats and its towers of Babel were built of bricks coated in oil. Sargon the Great and Moses the Prince are said to have been sent floating down the river as infants in cradles sealed with oil.

Just as history credits Columbus for "discovering" America, histories of oil in both North America and the Middle East give white men the credit for discovering the substance whenever and wherever they successfully laid down a pipe to pump it out of the ground. While Daniel Yergin's book, *The Prize,* provides the best and most important history of oil, it is nonetheless typical of this tendency: "Oil had, at last, been struck in Persia," when the British pumped it out of the ground on May 26, 1908. The time was 4:00 A.M., and the place was the Masjid-i-Suleiman No. 1 well, in present-day Iran. What the British and the European Americans can rightly be credited for is discovering methods for turning oil into a substance that could be extracted out of the ground in large enough quantities to service external markets—in other words, international trade.

Just as the British were successfully drilling for oil in Persia, California was in the midst of its great oil boom.

THE CALIFORNIA OIL BOOM

The "forest" that you see are oil derricks, all drilled in the 1920s.

—*Caption on photograph of Signal Hill, Long Beach, California, 1930*

Fortunately for people with an interest in tracing U.S. corporate histories, those companies that have been around for a century or more have a tendency to produce big glossy coffee table books and sophisticated websites complete with videos, interactive timelines, and old photographs that depict the trials and tribulations of their exciting creation stories. Each narrative combines equal parts truth, drama, and tall tale. The opening paragraph of Chevron's corporate history is a good example:

> Petroleum pioneers Demetrius Scofield and Frederick Taylor of the California Star Oil Works, a Chevron predecessor, took aim at Pico Canyon, a remote portion of the rugged Santa Susana Mountains. In September 1876, driller Alex Mentry succeeded in striking oil in Pico No. 4, despite rattlesnakes, wasps, mud and underbrush. The first successful oil well in California, Pico No. 4 launched California as an oil-producing state and demonstrated the spirit of innovation, ingenuity, optimism and risk-taking that has marked the company ever since.

The truth is that the first productive oil well in California was drilled in 1865, eleven years prior to Scofield and Taylor's discovery, and the company that found it was Union Matolle Company in California's Central Valley, not California Star Oil Works. In addition, Demetrius Scofield, in his three piece suits, starched white collars, and delicately groomed mustache, appears to have had more in common with President Bush's campaign finance "pioneers" (those who raised $100,000 and more for his 2000 campaign) than with pioneers of America's frontier age.

However, Scofield and Taylor did play a lead role in bringing the oil boom to California, and subsequently California's oil boom to the world, and in laying the groundwork for the other corporate founding fathers to emerge in California over the course of the next forty years. Before their company became Chevron, the California Star Oil Works was acquired by the Standard Oil Trust, which created Standard Oil of California, a company that remained one of the world's most important oil corporations for nearly a century.

Scofield and Taylor's discovery is officially considered part of a California "boomlet" that quickly fizzled at the end of the 1860s. It was followed by the California oil boom thirty years later. Neither oil boom would have taken place if California's early settlers had not learned from the Native tribes that the gooey substance seeping out of the ground made good sealant for wagon wheels. According to legend, Edward L. Doheny, an unsuccessful gold and silver prospector, and Charles A. Canfield, his mining partner, were in the downtown area of Los Angeles in 1892 when Doheny saw a cart with its wheels coated in tar. When he asked the owner of the cart where the substance came from, the man pointed to the northeast. Doheny and Canfield followed his finger and "discovered" the Los Angeles Oil Field near present day Dodger Stadium by "dipping the sharpened end of a eucalyptus tree" into the ground.[10] Theirs was the first well to strike oil in Southern California. The boom was on.

California's oil production grew rapidly and in staggering amounts—from less than half a million barrels per year in 1893 to twenty-four million in 1903. For the next twelve years, California led U.S. oil production to such a degree that in 1910, the state's oil output was larger than any foreign nation and accounted for 22 percent of total world production.[11]

In the 1920s, three new major oil fields also began producing. The largest was Signal Hill, whose development in 1921 involved unearthing the Sunnyside Cemetery on Willow Street. Relatives were "reimbursed" with royalty checks for oil drawn from the land that once held their deceased family members.

The success of California's oil boom is due in large part to the protection from foreign competition that the young industry received from the federal government. Like so many other business sectors in the United States and around the world, the oil industry's early development was made possible by tariffs. During the early years of the oil boom, there was widespread fear that the government would eliminate the tariffs placed on imported oil—allowing less expensive foreign oil to flow in and wipe out the newly budding domestic industry. Therefore, oil company executives came together and successfully lobbied the government not only to maintain the tariffs but to double their value. Protected from outside competition and able to develop, experiment, and grow, the oil industry blossomed. One could easily argue that none of the four companies profiled in this book would exist today without such industrial tariffs. These are the very same types of tariffs that the advocates of the Bush Agenda wish to eliminate all around the world today.

One by one, the corporate founding fathers staked their claims in California. In 1904, Warren Bechtel was drawn to San Francisco from Peabody, Kansas, to begin work in construction. In 1912, brothers Allan and Malcolm Loughhead used the wealth that they and their benefactors generated in the oil boom to found the Alco Hydro Aeroplane Company in Santa Barbara. The same year, the Loughhead brothers' future partner, Glenn Martin, got his airplane company off the ground in Los Angeles. Both companies benefited significantly from the oil industry's adoption of aerial surveillance and, of course, the abundance of fuel. In 1918, Erle Halliburton made his way from Memphis, Tennessee, to Los Angeles to work in the oil services industry.

CHEVRON CORPORATION

Chevron is the second largest oil company in the United States (after ExxonMobil), the sixth largest company in the country, the fifteenth largest in the world, and the fifty-second largest economy on the planet. In 2001, Chevron merged with longtime partner Texaco, and for four years went by the combined name "ChevronTexaco." When it purchased Texaco, Chevron also acquired the history of Texaco's founding father, "Buckskin Joe" Cullinan.

Buckskin Joe, the "rough-hewn forceful leader" of the Texas Fuel Company, stands in sharp contrast to Chevron's Scofield. As described by Daniel Yergin, Buckskin earned his name because "his aggressive, abrasive personality and his drive to get a job done reminded those who worked for him of the rough leather used for oil field gloves and shoes."

When Beaumont, Texas, became the home of the Texas oil boom on January 10, 1901, Buckskin Joe soon followed. The same year, he founded the Texas Fuel Company. In 1959, the company adopted the name Texaco. Four years after its 2001 merger with Chevron, Texaco's name was dropped from the corporate title, which reverted back to "Chevron" in May 2005.

An entire book could be devoted to the century-long series of mergers and acquisitions that led to the present day behemoth known as Chevron. However the most important transition may be the one most skimmed over in Chevron's own corporate history: the break-up of John D. Rockefeller's Standard Oil Trust in 1911.

If one person can be given credit for taking down one of the most powerful corporations in history, it is Ida Tarbell. In 2002, the U.S. Postal Service issued a stamp commemorating Tarbell exactly one hundred years after she published the first of her nineteen-part series, *History of the Standard Oil Company.* The stamp features a black-and-white photograph of Tarbell looking straight at the camera, in white blouse, bow tie, and a smart round hat atop her black hair. Tarbell is one of the most important journalists of the twentieth century, and certainly one of the century's most famous women. Her blistering two-year-long exposé revealed the destructive and illegal practices of both John D. Rockefeller and his Standard Oil Company and helped to bring about the monopoly's ultimate downfall.

Tarbell's series ran in *McClure* magazine from 1902 to 1904. As she explains in the preface to the collected edition, she wrote the series "in order that [*McClure's*] readers might have a clear and succinct notion of the process by which a particular industry passes from the control of the many to that of the few." Standard Oil was chosen "for obvious reasons. . . . It is the most perfectly developed trust in existence; that is, it satisfies most nearly the trust ideal of entire control of the commodity in which it deals." Tarbell brought the full weight of public knowledge to bear on the monopoly practices that Standard used to take over and control the U.S. oil industry. She saved some of her harshest criticism for Mr. Rockefeller himself: "Now, it takes time to secure and to keep that which the public has decided is not for the general good that you have. It takes time and caution to perfect anything which must be concealed. It takes time to crush men who are pursuing legitimate trade. But one of Mr. Rockefeller's most impressive characteristics is patience."

The public was not only convinced by Tarbell's words, it was moved to action. Demands that the government reign in Standard's power and break up the monopoly spread quickly across the country. An onslaught of suits and cases were brought against the company, including a 1906 federal suit brought by the Roosevelt administration charging Standard under the Sherman Antitrust Act of 1890. In 1909, the federal court ruled in favor of the government and ordered the dissolution of Standard Oil, and in May 1911, the U.S. Supreme Court upheld the ruling. Standard Oil was dissolved into some thirty-five smaller companies, dubbed "Baby Standards," including Standard Oil of California (SoCal).

While Chevron's history begins in California, it certainly does not end there. As the hunt for oil spread first beyond California's border and then America's, Chevron went with it. The quest has taken Chevron to virtually every corner of the globe, where it has worked with some of the world's most notorious and brutal governments and dictators, including Saddam Hussein. In fact, Chevron has maintained one of the longest and most profitable U.S. corporate histories with Iraq—a relationship that has grown significantly stronger since the 2003 invasion (as discussed in later chapters).

Chevron's vice president and general council, Charles James, served in the Departments of Justice of both George W. and George H.W. Bush. However, Chevron has the far more important honor of being the only oil company in the world that can count the former U.S. national security adviser and current secretary of state as a former director: Condoleezza Rice served on Chevron's board from 1991 to 2001. It is well known that Rice was so important to Chevron that it named an oil tanker in her honor. Less well publicized is that she served as head of the Committee on Public Policy and that her tenure coincides with a time when Chevron made disastrous policy decisions for which it is still being held to account. Chevron's activities in Nigeria provide a crucial example of these policy choices.

Bowoto v. ChevronTexaco

Nigeria is Africa's largest oil producer and the world's sixth largest exporter of oil. Like Iraq, Nigeria receives more than 95 percent of all government income from the sale of oil. And as in Iraq, foreign governments and oil companies have spent decades designing military, economic, and political policies to control that oil. Chevron has operated in Nigeria since the 1960s and today extracts some half a million barrels of oil per day.

Oil extraction is considered one of the most environmentally harmful industries in the world, and neither Nigeria nor Chevron offers an exception. For example, Chevron's operations include gas flaring from facilities located alongside populated villages. *Gas flaring* refers to burning flames of gas that shoot hundreds of feet into the sky from oil facilities, releasing a fiery cocktail of toxic substances into the air. Such flaring is virtually banned in the United States, but in Nigeria, the flares light the sky day and night, causing massive air pollution and severe illness.

Chevron has been accused of frequent oil spills from its operations and of using dredging techniques, which have salinized the fresh water supply, destroyed riverbeds and the natural ecosystem, and caused erosion such that several villages are on the verge of collapsing into the ocean. In other areas, silt causes the water level to drop too low for boats to travel or fish to survive, leaving many Nigerians without fresh water supplies or a means to support themselves.[12]

The demands of communities across Nigeria have been small in comparison to the damage done. Generally, they ask Chevron to clean up its mess, stop gas flaring, and to provide water wells, jobs, and electricity in the communities in which the company operates. Rarely are these demands met, and protests are thus as common in Nigeria as the gas flares.

In 1999, a case was brought against Chevron for the 1998 deaths of protestors at one of the company's Nigerian facilities and for the 1999 destruction of two Nigerian villages by soldiers in Chevron helicopters and boats. The charges were brought by Nigerian plaintiffs under the U.S. Alien Tort Claims Act, which permits suits in U.S. courts against individuals or corporations that commit international human rights violations anywhere in the world, if that person or corporation resides in or visits the United States. The case is set to go to trial in October 2006.

According to the plaintiffs' attorneys, a group of Nigerians went to Chevron's offshore Parabe Platform to demand that company officials meet with community elders on shore.[13] They demanded that Chevron clean up its operations and support the communities in which it operates by offering local people jobs and providing assistance with vital public services such as water, electricity, and schools. The local people claim not to have carried weapons, whereas Chevron's security officers were armed. After three days, when negotiations seemed to be progressing, the protestors agreed to leave the platform the next morning and informed Chevron of their intentions. In the early morning of May 28, 1998, the day the protestors planned to leave, Chevron-leased helicopters carrying soldiers and Chevron representatives approached the platform. The soldiers opened fire on the protestors before the helicopters landed. Two protestors were killed and others wounded, one of whom was shot and then bayoneted. According to plaintiffs, the leader of the protest was taken away by the soldiers and later tortured when he refused to sign a confession stating that he was "a pirate." Chevron representatives in Nigeria later admitted to reporters that Chevron's helicopters were used and that the company's head of security was on board during the attacks.[14]

Seven months later, on January 4, 1999, in apparent retaliation for the Parabe incident, plaintiffs claim that Chevron-leased helicopters carrying Chevron representatives flew over the fishing villages of Opia and Ikenyan and opened fire.[15] The helicopters were followed by Chevron-leased boats filled with soldiers who attacked the villages. The assaults left at least seven people dead and both villages almost completely burned to the ground. Many people were injured, and some remain missing to this day. Virtually everyone from these villages lost a home, boat, or other possessions in the fires.

Although Chevron does not deny that the killings and destruction took place, the company argues that its Nigerian subsidiary, not the parent company, is to blame and that the Nigerians were not peaceful, but were "kidnappers and extortionists who held 175 people hostage for three days while [Chevron Nigeria] vainly tried to negotiate with them." Chevron says the attacks on the villages were "in response to a violent insurrection" in which Chevron "was literally caught in the crossfire."[16]

Chevron has tried several arguments (all of which have failed) to have the case heard only in Nigeria. The most disturbing argument, in my mind, is the company's claim that the Nigerians cannot sue because the shooting of unarmed protestors who are trespassing does not violate international law. Rather, the company argues, such actions are only illegal under domestic Nigerian law and therefore should be argued in Nigerian courts against Chevron's Nigerian subsidiary. Fortunately, the court disagreed, finding that "torture and summary execution are two categories of conduct that unquestionably constitute a violation of most any norms of international law."[17]

In the most recent ruling, in March 2004, a federal judge in San Francisco found that Chevron can be held liable for acts of its Nigerian subsidiary on the grounds that it aided and abetted the subsidiary in its actions and that the subsidiary acted as Chevron's agent, although the judge dismissed the claims that Chevron could be held directly liable for these acts.

In 2003, undeterred by the lawsuits, protests, documentation of gross human rights abuses and environmental destruction, the World Bank rewarded Chevron with $75 million in investment insurance and $50 million in direct financial support for a gas pipeline that the company plans to build from Nigeria to Benin, Ghana, and Togo. The completion date is sometime in 2006.

Protest at "Slave" Terminal

The photograph published by the *Associated Press* on July 17, 2002, conveys an image that is nothing short of joyous: Some twenty-five women dancing, legs caught in midair, and arms lifted above their waists. They are smiling so broadly that you can see their teeth and almost hear their laughter. A blast of vibrant color jumps off African skirts, shirts, dresses, and headscarves in every conceivable hue. In stark contrast to this scene is the multimillion-dollar Chevron oil export terminal and dockyard sitting dormant directly behind them. The women are celebrating their victory after ten days of protest at the Chevron facility at Escravos terminal in the southern Delta of Nigeria. More than 650 women participated in the protest, some as old as ninety, and many carrying babies and toddlers. Together, they brought the facility, which normally exports half a million barrels of oil a day, to a complete standstill. They threatened to strip naked in a traditional gesture of shaming men if a satisfactory deal was not reached. Fortunately, it did not come to that. The women called an end to their protest when Chevron agreed, in writing, to meet their demands to hire twenty-five local villagers, build schools, and provide electricity and running water to their villages, which sit in the shadow of the terminal.

A giant gas flare greets visitors as they approach the Escravos terminal. *Escravos* means "slave" in Portuguese. The terminal earned its name in the seventeenth century as a slave collection point sitting at the southern tip of Nigeria on the Gulf of Guinea.[18] The protest at Escravos was just one of a wave of protests led by women across Nigeria in July and August of 2002 that were specifically targeted at Chevron facilities. These protests, in each instance peaceful, lasted until Chevron agreed to meet the women's demands for local jobs, clean drinking water, electricity, schools, and clinics. When their demands were met, the women left as peacefully as they arrived.

After forty-five years of operating in Nigeria, Dick Filgate, an executive at Chevron's San Ramon, California headquarters told the BBC, "We now have a different philosophy, and that is to do more with communities."[19] These are rare victories in a country where public protest against oil facilities are common, but the result is much more often violent repression and even murder, as in the Parabe Platform incident described earlier. What set these protests apart was that the participants were women.

Dennis Ojogor, an unemployed mechanical engineer, explained that the women's protest was a last-ditch effort after the men's protests had all ended in failure: "It used to be the men who did the protesting.

But the police and soldiers would use guns, chains and whips to drive us out," Ojogor said. "So now it is the women who have taken action. They cannot be touched."[20] Ojogor's sentiments were tragically premature.

A few weeks later, on August 8, 2002, over 3,000 Isekiri, Ijaw, and Urhobo women protested at the gates of the operational headquarters of both Chevron Nigeria Ltd. and Shell Petroleum Development Company. Chevron Nigeria is a joint venture between the state-owned Nigerian National Petroleum Company and Chevron; it is through this subsidiary that Chevron conducts all of its Nigerian operations. An Amnesty International report provides a detailed account of what took place. When the women arrived at the gates of the company headquarters, they were met by mobile police and soldiers who had emerged from inside of the corporate compound. The women sat down in front of the gates, peaceful and unarmed. According to Amnesty, with neither provocation nor warning, the security forces attacked. They threw tear gas, shot their guns in the air, and beat the women where they sat. Elisabeth Ebido, a forty-five-year-old Isekiri community leader, said she was beaten repeatedly with the back of a gun by four members of the security force leaving deep wounds on her arms and legs.

The women were brutally dispersed without being given the opportunity to speak with a single representative of either oil company or the government. In a letter to Amnesty International, Chevron Nigeria Ltd. wrote, "Chevron does not have a statement relating to the issue of a few women who came to the front of our office on 8 August. . . . We were not aware of the repression by a combined force of the mobile police and the army as you referenced."

Such stories are not limited to Chevron's activities. When the company purchased Texaco, it not only gained the legacy of "Buckskin Joe"—it also acquired Texaco's lawsuits.

Aguinda v. Chevron

Carmen Perez traveled for two days by bus from La Primavera in the Amazonian jungle of Ecuador to Quito, the nation's capital. In Quito, she boarded a plane and traveled for sixteen hours until she reached

San Francisco. From there, she rode in a minivan full of American human rights and environmental activists for one hour to San Ramon, a thoroughly unremarkable California suburb dotted with strip malls, industrial parks, and track housing. Carmen probably noticed the giant Whole Foods and Borders Bookstore before she saw the small, unmarked guardhouse on the other side of Bollinger Road. When she noticed that the guardhouse sat on "ChevronTexaco Way," she would have realized that she had finally reached her destination.

Chevron's headquarters is not visible from Bollinger Road, and there is no sign in front marking its existence. The building lies back beyond leafy trees, rolling green lawns, and a black iron fence that disappears off into the distance—giving the distinct impression that if you were to follow the fence you would end up right back where you began, in front of the small unmarked guardhouse. This is where I met Carmen. We stood on a freshly cut, watered lawn in front of the guardhouse at 7:00 A.M.

We were joined by Humberto Piaguaje, a leader of the Secoya people who had also made the long journey from Ecuador, as well as others from Burma, Mexico, and from across the Bay Area, including nearby Richmond and San Ramon, who had come to make the impacts of Chevron's activities on their community known. A few members of the local media were filming the police officers blocking the sidewalk to the distant and as yet unseen corporate headquarters. We were all gathered here for the 2005 Chevron annual shareholders meeting.

Carmen and Humberto had been specifically invited by a small group of shareholders who were offering a resolution at the meeting for an economic assessment of Chevron's activities in Ecuador. Carmen is the elected community delegate and U.S. spokesperson to the *Asemblea de Afectados por las Operaciones Petroleras de Chevron,* the Assembly of People Affected by the Petroleum Operations of Chevron. Her community is involved in a historic $6 billion lawsuit against Chevron brought in May 2003 by five indigenous groups and eighty Ecuadorian communities. The lawsuit, *Aguinda v. Chevron,* demands recompense for the destruction of their homes, health, environment, and livelihoods.[21]

Carmen is a health-care worker in a region believed to house the worst oil-related environmental disaster in the world. The amount of crude oil that has been dumped in and around her home in the Sucumbios Province is alleged to be roughly thirty times more than the amount spilled in the Exxon Valdez disaster off the coast of Alaska in 1989. Carmen is the president of the Amazon Health Foundation in an area where dozens of communities suffer from severe health crises, including shockingly high incidences of leukemia, lymphoma, cervical, stomach, larynx, liver, and bile duct cancer.[22] She is the vice president of the Parents' Association in an area where children are regularly born missing fingers, limbs, and internal organs. This is an area where an alleged 627 open toxic waste pits and antiquated waste facilities pollute rivers and streams that more than 30,000 people depend on for drinking, bathing, and fishing. She is the mother of six children, each of whom suffers from skin and throat infections from drinking and bathing in contaminated water.

Ecuador began receiving IMF loans in 1961. Although details of these loans have not been made public, it is safe to assume that Ecuador's status as an IMF loan recipient helped pave the way for Texaco's entrance into the country. Texaco received a concession for an area three times the size of Manhattan. From 1964 to 1992, the company built and operated oil exploration and production facilities in the northern region of the Ecuadorian Amazon. Indigenous communities were removed from their land to make way for the oil facilities, as were more than one million hectares of ancient rainforest. According to the suit, rather than install the standard environmental controls of the time for reinjecting toxic drilling waters back into the ground, Texaco dumped 18.5 billion gallons of toxic waste directly into the rainforest. The result is an exploding health crisis among the region's indigenous and farmer communities. Before Texaco came to Ecuador, Humberto's tribe, the Secoya, numbered in the thousands. Today, because of contamination, forest loss, and displacement, the tribe has dwindled to just 350 members.

When *New York Times* columnist Bob Herbert contacted Chevron in October 2005 to ask about the case, he was told by a spokesperson that "the billions of gallons of waste that was dumped 'wasn't necessarily toxic.' "[23] "We've done inspections," the spokesperson told Herbert. "We've done a deep scientific analysis, and that analysis has shown no harmful impacts from the operations. There just aren't any."

But the spokesman's contention, which is also Chevron's main defense in the case as it makes it way through Ecuadorian court, is looking more and more dubious. Chevron claims that Texaco remediated many of its former sites in the mid-1990s. However 98 percent of the water samples taken by Chevron's own scientists and presented as evidence at court from eighteen of its former well sites contained toxins at such high levels that they violate Ecuadorian laws—laws that are weaker than industry standards. For example, at "Sacha 6," a former Texaco well site reportedly remediated by the company in 1996, every one of the fourteen water samples presented by Chevron to the court was in violation of Ecuadorian law.[24] None of the claims made by the plaintiffs have been dismissed by the court and a decision in the trial is expected by early 2007.

A few weeks prior to Chevron's 2005 shareholders meeting, the *New York Times* published an article with the headline, "Big Oil's Burden of Too Much Cash." The article began, "Flush with cash, the world's giant oil companies find themselves in a paradoxical position—they are making more money than they can comfortably spend."[25] Now merged with Chevron, the company's profits had nearly doubled from $7.2 billion in 2003 to $13.3 billion in 2004. Chevron CEO David O'Reilly personally received over $10 million in total compensation for the year. Carmen had plenty of suggestions for how the company could spend that money.

Chevron's shareholders probably do not remember Carmen. Though they saw her at the meeting, they never heard about her journey or learned why she was there. Chevron CEO O'Reilly did not allow her to speak. Ten minutes before the scheduled close of the meeting, O'Reilly abruptly shut down the microphone and stopped the meeting before Carmen had a chance to begin. Denied the ability to speak to the shareholders and members of the board, Carmen spoke to us outside on Chevron's front lawn.

HALLIBURTON CORPORATION

Halliburton is the largest oil-and-gas services company in the world. It employs more than 100,000 people in over 120 countries, with over 7,000 clients. In 2002, Halliburton split into two parts: Kellogg Brown & Root (KBR), its engineering and construction arm, and Halliburton Energy Services Group, which performs virtually every conceivable service to the energy industry and provides nearly two-thirds of Halliburton's total revenue and more than 80 percent of its operating income.

Halliburton grew through mergers, acquisitions, and friends in high places, but its origins trace back to one "simple man, fiercely struggling to escape poverty, doggedly pursuing his piece of America's manifest destiny"—Erle Palmer Halliburton. Just as California reached the height of its oil boom, Erle Halliburton and his wife Vida arrived in Los Angeles. In 1918, Erle took a job driving a truck for the Perkins Oil Well Cementing Company, owned by Almond Perkins. Almond Perkins had patented a new method for oil well cementing. Erle liked the idea so much that he stole it and headed off for Texas to establish his own company, the New Method Oil Well Cementing Company. Perkins caught up with Halliburton and sued. Erle settled the suit but kept the method and made his early fortune from it. In 1924, Erle and Vida moved to Oklahoma and incorporated as the Halliburton Oil Well Cementing Company.[26]

The Texas oil boom that fueled Halliburton's rise also created the market for Brown & Root. In the early 1900s, brothers George and Herman Brown were building roads in Texas. Their financial backing came from Herman's brother-in-law, Dan Root. Dan had little involvement in the company, but his name stuck. In 1962, Halliburton purchased Brown & Root, allowing it to function as a nearly autonomous construction arm. Brown & Root became one of the nation's largest construction companies, due in large part to its exceptionally close and mutually beneficial relationship with former Texas senator and U.S. president, Lyndon Johnson. In fact, for most of its history, Brown & Root was considered the "Democrat's construction company," while Bechtel was the company of the Republicans.

While Dick Cheney, a solidly conservative Republican who was the CEO of Halliburton from 1995 to 2000, slowly changed the political bent of the company, he was not deterred from profiting quite handsomely from the Democratic administration of Bill Clinton. Cheney's success may ultimately lie in the fact that while the White House was controlled by a Democrat, the Congress, and even more important, the House Appropriations Committee, was controlled by Republicans. In addition, Cheney transformed Halliburton into a "virtual Pentagon," moving not only a sizeable amount of the federal budget but also a large number of Pentagon staff to the company. As retired Air Force Colonel Sam Gardiner once noted, Cheney "doesn't see the difference between public and private interest."[27] Cheney nearly doubled Halliburton's U.S. government contracts, from $1.2 billion in 1995 to $2.3 billion at the time of his departure. Cheney also increased the amount of U.S. taxpayer loans and loan guarantees Halliburton received from the U.S. Export-Import Bank and the Overseas Private Investment Corporation, which ballooned from $100 million in the five years before Cheney's arrival to about $1.5 billion on his watch.

Cheney brought several former Pentagon officials and friends of the Bush Family with him to Halliburton:

- **Joe Lopez**, a retired four-star admiral and former aide to Defense Secretary Cheney, came to Halliburton in 1999 and became senior vice president of government operations at KBR.
- **Dave Gribbin**, former assistant to Congressman Cheney and Cheney's Chief of Staff at the Pentagon, was Halliburton's vice president for government relations and its chief lobbyist from 1996 to 2000 when he returned to the White House with Vice President Cheney.
- **Charles Dominy**, a retired three-star general and a former commander at the U.S. Army Corps of Engineers, began working for Halliburton in 1995 and replaced Gribbin as Halliburton's chief lobbyist in 2001.

- **Lawrence Eagleburger**, former president of Kissinger Associates, George H.W. Bush's secretary of state, and a former board member of Dresser Industries, served on Halliburton's board from 1998 to 2003.
- **Ray Hunt** of Dallas-based Hunt Oil Company, who provided financial support to the presidential candidacies of both George Bushes and was appointed to George W. Bush's Foreign Intelligence Advisory Board, joined Halliburton's board in 1998 and continues to serve today.

In addition, Halliburton has benefited from the services of at least three men with important connections to the current Bush administration:

- **C.J. "Pete" Silas**, who served on George W. Bush's Transition Energy Advisory Team, joined Halliburton's board in 1993 and remained until 2005.
- **Kenneth Derr**, former Chevron CEO, joined the board in 2001 and remains today.
- **Kirk Van Tine**, is a recent addition to Halliburton's team as a registered lobbyist. He was George W. Bush's general counsel and then Deputy Secretary of the Department of Transportation and a former partner with Baker Botts Law Firm.

In 1998, Halliburton purchased Dresser Industries, adding more than just a "major provider of integrated services and project management for the oil industry" to its repertoire. This purchase also bought a key connection to the Bush family, as Kevin Phillips documents in *American Dynasty: Aristocracy, Fortune, and the Politics of Deceit in the House of Bush.* Former Senator Prescott Bush, George W. Bush's grandfather, ran the W.A. Harriman Company, which financed the reorganization of Dresser Industries in the late 1920s. Prescott Bush then represented W.A. Harriman on Dresser's board for more than twenty years, from 1930 to 1952. When George H.W. Bush needed a job in 1948, he went to Dresser. In 1988, Dresser Industries purchased

M.W. Kellogg, a company that creates technology for petroleum refining and petrochemical processing and performs engineering and construction services.

When Cheney purchased Dresser Industries, Halliburton inherited asbestos claims ultimately involving more than 400,000 people. In January 2005, Halliburton made a $4.7 billion payment to settle these cases. To protect its assets from the asbestos litigation, Halliburton reorganized by splitting its operations into Halliburton Energy Services Group and KBR, and then placed KBR under bankruptcy protection.

President George W. Bush and Vice President Cheney's relationship with Dresser Industries and its asbestos troubles might shed some light on the president's odd decision to use his 2005 State of the Union Address to make an appeal for the embattled asbestos industry. In the opening minutes of his speech, even before he mentioned the "purple fingers" of Iraqis who had just voted in national elections, President Bush told Congress, "Justice is distorted, and our economy is held back by irresponsible class actions and frivolous asbestos claims." The U.S. government classifies asbestos as a known carcinogen. It is banned in thirty countries and few people consider it a topic that wins votes in America's heartland. It seems the president may have been stumping for an old family ally, a company from which the vice president also continues to receive anywhere from $180,000 to $1 million annually in deferred compensation.

Halliburton in Nigeria

> The problem is that the good Lord didn't see fit to always put oil and gas resources where there are democratic governments.
>
> —*Dick Cheney, Halliburton CEO, 1996*[28]

Halliburton has no fewer than thirteen offices in Nigeria and actively participates in both oil and natural gas projects across the country. Chevron has regularly hired Halliburton to build its Nigerian oil facilities. For example, in April 2005, KBR agreed to build Chevron's Agbami offshore drilling facility, about 220 miles off Lagos in the Gulf of Guinea. Halliburton reports that the Agbami field is one of the ten largest discoveries of oil in the world in the past decade. The deal was signed with KBR rather than the company's Nigerian affiliate because the civilian government of President Olusegun Obasanjo had recently approved a decision by the Nigerian House of Representatives to ban Halliburton Energy Services Nigeria from bidding on all new contracts.

The ban was implemented in September 2004 after Halliburton Nigeria was found negligent for allowing two highly sensitive radioactive devices to disappear from its Nigerian operations, which led to widespread fears of a potential terrorist attack. The devices, used to take measurements in oil wells, mysteriously reemerged in Germany at a steel recycling plant in Bavaria. Halliburton refused to explain how the devices ended up in Bavaria and then would not return them to Nigeria, sending them instead to the United States.[29]

The Nigerian government also cited Halliburton's refusal to cooperate in a $180 million bribery investigation as a reason for implementing its ban on the company. In 2003, Halliburton admitted that its employees paid $2.4 million in bribes to Nigerian government officials in 2001 and 2002 to "obtain favorable tax treatment," according to the company's SEC filing.[30] Halliburton fired several of its employees as a result. The SEC is currently conducting its own investigation of the case.

The new bribery investigation is potentially even more damning to Halliburton not only because the amount of the alleged bribe is seventy-five times larger but also because Vice President Cheney may ultimately be tried in the case. The case involves Africa's largest construction project, the $12 billion Bonny Island liquefied natural gas complex off the coast of Nigeria. Halliburton's KBR is the lead company in a consortium working on the project under the name TSKJ. KBR's TSKJ partners are Technip SA of France, Snamprogetti SpA of Italy, and Japan Gasoline Corp. of Japan. TSKJ has won all the major contracts to build the Bonny Island complex, which will be among the largest plants of its kind in the world on its completion in 2007.

The U.S. Department of Justice, the SEC, a French magistrate, and the Nigerian government are each conducting criminal investigations into Halliburton's admission that its employees "discussed bribing public officials in Nigeria in order to secure a multibillion-dollar contract there" and "may" have actually paid $180 million in bribes to Nigerian government officials from 1995 to 2002 to win an $8.1 billion contract.[31]

The Nigerian House of Representatives voted in early September 2004 to summon Halliburton CEO David Lesar to respond to allegations that his company bribed government officials. The Nigerian parliament also issued a report accusing Halliburton of playing "hide-and-seek games" with Nigerian officials rather than assisting in the case.[32] The French magistrate, Reynaud Van Ruymbeke, has signaled his desire to subpoena Vice President Cheney as a witness in his investigation into the bribery charges.[33] Ruymbeke has the authority to charge Cheney under an international convention against "the corruption of foreign public officials in commercial negotiations" adopted in 1997 by the Organization for Economic Cooperation and Development (OECD) and implemented as law in France in 2000. Both the United States and France are members of the OECD and therefore the convention applies equally to both nations. Each investigation into the scandal remains ongoing.

In addition to facing bribery charges, Halliburton, like Chevron, has been accused by Nigerians of using murder to protect its oil interests. Only one Nigerian, however, has had his accusation recounted in the American media. Oronto Douglas is Nigeria's leading environmental and human rights lawyer and the founder of Nigeria's Earth Rights International. Among his many other credentials, Douglas served as a defense lawyer for famed human rights activist Ken Saro-Wiwa. Saro-Wiwa was ultimately hanged on November 10, 1995, with eight other Nigerian activists for their leadership roles in the nonviolent struggle of the Ogoni people against Shell Oil Corporation and the Nigerian military dictatorship of Sani Abacha. The "Ogoni Nine," as they are called, were found guilty of what have since been widely proven to be false charges of murder in a "kangaroo court." Though Douglas was arrested and tortured for his activism in the 1990s, he remains publicly active in his work on behalf of the people of Nigeria.

I was fortunate enough to meet and work with Douglas on several occasions and in as many countries. Douglas is short, boisterous, and kind, and he always seems to be smiling. Douglas' face is worn and scarred, making him appear older then his thirty-nine years. His energy, however, is that of a much younger man, and his optimism is that of someone who has learned the hard way that pessimism is a luxury.

In August 2000, Douglas told reporter Doug Ireland about the alleged killing of Gidikumo Sule, a young man from the Opuama village in the Niger Delta, at the hands of the Nigerian Mobile Police in July 1997. According to Douglas, Sule participated in the seizure of a Halliburton barge in protest of the company's failure to meet its own commitment to employ local youths. Douglas told Ireland that Sule was unarmed when he was killed. Douglas explained, "The Mobile Police are paid for by the oil companies, both under the military dictatorship of General Abacha we had then and the civilian dictatorship we have now [in 2000], and deploying them is always done at the oil companies' request. We call them the 'Kill and Go' squads, because they can kill and go away with no questions asked. At Opuama, the order to open fire was given by Halliburton officials. Their lives were not threatened . . ." Douglas told Ireland that he was also looking into incidents in which Halliburton officials ordered what he calls "brutal repression" of peaceful protesters near Warri and Gbaruamata leading to the "serious injury" of four people.[34]

Since speaking with Ireland, Douglas became the state information minister of the Obasanjo government—the first democratic government in Nigeria after sixteen years of corrupt and brutal dictators. Douglas has not pursued his accusations against Halliburton. For its part, Halliburton has never commented on the charges.

Halliburton in Iran

As comptroller of New York City, William Thompson is one of the most successful and noticeable advocates for a newly emerging form of activism—pension holder action. In total, U.S. state-run pension investments amount to approximately *$7 trillion.* The owners of this stock are increasingly using their influence to demand fundamental change in the behavior of U.S. corporations. William Thompson manages the largest state pension, New York City's $82 billion in pension funds—a sum greater than New York City's entire budget of $54 billion in 2005. When New York City's police and firefighters discovered that their pension funds were being used to support the government of Iran, they turned to Thompson for help.

Iran is one of the three points on President Bush's "Axis of Evil" and a country that the president has charged with supporting acts of terrorism against the United States. New York City's pensions are heavily invested in Halliburton, which, among other companies, has been working in Iran for over a decade and most likely in direct conflict with U.S. law.

U.S. companies are barred from doing business with Iran under the 1996 Iran-Libya Sanctions Act. However there is a loophole. The law does not ban foreign subsidiaries of U.S. corporations from working with Iran, as long as the subsidiaries do not employ U.S. citizens and are not simply a front for the parent company. The U.S. Department of Justice, a federal grand jury in Texas, and the SEC have each launched formal investigations to determine if Halliburton's work in Iran is in conflict with the Act.

In 1998, while he was CEO of Halliburton, Cheney personally lobbied the Senate, seeking special relief for Halliburton from the Iran-Libya Sanctions Act. Two years later, Cheney publicly argued that American companies should be allowed "to do the same thing that most other firms around the world are able to do now, and that is to be active in Iran. We're kept out of there primarily by our own government, which has made a decision that U.S. firms should not be allowed to invest."[35] If nothing else, Cheney's opposition to sanctions was consistent. A decade earlier, while serving in the House, Cheney twice voted to oppose sanctions against the apartheid government of South Africa.

While Cheney lobbied against sanctions, a Halliburton subsidiary was actively at work in Iran. *The Middle East Economic Digest* reported in 2000 that oil "industry suppliers such as Halliburton are already openly selling to Tehran through subsidiaries." In 2004, CBS News estimated that Halliburton sold about $40 million a year worth of oil field services to the Iranian government through its subsidiary, Halliburton Products and Services, Ltd., which is registered in the Cayman Islands.

In December 2004, Thompson submitted a shareholder's resolution calling on Halliburton to review and justify its Iranian operations. Halliburton tried to keep the resolution from ever reaching its shareholders, but the SEC denied its request, drawing even more attention to Halliburton's Iranian activities—including the attention of the CBS news program *60 Minutes.*

In January 2004, *60 Minutes* correspondent Lesley Stahl traveled to the Cayman Islands to visit Halliburton Products and Services. She found a mailbox, but no office or employees. She was told that mail for the subsidiary was rerouted to Halliburton headquarters in Houston. Halliburton responded that, while registered in the Cayman Islands, the subsidiary was run out of Dubai. So, *60 Minutes* went to Dubai, where they found the subsidiary sharing an office, a phone, and a fax line with none other than Halliburton—the parent company, directly under the control of executives in Houston.

A federal grand jury in the Southern District of Texas subsequently opened a criminal investigation and issued a subpoena requesting company documents. At the same time, Halliburton announced plans to start work with two Iranian companies, the Oriental Kish and Pars Oil and Gas Company, on a natural gas project in Southern Iran. This announcement brought even more opposition from shareholders, members of Congress, the public, and the media. Finally, in January 2005, citing business conditions in Iran, Halliburton announced it would not enter into any new contracts there. The announcement did not go far enough for William Thompson, who asked for and subsequently received a commitment from Halliburton in writing. On March 23, 2005, Halliburton submitted a letter to Thompson stating, "Halliburton will take appropriate corporate action to cause its subsidiaries to not bid for any new work in Iran." However, Halliburton has not agreed to end its current work in Iran, insisting that it will maintain "existing contracts and commitments which the subsidiaries have previously undertaken" there.

THE BECHTEL GROUP

> We are not in the construction and engineering business. We are in the business of making money.
>
> —*Steve Bechtel Sr., president, Bechtel Company*[36]

Today, Bechtel is the largest engineering and construction firm in the United States and one of the largest in the world, with more than 22,000 projects in 140 nations on all seven continents. Its projects include petroleum and chemical plants, nuclear power and weapons facilities, oil pipelines, mining and metal projects, water privatization, and more.

But back in 1891, "The Bechtel Group" had a better chance of becoming a pop music sensation than a construction company. At age nineteen, Warren Bechtel left his home and his parents in Peabody, Kansas, to hit the road with his slide trombone and a "ladies dance band." The group toured dance halls across the Midwest, but apparently with little success, because Warren returned home within a year to begin looking for steadier work. It would take another fifteen years, one wife, three sons, a move to San Francisco, and the Great San Francisco Earthquake of 1906 to jolt Warren into the business that would define his family for generations to come. After the quake, Warren purchased a Model 20 Marion, a steam shovel originally developed to dig the Panama Canal, splashed W.A. BECHTEL CO. across the cab, and proclaimed himself in business.[37]

The Bechtel Group was, is, and likely always will be a private family-owned and family-run business. It has already passed from father to son through four generations of Bechtels. When Warren died in 1933, control of the company went to Steve, his middle son. Steve passed it to Steve Jr., who turned it over to Riley—which brings us to today.

The War on Terror has been very good for the Bechtel men and their company. *Forbes* magazine estimated both Steve Jr. and Riley Bechtel's personal wealth in 2005 at $2.4 billion apiece. They tied as the 109th wealthiest men in the United States and the 258th wealthiest in the world. Company revenue has grown steadily over the past three years from $11.6 billion in 2002 to $16.3 billion in 2003 and to $17.4 billion in 2004.

Bechtel's connections to the current Bush administration include:

- **Riley Bechtel**, whom President Bush appointed in 2003 to the President's Export Council, which advises the president on international trade issues;
- **Ross Connelly**, an employee of Bechtel for over two decades, who has served for more than four years as president and CEO of the U.S. government's Overseas Private Investment Corporation, which provides U.S. taxpayer assistance to U.S. corporations operating overseas;
- **Retired General Jack Sheehan**, a senior Bechtel vice president who served on the Defense Policy Board advising Donald Rumsfeld; and
- **Daniel Chao**, a Bechtel vice president appointed in 2002 to the advisory committee for the U.S. government's Export-Import Bank, which provides loans, loan guarantees, and other U.S. taxpayer financing for U.S. companies operating abroad.

For more than one hundred years, Bechtel has made its home in San Francisco where, in recent decades, it has been the subject of constant public protests. These protests grew increasingly large and media grabbing as opposition to Bechtel's involvement in the Iraq War spread. In 2004, Bechtel announced that it was moving its headquarters to Bethesda, Maryland, in order to be closer to its largest client—the federal government. Almost two years later, however, while Bechtel has greatly enlarged its operations in Bethesda, its corporate headquarters remain in San Francisco.

Bechtel owes much of today's profits, as well as profits of yesterday, to its century-long connection to the Republican Party. That connection runs so deep that there is an entire book devoted to the topic, *Friends in High Places: The Bechtel Story* by Laton McCartney. Bechtel has worked hand in hand with the U.S. government to write policies that have ensured the company's growth and profits. A prime example is Bechtel's move into the nuclear industry.

Bechtel Goes Nuclear

> Nobody doubted that nuclear energy could work. The real question was, could anyone make a profit in it?
>
> —Bechtel 1898–1998: Building a Century[38]

Today, Bechtel is one of the world's largest purveyors of nuclear power, having a hand in the design or construction of forty-five nuclear power plants in twenty-two states in the United States and more than 150 plants worldwide. The company estimates that it has built 40 percent of the U.S. nuclear capacity and 50 percent of nuclear power plants in the developing world.

Bechtel played its part from the very beginning of the nuclear age. It built heavy water storage plants for the Manhattan Project, the program that developed the atomic bomb, and it helped build the "doomsday town," a full-sized model city built in the Nevada desert to test the impact of a nuclear bomb on a typical mid-sized American city. In 1949, Bechtel built the first nuclear reactor test station in Idaho Falls, Idaho.

Bechtel was most interested, however, in making nuclear power both commercial and profitable. Arguably, the world might not even have commercial nuclear power if it were not for Bechtel. One of Bechtel's first opponents in this endeavor was President Truman, who argued that nuclear energy "was too important a development to be made the subject of profit-seeking." Bechtel responded by supporting Dwight Eisenhower's successful 1953 presidential bid and was quickly rewarded for its efforts. In 1954, President Eisenhower passed the Atomic Energy Act, allowing the Atomic Energy Commission (AEC) to issue licenses to private companies to build and operate nuclear power stations. That same year, Bechtel founded the National Power Group to invest in and support research to demonstrate the potential for commercial uses of nuclear power. The group came with a $1 million membership fee. In 1957, President Eisenhower named John McCone head of the AEC. Twenty years earlier, Bechtel and McCone had founded the Bechtel-McCone Corporation, which in 1938 brought on Ralph Parsons, forming the Bechtel-McCone-Parsons Corporation, until the three eventually split amicably during and after World War II.

During McCone's confirmation hearing, Ralph Casey of the General Accounting Office discussed the profiteering of Bechtel-McCone and other companies during the war. Casey declared, "At no time in the history of American business, whether in wartime or in peacetime, have so many men made so much money with so little risk, and all at the expense of taxpayers, not only of this generation but of generations to come."[39]

McCone would prove to be an incredible ace in the hole for Bechtel. His first action as head of the AEC was to recommend that federal subsidies be paid to utilities for the construction of prototype nuclear plants. He then appointed three executives from Standard Oil of California to study his recommendation. Not surprisingly, their study called for a "vigorous [government supported] development program." Bechtel became one of the largest beneficiaries of this program. McCone also recommended that the federal government stop buying uranium from foreign sources and instead obtain it exclusively from U.S. companies. Finally, in 1969, President Nixon reversed two decades of U.S. policy by allowing commercial businesses to produce and sell enriched uranium, including plutonium. Weapons-grade nuclear material was now in private hands and for sale.

That same year, Bechtel built India's first commercial nuclear power plant in Tarapur, sixty miles north of Bombay. Clifford Beck, the director of the Government Liaison-Regulation Office of the U.S. Atomic Energy Commission, was dispatched to India in 1972 to report on the plant. Beck found a facility beset with fundamental structural and engineering problems. He reported "substantial" leaks in fuel shipments, haphazard storage of radioactive materials, and a high amount of local contamination.[40] Although the plant was originally designed for operation by only 250 workers at a time, Beck discovered that, in the three years since the facility had been completed, more than 1,300 employees had received their maximum allowable doses of radiation and had been replaced with new employees. In the 1976 maiden issue of *Mother Jones* magazine, Paul Jacobs interviewed Beck and reported that, when Beck returned to AEC headquarters in Maryland, he told his colleagues, "Tarapur is a prime candidate for disaster."

Jacobs also reported that further studies of Tarapur revealed radioactivity along the Arabian Sea's shoreline up to forty kilometers from the plant and in the bodies of local people who ate fish. Indian physicists charged Bechtel with building a faulty facility. In 1973, nearly one year after Beck's initial report, Bechtel sent one of its own engineers to investigate the plant. The engineer uncovered fundamental problems with both design and construction. The amount of waste generated exceeded the plant's capacity for proper disposal. The network of pipes, pumps, and valves leaked so badly that the engineer described it as "a sieve." A full 3,000 to 4,000 gallons of radioactive waste leaked daily from one section of the plant alone. Up to 1,500 workers had suffered extremely high radiation doses, and during one refueling outage in May 1973, 400 people suffered severe exposure.

The problems at Tarapur raised red flags all across the United States because Bechtel had produced all of its reactors in "cookie-cutter" fashion. A design flaw in one meant a design flaw in all. According to one report, Bechtel may in fact have "botched plant construction all over the United States."[41] For example, Bechtel's San Onofre nuclear generating station, which lies about 450 miles from my home, was installed *backward* and two miles from a fault line reportedly overdue for an earthquake. In the words of Mark Massara of the local Sierra Club, San Onofre is "an unequivocal environmental and economic disaster."[42]

Back in India, the Indian government used Bechtel's Tarapur plutonium to join the nuclear weapons club, setting off its first detonation in May 1974. This detonation led the U.S. government to regulate strictly the global distribution of uranium enrichment technology. However, after President Reagan took office in 1981 and appointed Bechtel's president, George Shultz, secretary of state and Kenneth Davis, Bechtel's vice president of nuclear development, undersecretary of energy, U. S. government policy was reversed and U.S. nuclear fuel and technology were once again made available worldwide. Bechtel quickly made lucrative deals with Japan, Brazil, China, Mexico, and Argentina, which included technology for making weapons-grade plutonium. Reagan, reportedly through Shultz's coaxing, reduced the restrictions for licensing international nuclear plant construction as well.[43] Bechtel was an immediate beneficiary, receiving generous U.S. government backing to bring commercial nuclear power to the world.

Bechtel in Kentucky

In 1997, the Bechtel-Jacobs Engineering Co., a joint venture between Bechtel and Jacobs Environmental Management Company, won a $2.5 billion five-year contract to manage cleanup operations in three government-owned uranium enrichment sites, including the Paducah, Kentucky, and Oak Ridge, Tennessee, Gaseous Diffusion Plants.

Under intense criticism of its work, Bechtel-Jacobs' Paducah and Portsmouth contracts were allowed to expire on May 31, 2005, and the company was replaced. In 1999, the U.S. Department of Energy conducted an investigation into Bechtel-Jacobs' performance at the plant, which produces enriched uranium for nuclear weapons, submarines, and commercial power plants. The agency concluded that "The current radiation protection program and some elements of worker safety programs do not exhibit the required levels of discipline and formality. . . . Further, there has been little progress in reducing or mitigating site hazards or sources of environmental contamination. Weaknesses in hazard controls are evident . . . oversight has not been sufficient, and communication with stakeholders and workers has not been comprehensive and responsive to stakeholder needs."[44]

Kentucky's congressional delegation expressed its concern. U.S. Senators Jim Bunning and Mitch McConnell, along with U.S. Congressman Ed Whitfield, told a local Kentucky newspaper that Bechtel-Jacobs was spending too much money "on paperwork and too little on cleanup."[45] Local residents were also worried. Ronald Lamb, who lived two miles from the Paducah facility, said of Bechtel, "My father died of cancer, my next door neighbor died with cancers behind both eyes. Seventeen people have died of cancer in the thirty houses on the next street and they are still studying what to do?"[46]

Criticism is not limited to Bechtel-Jacobs' negligence at the Paducah facility. On March 29, 2000, former Bechtel-Jacobs employee Pamela Gillis Watson told the U.S. Senate Government Affairs Committee, "My own experiences over the past two years as an employee of the Bechtel-Jacobs Company have solidified my belief that the Department of Energy and its contractors [Bechtel-Jacobs]—past and current—have perpetrated a massive betrayal of the workers, the public, and the environment."[47] Watson was an employee at the Oak Ridge, Tennessee, Gaseous Diffusion Plant for ten years. She told the Committee that "Bechtel-Jacobs' management has gutted medical services. . . . I was told that Bechtel Jacobs even laid off the K-25 occupational medicine physician, who was one of the only advocates for the sick workers."

In her testimony, Watson recounted that, in 1998, she began reporting a number of health and safety violations, instances of fraud, waste, and abuse, and information security concerns about the diffusion plant to Bechtel-Jacobs management. Watson explained that at first she was ignored. When she persisted, she was told by her supervisor to "lay low" because "management was going to try to get her." After she took her complaints directly to the Department of Energy, she was told by a senior manager that he could no longer "do anything to help" her because her complaints had created a stir among senior management. In October 1999, she filed another health and safety complaint with the Department of Energy and was fired the following month. Watson says that her concerns were proved correct when the complaint was judged valid by the company and then resolved years later.

After providing her testimony, Watson disappeared from the public eye. I tried in vain to reach her, but I was able to talk with a Bechtel-Jacobs spokesperson in Oak Ridge who told me that "the matter was resolved in 2000 to the mutual satisfaction of the parties. Ms. Watson withdrew her complaint."[48] Dennis Hill, Bechtel-Jacobs' Media Relations spokesman, told me that when its contract expires the company will not only leave Oak Ridge but also will cease to exist altogether—having been created for the sole purpose of administering its Gaseous Diffusion Plant contracts.

Bechtel v. the Cochabamba Water Warriors

Cochabamba is the third largest city in Bolivia with 850,000 residents. In late 1999, the World Bank required that Bolivia privatize Cochabamba's water in return for reduction of its debts. Bechtel—one of the top ten water privatization companies in the world—won the contract through a Bolivian subsidiary, *Aguas del Tunari.* Bechtel has interests in over two hundred water and wastewater treatment plants worldwide and has privatized systems in countries as diverse as Estonia, Ecuador, the Philippines, and Bulgaria.

Immediately after Bechtel took over the Cochabamba water system, and before any of the promised investments in infrastructure were made to improve or expand services, the company raised the price of water. The average water bill increased by 100 percent, while some families received bills that were a full 300 percent higher. Families earning the minimum wage of $60 per month suddenly faced water bills of $20 per month. As a result, many were simply forced to do without running water. Furthermore, not only did Bechtel increase prices, but the same law that privatized the water system also privatized any collected water, including rainwater collected in barrels and used for washing and irrigating plots of land and water collected via aqueducts used to irrigate farms.

In December 2000, I flew to Cochabamba as part of a delegation of international trade and globalization specialists. My hosts were Marcela and Oscar Olivera, brother and sister members of *La Coordinadora de Defensa del Agua y de la Vida,* the Coalition in Defense of Water and Life, a network of citizens and organizations that formed to demand fairness in the delivery of the city's water. They told me what happened next.

La Coordinadora first went to Bechtel and demanded a reduction in prices. Bechtel told them to talk to the government. Then they went to the government and were told to talk to the World Bank. They went to the World Bank and were told to talk to Bechtel. *La Coordinadora* held a public referendum to determine the best next steps. The majority of the people voted for the cancellation of the contract with Bechtel. When this demand was met with silence from government officials, the citizens went on a citywide strike, blocked roads, marched, and sat down in the streets. Bechtel waited while the Bolivian government defended Bechtel's right to privatize by sending armed military troops into the streets to disperse the crowds. At least one seventeen-year-old boy was shot and killed and hundreds more were injured, but the people persisted.

Marcela described how her brother was among those taken by the government from their homes, blindfolded, hands bound behind their backs, and placed onboard a helicopter. Fully expecting to be assassinated, they were relieved when the blindfolds were removed to find themselves instead in a prison camp in the jungle. I met indigenous women from rural Bolivia who had marched for miles into the city of Cochabamba to sit down in front of *Aguas Del Tunari* and the Bolivian government in protest. They sat for days and nights on end without moving. I saw a ninety-year-old woman who suffered deep flesh wounds from police beatings in response to her peaceful sit-in. These women told stories of the quiet support they received from the wealthy women living in the city who, under the cover of night, brought home-cooked meals, warm blankets, and water to help ensure the success of the protest.

On April 10, 2000, the Bolivian government canceled Bechtel's contract. Bechtel left but then immediately launched a $25 million lawsuit against the Bolivian government for profits that it claims it would have earned through the privatization of Cochabamba's water. It is suing through a foreign investment agreement much like the Multilateral Agreement on Investment, a bilateral investment treaty between Bolivia and the Netherlands. The United States does not have such a treaty with Bolivia, so Bechtel established a holding company in the Netherlands in order to enact the suit. The suit is still pending at the International Centre for the Settlement of Investment Disputes, which happens to be housed at none other than the World Bank.

The people of Cochabamba have established a new publicly run water system. The new company has had considerably greater success than both Bechtel's system and the one that predated it. While not perfect, it provides water more universally, equitably, and of a higher quality than Bechtel's effort.

LOCKHEED MARTIN CORPORATION

Following Lockheed Martin's corporate history is a little like taking a voyage through humankind's evolution from innocence to increasing self-destruction. Lockheed Martin began with a miraculous invention—one of the first airplanes in the world, designed and flown in 1909 by Glenn Martin in Santa Ana, California. Just six years later, Martin was costarring alongside silent movie legend Mary Pickford and his Model T plane in *A Girl of Yesterday*. For one year, he even merged his company with the Wright brothers to form Wright-Martin Aircraft, but returned to his own company in 1917. Martin Marietta was established in 1961 when the Glenn L. Martin Company merged with American-Marietta Corp., a supplier of building and road construction materials.

The Loughhead brothers, Allan and Malcolm, formed the Alco Hydro Aeroplane Company in 1912. One year later, they designed and flew their own wood and fabric seaplane over the San Francisco Bay. In 1926, they decided it was best to have a name customers understood, so they set up shop in Hollywood as the Lockheed Aircraft Company, using the phonetic spelling of their last name. From there, one can trace through these two companies just about every significant moment in flight history—from the Lindberghs to Amelia Earhart to the first walk on the moon. One can also trace virtually every major advance in weapons technology—from airplanes to fighter jets, ships to submarines, and ballistic to nuclear weapons.

Lockheed and Martin Marietta merged in 1995 to form Lockheed Martin, which has become the largest military contractor in the United States and the largest arms exporting company in the world. With more than 135,000 employees worldwide, Lockheed Martin produces an astonishing array of weapons, but its specialty is the fighter jet. In 2001, it was awarded the world's largest weapons contract to date, a $200 billion deal to build the Joint Strike Fighter, a "next-generation" combat jet "designed to set new standards for lethality," according to Lockheed. Lockheed describes its F/A-22 Raptor as the herald of the "era of U.S. air dominance against all ground and air-based threats." The company also makes the Patriot Advanced Capability 3 missile, which "uses hit-to-kill technology to destroy its targets, and was selected principally for the extremely high lethality the missile delivers" in the 2003 Iraq War (an earlier generation of the Patriot missile hit only one out of ten of its targets in the first Gulf War,[49] a fact rarely noted on the *CNN* newscasts watched by most Americans at the time). Lockheed designs the majority of the Terminal High Altitude Area Defense (THAAD) missile defense system, known as "Star Wars." The program has failed on virtually all of its trial runs at a cost of some $4 billion to U.S. taxpayers. Lockheed also produces a multiple warhead long-range nuclear missile called the Trident II Submarine Launched Ballistic Missile.

The War on Terror has been extremely profitable for Lockheed Martin. In the past five years, Lockheed's stocks have more than tripled in value while it's sales increased by almost $13 billion. Lockheed Martin spokesman Thomas Jurkowsky told the *New York Times* that the company's success came from the "changed geopolitical landscape" in which Lockheed Martin helped the Pentagon "meet the demands that have been placed on it by providing a broad range of advanced technologies and capabilities."[50] Lockheed's 2005 earnings were just over $37 billion, while in 2004, it earned nearly $35.5 billion. The Arms Trade Resource Center determined that almost 80 percent of Lockheed's 2004 earnings were paid for by U.S. taxpayers. In fact, if you paid your taxes last year, you effectively handed $159 of it over to Lockheed. Moreover, the Resource Center found that in

2002 Lockheed paid so few taxes that it was effectively taxed at just 7.7 percent, compared to the average American's tax rate of about 20 to 35 percent.[51]

To understand why Lockheed has had such dramatic success since 2000, it is useful to describe the company's umbilical attachment to the Bush administration. Lynne Cheney, the vice president's wife, served on Lockheed's board of directors from 1994 to 2000. Bruce Jackson, a former vice president of Lockheed, served in the administration of George H.W. Bush, drafted the 2000 Republican Party foreign policy platform, cofounded the Committee for the Liberation of Iraq, and is a current director of PNAC. Until he was elected Governor of Mississippi in 2003, Haley Barbour was a Lockheed lobbyist and the former chairman of the Republican National Committee. In addition to these three, at least thirteen of the company's current and past executives and directors have served in the administration of President George W. Bush:

- **E.C. Pete Aldridge Jr.** serves as chair of President Bush's Commission on the Implementation of the U.S. Space Exploration Vision. He served as the U.S. undersecretary of defense from May 2001 to May 2003, where he conducted major weapon system acquisitions and research. He currently sits on Lockheed Martin's board of directors.
- **Everet Beckner**, a former Lockheed Martin vice president, served as the deputy administrator for defense programs at the Department of Energy from February 2002 to April 2005.
- **James O. Ellis Jr.** retired in July 2004 as admiral and commander of U.S. Strategic Command, Offutt Air Force Base, Nebraska. He was responsible for the global command and control of U.S. strategic forces or USSTRATCOM. He currently sits on Lockheed Martin's board of directors.
- **Gordon England**, a former vice president of Lockheed Martin from 1993 to 1995, is the acting U.S. deputy defense secretary (awaiting Congressional approval) and current secretary of the navy.

- **Stephen J. Hadley** is the assistant to the president for national security affairs. From 1977 to 2001, Hadley was a partner at the law firm of Shea & Gardner, which represented Lockheed Martin.
- **Michael P. Jackson**, a former chief operating officer (COO) of Lockheed Martin, is the deputy secretary of the U.S. Department of Homeland Security.
- **Norman Mineta**, a former vice president at Lockheed Martin from 1995 to 2000, is the U.S. secretary of transportation. He helped create the Transportation Security Administration, which has awarded at least seven contracts to Lockheed Martin worth some $626 million.
- **Anthony Principi**, a former senior vice president of Lockheed Martin, was named by President Bush as chairman of the Defense Base Closure and Realignment Commission. He served as secretary of veterans affairs from 2001 to 2005.
- **Otto Reich**, former Lockheed Martin lobbyist, was secretary of state for western hemisphere affairs from 2003 to 2004.
- **James M. Loy**, a current member of Lockheed Martin's board of directors, served as the first deputy secretary of homeland security from 2003 until 2005. Prior to this position, Loy served in a variety of capacities at the Transportation Security Administration, including undersecretary for security.
- **Joseph Ralston**, a current member of Lockheed Martin's board of directors, completed a thirty-seven-year career in the U.S. Air Force in 2003, culminating with the position of commander of U.S. European Command and Supreme Allied Commander Europe, NATO. In this post, Ralston oversaw training assistance for Operation Enduring Freedom and commanded daily missions in the War on Terrorism.
- **Peter Teets** was the undersecretary of the U.S. Air Force from 2001 until 2005. Prior to his resignation, Teets served briefly as acting secretary of the air force. He began working at Martin Marietta in 1963 and served as COO of Lockheed Martin from 1997 to 1999.

- **Michael Wynne**, a former vice president of Lockheed Martin from 1994 to 1997, is undersecretary of defense for acquisition, technology, and logistics.

Many of these men have served in acquisition, design, and research roles for both the federal government and Lockheed Martin, leading directly to contracts and profits for the company.

10,000 Kentucky Workers v. Lockheed Martin

Jim and Terri Hutto made a disturbing discovery in the summer of 1999. Someone had buried three radioactive barrels in their backyard. A Department of Emergency Services worker who came to test the barrels found high levels of radioactive contamination not only on the drums but also on Jim's hands and shoes.[52] Jim and Terri live in Paducah, Kentucky, just three miles from the Paducah Gaseous Diffusion Plant. Senator Chuck Grassley has called the plant "the site of some of the worst environmental contamination anywhere in the country."[53]

Martin Marietta operated the Paducah plant for fourteen years before Lockheed Martin took over from 1984 to 1999. Workers at the plant made a discovery similar to Jim and Terri's when, one day, they noticed a tarlike substance in the tracks left by a truck near the plant. After poking around, they discovered more ooze, which led them to "a burial ground for radioactive debris just north of the plant. The waste was barely hidden under a thin layer of soil in a grassy lot."[54] Two lawsuits have been brought against Lockheed Martin for its operation of the facility. The first involves 10,000 former workers of the plant in a $10 billion case against Lockheed Martin and each of the additional companies that ran the plant during its forty-seven years in operation. The companies have been charged with "egregious health physics violations involving workers" at the facility.

The second lawsuit is against Lockheed Martin alone. The case was originally brought by the Natural Resources Defense Council (NRDC), an NRDC scientist, and three Paducah plant employees. After the case grew to $1 billion, which made it one of the nation's largest environmental whistleblower lawsuits in history, the NRDC retained an outside firm, Egan & Associates, P.C. The case was brought in 1999 under the whistleblower provisions of the False Claims Act, which allows private parties to sue on behalf of the federal government if they believe the defendant submitted false claims for federal funds. In this case, Egan & Associates argues that Lockheed obtained approximately $1 billion from the U.S. government fraudulently, because it disposed of radioactive waste illegally and violated federal regulations concerning worker health and safety.

The reports of worker exposure to radioactive material are shocking. For example, workers from the plant told the NRDC that they worked in "a black fog, breathing in visible clouds of uranium and plutonium contaminated dust."[55] The contaminated dust was so thick that it coated the floor, their skin, and even their teeth. Al Puckett, a retired Paducah worker, told the *Washington Post* in 1999 that workers would brush black powder or green uranium dust off their food at lunchtime: "They told us you could eat this stuff and it wouldn't hurt you." To dramatize the point, he said, "Some supervisors 'salted' their bread with green uranium dust." At the end of the day, the dust went home with the workers and "we frequently discovered that our bed linens would be green or black in the morning, from dust that apparently absorbed into our skin," Jenkins said.[56]

Jenkins' description is from the 1970s, before Lockheed's time. While conditions certainly improved in the 1980s, more stringent rules were not formally adopted until 1989. Plaintiffs in the case say that conditions did not improve until the 1990s, and that even then they did not improve nearly enough.

According to the *Washington Post,* Charles Deuschle, a health physics technician who began working at Paducah in 1992 and is an employee in the lawsuit against Lockheed, said that surveys discovered radioactive contamination in the plant's cafeteria in the 1990s: "I saw conditions that would never have been tolerated in any other nuclear location where I have worked." Internal plant surveys included in the suit found high levels of radiation on street surfaces, manhole covers, loading docks, and in locker rooms as recently as 1996. Workers were not even told about the risks associated with plutonium at the plant until around 1990, when Lockheed (then Martin Marietta) officials held a meeting with union officials. However, one union member told the *Post* that the company's focus was exclusively on cleanup, not on worker health and safety.

The Department of Energy (DOE) conducted an audit of safety practices at the plant in 1990 and found scores of deficiencies in radiation monitoring and worker protection. According to the *Post*, the audit team found that "Paducah failed to properly monitor radiation to workers' internal organs—even though plant managers had been repeatedly warned to do so. Radiation-measuring equipment was either missing or not properly calibrated, the report said, and workers weren't being tested for the kinds of radiation known to exist at Paducah."

In 1988, a county health inspector found technetium and chemical carcinogens from the plant in a farmer's well. Lockheed posted creeks and ditches with warning signs in the early 1990s, but the signs did not refer to plutonium or radioactivity. Instead, some warned of possible contamination with cancer-causing chemicals; others merely cautioned against eating local fish. Finally, the 1990 DOE audit cited inadequate controls over waste disposal and a system for tracking contamination that amounted to "word of mouth" between managers.[57]

Thus, charges against Lockheed in the second lawsuit include illegal storage, dumping, and disposal of radioactive waste, false reporting of plutonium and other contamination, and illegal placement of radioactive metals into commerce. The suit says that contaminated waste streamed out of the plant for years, exposing workers to dangerous levels of radiation. Some waste was even dumped in woods and abandoned buildings in a nearby state wildlife area.

The George W. Bush administration sat on the case against Lockheed for three full years. Finally, in both anger and frustration, Senator Chuck Grassley of Iowa wrote to Energy Secretary Spencer Abraham in August 2002 to demand an explanation. The senator made clear his concern that the reason why the case was collecting dust was because of the close relationship between the Bush administration and Lockheed

Martin; he demanded to know "whether any relationships with Lockheed Martin may be slowing, or stalling, your department's pace in reaching a resolution in this matter expeditiously." He pointed out that "the previous Energy Secretary personally apologized to the Paducah workers, so why the delay?"

One year later, the Energy Department agreed to move the case forward, allowing the Justice Department to join the case against Lockheed on the charges that the company submitted false statements to the Energy Department and government regulators regarding radioactive waste; that wastes were improperly disposed of in landfills at the Paducah plant; and that Lockheed's storage and disposal of the wastes violated the Resource Conservation and Recovery Act. However, the Bush administration would not intervene on the allegations that Lockheed improperly exposed workers to radiation hazards. Both cases are pending.

Lockheed Martin Faces Antiwar Protests in Sunnyvale, California

They started to arrive at 5:30 A.M. on April 22, 2003. On this gray, cold, rainy Northern California morning, the signs welcoming them to "Sunnyvale" only added insult to injury. But the U.S. military war veterans, former Lockheed Martin employees, students, scientists, and concerned citizens were not easily dissuaded. At the appointed time, they fanned out across the three entrances and sat down in the roads. The Lockheed Martin facility in Sunnyvale, California, was to be shut down for the day. Over the course of the day, some 600 people would participate in the peaceful protest. Shahrzad Rose Roome told reporters, "We are shutting down Lockheed Martin in order to stop the production and use of illegal weapons of mass destruction. Weapons being used against the Iraqi people in an unjust, illegal, and amoral U.S. invasion. To stop the war, we must stop the military-industrial complex driving the war."[58]

The protest was led by a self-appointed citizens weapons inspection team, who announced that they "were unable to find weapons of mass destruction in Iraq, but we did find them right here in Sunnyvale." The members of the informal team explained that the "Sunnyvale facility produces the Trident II nuclear missile and the Star Wars Ballistic Missile Defense System." In addition, the team noted Lockheed's manufacture of the key tactical strike weapons used in the 2003 war against Iraq, including air to surface and cruise missiles such as the AGM-142 and HAVE LITE air to ground missiles, AUP and BLU-109 conventional "bunker buster" missiles, and the JASSM long range conventional air to ground missiles.

Local residents expressed outrage against the toxic contamination that Lockheed had brought to Sunnyvale. They explained that Lockheed was found guilty by the California State Health Services Department of illegal storage and treatment of hazardous wastes at the Sunnyvale location, including toxic leaks and removal of hazardous waste pipelines, for which the company was forced to pay $1.3 million in fines. They also discussed reports of massive water contamination reaching storm drains and toxic plumes in shallow and deep groundwater under the site.

The protestors held "die-ins," lying across the road to create a visual representation of the causalities caused by Lockheed's weapons of war—deaths which are rarely, if ever, shown on television. The Brass Liberation Orchestra played music while participants handed out literature that explained the protest to Lockheed Martin employees and Sunnyvale residents. The day ended when some fifty of the protestors blocking the roads were arrested and the organizers of the event, the Bay Area's Direct Action to Stop the War, knew that their message had been heard: In order to stop the war, the war machine would have to be "unplugged at its roots."

~

Understanding the Bush Agenda requires understanding the corporations behind and within it. Bechtel, Chevron, Halliburton, and Lockheed Martin are intimately entwined with the Bush administration and its policies. Like the members of the Bush administration, the companies owe all or much of their fortunes and longevity to the oil industry. However, whether operating in Nigeria, Ecuador, India, Bolivia, Kentucky, California, or elsewhere, the histories and activities of each company proves that, while corporate globalization policy benefits some, it is extremely costly for countless numbers of people and communities in the United States and around the world.

Of course, the influence of each company is not limited to the current Bush administration, or even to the U.S. government. The histories of Bechtel, Chevron, Halliburton, and Lockheed Martin in the Middle East generally and Iraq in particular, are the focus of the following chapter. Each company has worked in or with Iraq for decades, successfully lobbying the Reagan and George H. W. Bush administrations to expand economic relations with Iraq and the regime of Saddam Hussein. Executives with each company then joined with the George W. Bush administration in the drumbeat for war against Hussein and the 2003 invasion of Iraq, and have since profited greatly from the invasion.

FIVE

"A MUTUAL SEDUCTION"

TURNING TOWARD IRAQ

Iraq possesses huge reserves of oil and gas—reserves I'd love Chevron to have access to.

—*Kenneth T. Derr, CEO, Chevron, 1998*[1]

We hope Iraq will be the first domino and that Libya and Iran will follow. We don't like being kept out of markets because it gives our competitors an unfair advantage.

—*John Gibson, chief executive, Halliburton Energy Service Group, 2003*[2]

Just as California's oil boom was giving birth to Bechtel, Chevron, Halliburton, and Lockheed Martin, another great twentieth-century oil boom had begun on the other side of the world. Within a few decades, all four companies made their way to the Middle East, either working directly in the oil industry or, as in the case of Lockheed Martin, because of it.

Most significant to the Bush Agenda is the involvement of these corporations in Iraq. Over the course of several administrations, each company has used its influence in the U.S. government first to increase economic engagement with Iraq and then, when Saddam Hussein no longer played ball, to advocate for war against Iraq. Along the way, the corporations' most active allies were the administrations of Ronald Reagan, George H. W. Bush, and George W. Bush.

One constant in U.S. economic engagement with Iraq over the past twenty-five years has been the ability of U.S. corporations to influence U.S. policy to their own extreme benefit. In Iraq, many of these corporations helped to arm Hussein, lobbied for war against him, and are now profiting from his removal. In the 1990s and in the lead-up to and following the March 2003 invasion of Iraq, Bechtel, Chevron, Halliburton, and Lockheed Martin were part of a chorus of corporations desiring increased and more secure access to Iraqi profits. Their spokespeople began to advocate loudly for an invasion of Iraq and the ouster of Saddam Hussein.

The *Pax Americana* brotherhood of the Bush administration would do well to read up on not only their Roman history but also their Iraqi history—for it is a story of overthrown empires and ejected outside interests, with the worst repercussions felt by those intent on walking away with the nation's oil wealth. Iraq's history is intimately intertwined with oil, power, and war.

THE PURSUIT OF OIL IN THE MIDDLE EAST

U.S. corporations were vying for Iraqi oil before Iraq even existed as a country. They would first have to overcome several obstacles, however, before finally entering the region in 1912. The first obstacle was the Ottoman Empire. Beginning in 1638, the provinces of Basra and Baghdad were under Ottoman control. But after World War I, the British ousted the Ottomans and laid a new imperial claim on the region—and to the victors go the spoils of war. The British acquired a mandate over Mesopotamia, out of which they fashioned a country: Iraq. The British imposed their own model of government on the new country, instituting a constitutional monarchy. They then chose a royal family over whom they could maintain both military and economic control. As Iraqi scholar Dilip Hiro explains in *Iraq: In the Eye of the Storm,* "Political stability in the area was required not only by the prospect for oil but also for the defense of the Persian Gulf. . . . The British did not want to rule the region directly; that would cost too much. Rather what Churchill, then the head of the Colonial Office, wanted was an Arab government, with a constitutional monarch, that would be 'supported' by Britain."

The U.S. government and its oil companies were excluded from the meetings where the oil was divvied up because, in the words of one British oil historian, "she had not fought in the Middle East nor had she declared war on Turkey."[3] Their exclusion, however, led the Americans to "whip up a storm of public indignation. As usual the British pusillanimously gave way to the Americans and a new 'shareholding' agreement was reached." American oil companies were given a small stake in the newly established Iraq Petroleum Company (IPC). The British controlled 50 percent of the IPC; the French 24 percent; Calouste Gulbenkian—"Mr. Five Percent," the leading dealmaker between the corporations of the West and the governments of the Middle East—got his standard cut; and the four American companies working as the Near East Development Corporation got exactly 16.666 percent to divvy up between them. The U.S. companies were Standard Oil of New Jersey (25%) and Standard Oil of New York (25%), which later became ExxonMobil; Gulf Oil Corporation (16.666%) and Standard of Indiana (16.666%), which together became BP; and the Atlantic Refining Company (16.666 %), which later became Sunoco.

The Iraqis, however, wanted the British out. In 1932, in a situation remarkably similar to that of present-day Iraq and the United States, the British granted Iraq *nominal* independence while British troops remained stationed in the country. British officials maintained posts in all levels of the Iraqi government, and both the British government and British companies exercised control over key sectors of the Iraqi economy.

At the same time, Chevron and Bechtel were getting their feet wet in the Middle East in Bahrain and Saudi Arabia. The same year that Iraq declared independence from the British, American oil companies did the same in Saudi Arabia. The first American oil company to break the British monopoly over Mid-East oil was Standard Oil of California (Chevron), when it struck oil in Bahrain in 1932. One year later, it was awarded the big prize, an exclusive fifty-year concession with Saudi Arabia. A "concession" is when a company obtains the rights from a government to explore for, own, and produce oil in a given area. The Saudis had so much oil that Standard of California needed help with its concession, so it partnered with Texas Oil Company (Texaco) on global marketing and with Bechtel on infrastructure.

Both Brown & Root and Lockheed began operating in the Middle East in the 1950s. Brown & Root sold oil well services to Saudi Arabia, Libya, and Iran, while Lockheed sold fighter jets. Lockheed Aircraft Corporation produced 715 C-130 Hercules transport planes for military services in several countries, including Iran and Pakistan, from 1955 to 1964.[4] These sales marked the beginning of a long and highly profitable relationship between Lockheed and Iran that continued for years until coming to an abrupt halt with the 1979 Iranian revolution. Thereupon Lockheed moved along and sold its wares to Iraq, Syria, and Saudi Arabia.

THE DEMAND FOR OIL SOVEREIGNTY

Independence in Iraq is marked on July 14, 1958, the day the royal family was murdered, the British were evicted, and the republican government of General Abdul Karim Qasim was established. "Liberation Monument" in Baghdad's Liberation Square commemorates this day. Qasim was immediately unpopular among Western governments and companies, and not just because of his revolution or his subsequent turn toward Moscow. Qasim really upset the West by taking control of Iraq's oil sector, encouraging other governments to do the same, and hosting the founding meeting of the Organization of Petroleum Exporting Countries (OPEC). I discussed OPEC and the 1970s oil crises in chapter 3, but I revisit the topics in this chapter to provide an Iraqi historical perspective.

Qasim set aside 70 percent of Iraq's oil revenue for development and established an independent development board to oversee the funds in 1959. In September of the following year, he convened a meeting in Baghdad with representatives of Iran, Kuwait, Saudi Arabia, and Venezuela, who collectively represented 80 percent of the world's crude oil exports. The result was an organization designed to unify the activities of oil exporting countries, just as the oil companies worked together to coordinate their activities.

OPEC was created as a counterweight to the long-standing hegemony of the so-called Seven Sisters. The name was coined by Italy's Enrico Mattei, who argued that the "Sette Sorelle," the world's seven largest oil companies, worked in such close association and through so many multiple joint ventures that they were, in fact, a cartel. Acting in concert, the companies set the terms that the exporting countries, which operated independently and often in conflict which one another, were forced to follow. Enrico Mattei's characterization was embraced the world over, and the Seven Sisters nomenclature stuck. The Sisters included the four members of Aramco, the Arabian American Oil Company: Exxon, Mobil, Chevron, and Texaco. It also included Gulf Oil, Royal Dutch/Shell, and British Petroleum. Through an array of mergers and acquisitions, the seven became just four by 2001: ExxonMobil, Chevron, Shell, and BP.

Qasim took on the companies and all but nationalized Iraq's oil. In 1961, he passed Public Law 80, which removed 99.5 percent of the Iraq Petroleum Company's concession territory. Then he established the new government-owned oil company, the Iraq National Oil Company.[5]

That same year, Brown & Root was commissioned by the Iraq Petroleum Company to lay two underwater pipelines from the Iraqi mainland to a new supertanker loading dock in the Gulf. According to Brown & Root's coffee table history book, *Offshore Pioneers,* "150 Brownbuilder welders, crewmen, and technicians" went to the Persian Gulf, "where they remained for several years on other marine pipeline jobs." The company thus established its presence in the Middle East, "where important contacts were made with other oil companies, such as Gulf Oil and British Petroleum."

Brown & Root's longevity in Iraq outlasted Qasim's. The CIA joined with the Iraqi Ba'ath Party in a successful military coup to overthrow Qasim in 1963.[6] Roger Morris, a White House National Security Council staff member at the time, wrote of the Iraq coup, "As in Iran in 1953, it was mostly American money and even American involvement on the ground" that led to the coup.[7] The newly established Iraqi government strengthened its ties with the West, particularly the United States. Public Law 80, Qasim's oil law, was never fully enforced, allowing the U.S. oil companies to retain much of their concession territory. However, the new government was highly unpopular with the people of Iraq and was soon replaced with a government that was less friendly to the West. With each successive Iraqi government, Iraq cut back its relations with the West until it ended all economic and diplomatic relations with the U.S. government in 1967, in retaliation to America's support of Israel in the Six-Day Arab-Israeli War.

Meanwhile, a larger shift was taking place among Third World countries—a demand for equality and redistribution of wealth and power. The world's increasing demand for oil, combined with the fact that the United States hit its oil production peak in 1970, marked a change in the global balance of power. Third World countries rejected the dominant oil concession system by which oil companies had been granted exclusive rights to explore for, own, and produce oil. Instead, they demanded greater ownership over their oil through participation agreements in which only partial ownership was granted to the oil companies and through full nationalization of their entire oil sectors. Libya nationalized its oil sector in 1971, and one year later, Iraq followed suit, nationalizing the Iraq Petroleum Company's remaining concession.

THE 1970s OIL CRISIS—TAKE ONE

The shift in power to the oil-exporting countries and away from the importers fully materialized in response to the 1973 Arab-Israeli war. Once again, the U.S. government backed Israel and the oil-exporting governments retaliated. In May 1973, King Faisal of Saudi Arabia told the American press, "America's complete support for Zionism and against the Arabs makes it extremely difficult for us to continue to supply the United States with oil, or even to remain friends with the United States." The U.S. oil companies, afraid of losing their access, chose sides. Texaco, Chevron, and Mobil went public and demanded that the U.S. government change its Middle East policy to oppose Israel.[8]

The Arab-Israeli war began on October 6, 1973. The next day, Iraq nationalized Exxon and Mobil's shares in the Basra Petroleum Company. One week later, OPEC oil ministers agreed to use the "oil weapon" by cutting exports and recommending an embargo against unfriendly states. Iraq led a successful charge for a full oil embargo against the United States by all of the Arab nations of OPEC. In December, OPEC set the price of oil at $11.65 per barrel ($51.44 in 2005 dollars), a *468 percent* increase over the price in 1970.[9] The embargo ended on March 18, 1974, with the cooling off of active hostilities between the Arab nations and Israel.

In the United States, this era is often referred to as the "1970s Oil Crisis." The Middle East was in turmoil, oil prices soared, U.S. consumers suffered, and U.S. oil companies made out like bandits. Rather than accept lower profits, American oil companies chose to raise oil prices and force all of the sacrifices onto the American public. The public, however, chose to fight back.

From the start of the oil crisis in October 1973 to February 1975, prices for gasoline in the United States increased 30 percent and for home heating by more than 40 percent. At the same time, oil company profits skyrocketed. The U.S. Treasury Department found that the twenty-two largest U.S. oil companies averaged higher earnings in 1973 than in any of the preceding ten years. In 1974, Standard Oil Company of Indiana's profits were up 90 percent; Exxon's were up 29 percent; Mobil's were up 22 percent, and Texaco's were up 23 percent.[10]

While the oil companies raked in their profits, the U.S. economy was in peril. In January 1974, Ford Motor Company laid off 62,000 workers and said more layoffs were on the way because of a sales slump induced by fuel shortages. A member of the Oil, Chemical, and Atomic Workers International Union told the *New York Times*, "If the oil companies have all that money to spend on executive raises, outside investments and political contributions, they must be making too much money."[11] Similarly, a February 3, 1975, story with the headline, "Attacks on Oil Industry Grow Fiercer," reported on the growing anger of people across American who "turned down their thermostats and reduced their speed on the highways to save costly fuel [and who] have focused their outrage on the national large producers and refiners."[12] There were no fewer than ten major lawsuits leveled against the largest oil companies during this time. Most of them, however, were ridden out by the companies until the plaintiffs were forced to drop the cases from lack of time and/or resources.

Still, the public demanded answers and turned to Congress. In 1974, seven executives of the nation's largest oil companies stood before the U.S. Senate and the nation and declared their innocence. They insisted that they had not used the 1970's oil crisis to their companies' advantage. For three days, the executives of Exxon, Texaco, Gulf, Mobil, Standard Oil of California, Standard Oil of Indiana, and Shell appeared before the U.S. Senate Subcommittee on Investigations. They were grilled by Democratic Committee Chairman, Senator Henry Jackson of Washington. The prevailing attitude was captured by Senator James B. Allen, Democrat of Alabama, when he asked the executives, "Would it be improper to ask if the oil companies are enjoying a feast in the midst of famine?"[13] Many senators and members of Congress are posing the same question today.

The 1970s also proved to be a profitable time for U.S. arms dealers trading with the Middle East. For example, Lockheed sold millions of dollars worth of planes and training expertise to the Shah of Iran. Lockheed's C-5A transport plane had been a controversial item for the U.S. Congress because of massive cost overruns in its production. In 1972, the Congress cut its purchases from 121 planes to 80 planes, causing the production line in Marietta, Georgia, to shut down. However, the Shah of Iran wanted these planes and agreed to fully fund the reopening of the production line and to purchase ten of the $55 million planes, including an initial outlay of $175 million.[14]

It seems that Iran also experienced some difficulty in its integration of more than $10 billion worth of U.S. military equipment purchased between 1972 and 1976, including at least six Lockheed P-3 sea reconnaissance aircraft. In 1976, Lockheed began a three-year $200 million program to train Iranian airmen in jobs, including supply management and inventory control. The program was designed to help reduce Iranian Air Force dependency on U.S. military civilian contractors and managers. Overall, the program was slated to include approximately 400 Lockheed employees, all stationed in Iran.[15]

Halliburton continued its work in Iraq and Iran throughout the 1970s. In 1973, Brown & Root received a $117 million contract to build two oil terminals in Iraq. By 1975, Brown & Root and other Halliburton divisions were working on projects in Iran worth $9.2 billion.[16]

THE 1970s OIL CRISIS—TAKE TWO

If the first energy shock seemed difficult, the world had no idea what was in store with the second oil crisis. In 1979, a popular revolution in Iran ousted the U.S.-installed government of Mohammad Reza Shah Pahlavi and replaced him with Ayatollah Ruhollah Khomeini. Iran had long been a leading ally of U.S. corporations and the U.S. government. In fact, relations between the United States and Iran were so good that President Carter and Mrs. Carter celebrated New Year's Eve 1977 in Tehran with the Shah and his wife.[17] Israel may have been the United States' main political and military ally, but Iran and Saudi Arabia provided the United States with oil. The overthrow of the Shah ended the cozy U.S. relationship with Iran. A virulently anti-American government took hold of the nation, and on November 15, 1979, Iran canceled all of its contracts with U.S. oil companies, including Standard Oil of New Jersey (ExxonMobil), Standard of California (Chevron), and Gulf (BP).

The second oil shock was far worse than the first for American consumers, although equally profitable for U.S. oil companies. The price per barrel of oil more than doubled from 1979 to 1981, rising from $40 to $95 per barrel (in 2005 dollars). Gas prices at the pump increased 150 percent. However, the effect of the Iranian oil embargo was only to reduce worldwide supply by 4 to 5 percent. Much more significant was the stockpiling of oil by the companies, which reduced supply even further and caused prices to skyrocket. In reality, imports of oil into the United States were 8 percent *higher* in 1979 than in 1978, while gasoline inventories were 6 percent *higher.* The combined net earnings of the five largest domestic oil companies, Exxon, Mobil, Socal (later Chevron), Gulf (later BP), and Texaco, increased by 70 percent between 1978 and 1979, from $6.6 billion to $11.2 billion, while 1980 was the most lucrative year in the industry's history up to that point.[18]

Joel Jacobson, the New Jersey State Energy Commissioner, wrote in the *New York Times* on January 11, 1981, "The blunt fact is that the gasoline shortage of 1979 was caused by an industry that manipulated production and marketing on the near side of the supply-demand equation; contrived artificial shortages via a host of innovative ruses; stimulated panic-buying, which has inured the motorist—so grateful for his gallon of gasoline—to the meteoric rise in prices and profited handsomely as a consequence." The high gas prices and long lines of the two crises marked the beginning of the end for the Carter administration.

The failure of an American-led coup in Iran and the subsequent 1979 oil crisis, whether real or contrived, brought many lessons. One was that a new ally was needed in the Middle East to replace Iran and ensure American access to oil. The second was that overthrowing governments was tricky business. Better to try to work with potential allies, no matter how unsavory. After the Iranian revolution, Iraq, with the second largest oil reserves in the world after Saudi Arabia, became the next best option. Saddam Hussein was certainly an unsavory partner, but he was also a potential ally.

"A MUTUAL SEDUCTION"— THE TURN TOWARD IRAQ

> We should pursue, and seek to facilitate, opportunities for U.S. firms to participate in the reconstruction of the Iraqi economy, particularly in the energy area.
>
> —*President George Bush, National Security Directive 26, October 2, 1989*

> Iraqi leader Saddam Hussein aggressively courted U.S. companies and government agencies. . . . It was a mutual seduction. The U.S. government bitterly opposed to Iran and its leader, Ayatollah Ruhollah Khomeini, first tilted toward Iraq to ensure that Iran did not win the war; later, it became equally interested in Iraqi oil and trade.
>
> —*Guy Gugliotta,* Washington Post, *September 16, 1990*[19]

With the Iranian revolution in 1979, the United States lost one of its most important allies in the Middle East. However, unlike George W. Bush's administration, the Reagan and Bush Sr. administrations sought primarily to *open* and *increase* trade between the United States and Iraq, rather than to control the country's *internal* economic policies. In fact, until the 1991 Gulf War, the Reagan and Bush Sr. administrations were so driven to increase economic ties with Hussein that they were willing to ignore both his brutal human rights atrocities and his support for international terrorism—despite vocal concerns from within their own administrations. Reagan and Bush supplied money, arms, commercial products, and other forms of trade to Iraq. U.S. corporations, particularly through the efforts of the U.S.-Iraqi Business Forum and Kissinger Associates, worked aggressively to define and expand this relationship to their extreme benefit.

The 1979 Iranian revolution and subsequent hostage crisis helped dispose one president, Jimmy Carter, and helped put two others in place, Saddam Hussein in 1979 and Ronald Reagan two years later. The same year that Hussein became president of Iraq, Carter made Iraq one of the first countries to appear on his newly established list of state sponsors of terrorism, due to its support of the Palestine Liberation Organization and other groups linked to terrorist activity. This led to formal U.S. economic sanctions against Iraq.

Iraq may have been a sponsor of state terrorism, but it was also one of the wealthiest countries in the Middle East, with hard currency reserves between $25 billion and $38 billion. The reason for this wealth, of course, was its oil. Iraq's official reserves of oil are currently stated as 112 billion barrels, the world's second largest deposits after Saudi Arabia. According to a 2003 report by the U.S. Department of Energy, however, Iraq's real reserves may be far greater—as much as 300 to 400 billion barrels.

On September 22, 1980, Saddam Hussein invaded Iran, igniting an eight-year war between the two nations. Hussein invaded to gain access to the country's primary waterway, its oil, and its regional power.[20] Hussein first claimed full sovereignty over the Shatt al-Arab River, the point where the Tigris and Euphrates rivers join together in the south of Iraq, near Basra. The Shatt carries the waters of these mighty rivers into the Persian Gulf and out to the Indian Ocean. The last 120 miles of this journey marks the boundary between Iraq and

Iran. The Shatt is a crucial bloodline to both countries, but it is Iraq's life blood. Except for the tiny opening that lies between the Umm al-Qasr port and the Shatt al-Arab, Iraq is a landlocked country, and the Shatt is its only direct avenue out of the Gulf to the ocean. In addition, a considerable amount of both countries' oil infrastructure—fields, pumping stations, refineries, pipelines, loading facilities, and storage tanks—are concentrated around and dependent on the Shatt. Just across Iraq's border in Iran, for example, lies Khuzestan, a region that holds 90 percent of Iran's oil. Hussein, flush with a million-man army, wanted to take control of at least this part of Iran and become a dominant Middle East player.

The United States was officially neutral throughout the eight years of the Iran-Iraq war. Unofficially, however, the Reagan administration was hoping to make Iraq "the new Iran"—or rather, what the pre-1979 revolution Iran was to the United States and its corporations. Reagan did not want Iran's revolution spreading across the Middle East, and he certainly did not want it spreading into the nation with the second largest oil reserves in the world. In addition, there was a significant amount of money to be made in Iraq. Thus, in May 1981, the State Department's William Eagleton met with Iraqi Deputy Prime Minister Tariq Aziz. Eagleton described the meeting in a telegram to the State Department: "I said the U.S. government supports the participation of American firms in projects designed to restore Iraq's oil facilities as rapidly as possible after the war." He added that the meeting "should be helpful to our position and that of U.S. business interests in Iraq."[21]

In March 1982, Reagan removed Iraq from the list of countries supporting terrorism, which rendered Iraq eligible for a broader range of trade and credits with the United States. Then, on November 26, 1984, just days after his reelection, Reagan restored full diplomatic relations with Iraq. This move came under extreme influence from U.S. business interests but against the advice of many within Reagan's administration.

U.S. business interests played a leading role in persuading the Reagan administration to open economic engagement with Iraq, to resist calls for sanctions, and to ensure that the U.S. taxpayer bore much of the economic risk. The business groups with the greatest influence were the U.S.-Iraq Business Forum and Kissinger Associates. These two groups used the power of their clients, and in the latter case, Henry Kissinger's personal influence, to ensure that U.S. economic policy toward Iraq would increasingly and consistently benefit both their own and their clients' bottom lines.

The U.S.-Iraq Business Forum

> The Administration looks "to those in the U.S.-Iraq Business Forum to help preserve and expand the overall U.S.-Iraqi relationship through its commercial side, as only the private sector can do."
>
> —*A. Peter Burleigh, deputy assistant secretary of state*[22]

Let's play a game. See if you can accurately determine the decade during which the following symposium took place. The title of the symposium is, "United States Commercial, Economic, and Strategic Interest in Iraq." Speakers on the first panel, entitled "The United States and Post-War Iraq," represent the White House National Security Council and the House Committee on Foreign Affairs. Attendees include representatives of the U.S. Departments of State, Commerce, Energy, and Agriculture; the U.S. Export-Import Bank; and several members of Congress. Corporate attendees include Bechtel, Chevron, Texaco, Brown & Root, Boeing, Pepsi, Ford, General Electric, General Motors, Exxon, and Mobil. Keynote speakers include the ambassador of Iraq and the Iraqi oil minister, while several representatives of the Iraqi Embassy participate in the day's events.

Give up? It was the 1980s, the same period during which Saddam Hussein was president of Iraq and the United States had on-again, off-again economic and diplomatic sanctions against the Iraqi regime. The exact date of the symposium was November 14, 1989, although this was the third of an annual series. The symposium followed the 1988 conclusion of the Iran-Iraq war. Peter Burleigh's earlier statement is quoted from the symposium's closing remarks. The event was organized by the U.S.-Iraq Business Forum, which played a key role in uniting the efforts of Saddam Hussein, the U.S. government and U.S. corporations. Its efforts were short-lived but highly profitable for those involved.

Marshall Wiley, a lifelong U.S. foreign service officer, founded the U.S.-Iraq Business Forum. Beginning in 1954, Wiley served as a foreign service officer in Israel, Yemen, Lebanon, Jordan, Egypt, Iraq, Saudi Arabia, and then Oman, where he was finally elevated to the role of ambassador for the last four years of his formal government service. Because Iraq severed diplomatic relations with the United States in 1967, the United States was not permitted to have an embassy in Iraq. On October 1, 1972, however, the U.S. State Department established a U.S. Interests Section at the Belgian Embassy in Baghdad. Wiley served at this desk from 1975 to 1977. After Reagan opened economic ties with Iraq in 1982, Wiley struck out into the world of international finance. His unofficial partner was Iraq's ambassador to the United States, Nizar Hamdoon.

The "bland, reassuring Wiley," as described by author Joe Conason, and the handsome, charismatic Hamdoon made an unlikely pair. After his death on July 4, 2003, an obituary described Nizar Hamdoon's "glory years as Iraqi ambassador in Washington, 1984–1987." Hamdoon "had a reach, a public presence, and an impact that fellow ambassadors could only envy." He was described by the *Washington Post* as a "master of the Washington scene," whose parties were attended by "policy-makers from the State Department and the White House, as well as academics, journalists, Jewish leaders, and members of Congress."[23] Leading newspapers printed glowing profiles and invited him to write his own columns.

Wiley established the U.S.-Iraq Business Forum shortly after economic relations were opened between the two countries, with what the *New York Times* termed the "aggressive encouragement of Nizar Hamdoon." Wiley founded the organization to serve as a mediator between the American private sector, the Iraqi government, and one of the world's few remaining unexploited and potentially lucrative markets. "I needed some assurance that the Iraqi Government would cooperate with such an organization if I created one," Wiley explained, "and it was agreed that they would."[24] The U.S.-Iraq Business

Forum was a trade association that eventually represented some sixty American companies, including Bechtel, Lockheed, Texaco, Exxon, and Mobil. Ray Hunt, who at the time was representing his oil company, Hunt Oil, was a member of the Forum. Today, Ray Hunt sits on the board of Halliburton.

According to *Financial Times* reporter Alan Friedman in his 1993 book, *Spider's Web: The Secret History of How the White House Illegally Armed Iraq*, the U.S.-Iraq Business Forum "played a large part in encouraging Administration officials to believe that assisting Saddam was an economic as well as a strategic interest of the United States." Hussein let it be known (through Hamdoon and Wiley) that if a company wanted to do business in Iraq, it had to join the Forum. A list of Forum members was reportedly given to the heads of each Iraqi Ministry for use when awarding contracts. Hamdoon made the point quite clearly when he told a Washington DC audience in 1985, "Our people in Baghdad will give priority—when there is a competition between two companies—to the one that is a member of the Forum."[25]

In 1989, Wiley articulated the feeling of many members of the business community and the administration of Reagan and George H.W. Bush when he laid out several reasons for courting Iraq in unusually blunt terms: "Iraq plays an important role now in bringing about stability in the Gulf region as an offset to Iran. This balance of power is important to us because of our interest further down the Gulf in the countries that are friendly to us and where we have substantial energy reserves that must be protected. So Iraq plays an important political and geopolitical role in addition to being an important trading partner for us, and we can't push all these considerations to the side and shut our eyes to them because of some perceived human rights violations."[26]

Opening the door was easy, but keeping it open required real work. Fortunately for American corporations, Wiley did not act alone. His close corporate ally was none other than Dr. Henry Kissinger. Kissinger Associates was the second doorway between American business and Iraq.

Kissinger Associates

> I think that in the modern world, if you don't understand the relationship between economics and politics, you cannot be a great statesman. You cannot do it with foreign policy and security knowledge alone.
>
> —*Henry Kissinger, April 20, 1986*[27]

Can one possibly tell a story of American politics from the last forty years without reserving a prominent place for the august Henry Kissinger? Kissinger served simultaneously as President Nixon and then President Ford's national security adviser and secretary of state from 1973 to 1975. He began as Nixon's national security adviser in 1969 and then continued as Ford's secretary of state until 1977. Kissinger made his fame in government, but he continues to earn his fortune in business.

While the twin oil shocks of the 1970s were bad for consumers and most importer governments, they were good for former government officials. America's corporate executives learned the hard way that doing business overseas, particularly in the Middle East, required unique expertise. If they wanted access to all those petro-dollars in the hands of those state-controlled oil industries, they would need people with economic know-how and close relationships to government leaders. Kissinger's strength, as explained by Anthony J. F. O'Reilly of the Heinz Corporation in 1986, "is analyzing people and their power base. He has a durable and great inventory of contacts. To say that he is a door opener sounds mildly disparaging, but it is helpful in countries with rusty hinges."[28] Countryrisk.com, a website that reviews country analysis consultants, describes Kissinger Associates as "the company that started it all."

Kissinger founded the company in 1982, the same year that President Reagan opened economic ties with Iraq, and it remains one of the world's preeminent providers of advice and political risk assessments to the largest corporate-multinationals. It has also maintained its status as one of the most secretive businesses in the United States. Its customers are required to sign strict confidentiality agreements committing not to disclose their discussions with Kissinger Associates or even their status as Kissinger clients. In fact, Kissinger resigned from his appointment on the 9/11 Commission in 2002 in order to keep his client list secret. The company has no known website and, the most recent comprehensive interview conducted with Kissinger about his business was almost twenty years ago, with Leslie Gelb of the *New York Times* in 1986.

Some information has found its way into the public domain, most notably during the 1989 Senate confirmation hearings of Lawrence Eagleburger to become undersecretary of state. Eagleburger, who served as president of Kissinger Associates from 1984 to 1989, told that committee that Kissinger clients typically pay a fixed annual fee for their top management to meet several times a year for discussions lasting a day or two on international political, economic, and security trends—ranging from Soviet affairs to the price of oil.[29] An October 2004 press release by APCO Worldwide, announcing a "strategic alliance" with Kissinger Associates, says that the company "provides strategic advisory and advocacy services to a select group of multinational companies. The firm provides advice regarding special projects, assists its clients to identify strategic partners and investment opportunities, and advises clients on government relations throughout the world."

The names of a few Kissinger clients have leaked out over the years, including Hunt Oil, H.J. Heinz, Arco (now BP Oil), American Express, Shearson Lehman, Union Carbide, Coca-Cola, ITT Corp., and engineering giant Fluor, which, in 2004, became the recipient of the third largest reconstruction contract in Iraq, worth $3.75 billion. Voicing his frustration after some of these leaks were made, Kissinger snapped, "Every smart-alecky newsman is asking my clients what I did for them, and I think it's an outrage."[30]

While Kissinger is chairman and sole owner of the company, he brought along his key government advisers to work alongside him. The next section describes how each Kissinger adviser played a lead role in opening economic ties with Iraq while in government and then turned around to cash in on those ties at Kissinger Associates. Other than Kissinger himself, Eagleburger may have made the most successful use of this particular revolving door.

Before joining Kissinger Associates, Eagleburger had already worked for the U.S. State Department from 1969 to 1983. After serving as President George H. W. Bush's deputy secretary of state during the first Gulf War, he became secretary of state in 1992. Today, he sits on the board of directors of Halliburton, the number one recipient of 2004 Iraqi reconstruction contracts, worth $11.4 billion.

Kissinger Associates' vice chairman was Lieutenant General Brent Scowcroft, national security adviser to Presidents Ford and Bush Sr. Scowcroft worked for Kissinger from 1982 to 1989. Another former managing director of Kissinger Associates is L. Paul Bremer, III, who, after leaving the State Department, worked for Kissinger from 1989 to 2000. Bremer left Kissinger to form his own crisis consulting practice at Marsh, Inc. and then went on to serve as U.S. administrator of the Coalition Provisional Authority in Iraq.

Rounding out the key players at the firm's inception was Managing Director Alan Stoga who served with Kissinger from 1984 to 1996, continuing to work as a director and consultant to the firm today. Stoga was the firm's most direct link to the U.S.-Iraq Business Forum. He appeared regularly at Forum events and traveled to Iraq with Wiley and Abboud on a Forum-sponsored trip in 1989, where they met directly with Saddam Hussein. Many Kissinger clients were known members of the U.S.-Iraq Business Forum, and they also became recipients of contracts with Iraq.

Enter the Corporations

Wiley and Kissinger were skilled at their jobs. It did not hurt that they had the nation's largest corporations behind them. The Reagan and Bush Sr. administrations willingly opened the door as far as they could to U.S.-Iraqi trade, which took three main forms: arms, oil, and agriculture. Though U.S. corporations desperately wanted to make a profit off of Iraq, they did not want to bear the risk. The U.S.-Iraq Business Forum and Kissinger Associates thus persuaded the Reagan administration to reverse a ban on U.S. commodity credits and loan guarantees to Iraq. This allowed the U.S. Export-Import Bank and the U.S. Overseas

Private Investment Corporation (OPIC) to start making loans to Iraq. The Export-Import Bank and OPIC are U.S. taxpayer-funded lending institutions that provide loans and credits either directly to U.S. corporations to help them operate abroad, or to foreign governments to help them purchase goods produced by U.S. companies. Both forms of lending were aggressively used to bring U.S. corporations and their goods into Iraq. U.S. taxpayers therefore bore a good deal of the financial cost of U.S. corporate activities with Iraq and Saddam Hussein.

In 1981, there was essentially no trade between the United States and Iraq. Then, between 1983 and 1989, annual trade between the two nations grew nearly sevenfold, from $571 million to $3.6 billion. This amount was expected to double in 1990, before Iraq invaded Kuwait. In 1989, Iraq became the United States' second-largest trading partner in the Middle East. Using loans from U.S.-based banks, many guaranteed by the federal government, Iraq purchased $5.2 billion in U.S. exports of food, technology, and industrial goods. The United States, in turn, bought $5.5 billion in Iraqi petroleum. In fact, from 1987 to the beginning of 1990, U.S. consumption of Iraqi oil increased from 80,000 to 675,000 barrels per day and then increased again to 1.1 million barrels per day in July 1990.[31]

From the beginning, it was difficult to distinguish between business and political activity in Iraq. A prime example is the Reagan administration's support of Bechtel and the Aqaba oil pipeline. In early 1983, Secretary of State George Shultz met with Iraq's foreign minister, Saadoun Hammadi, to discuss further opening U.S.-Iraqi relations. Hammadi told Shultz to "think of it as a long-term investment."[32] Shultz, who had only recently finished a nine-year stint as president and director of Bechtel (1974 to 1982), probably did not need to do much thinking. Shortly after the meeting with Hammadi, Shultz's State Department, often led by Undersecretary of State Lawrence Eagleburger, began lobbying directly for Bechtel in Iraq. After successfully expanding U.S. government ties to Iraq and opening Iraq's economy to U.S. businesses, Eagleburger departed to Kissinger Associates and Shultz went back to Bechtel, respectively, where they could benefit directly from the fruits of their government service.

In December 1983, the State Department began discussions with Bechtel on the construction of an oil pipeline to transport Iraq's oil out of the port of Aqaba, Jordan. Bechtel's P.T. Hart reported to his colleagues that the "U.S. Department of State and the National Security Council continue to be most interested in means of exporting safely Iraqi crude oil." A few days later, Reagan's newly appointed Middle East envoy, Donald Rumsfeld, flew to Baghdad to meet personally with President Saddam Hussein and Tariq Aziz, Iraq's deputy prime minister and foreign minister. Rumsfeld was the first senior American official to visit Iraq in six years.

Hussein, Aziz, and Rumsfeld spoke at length about Bechtel's proposed oil pipeline. According to a State Department cable reporting on the meeting, "Saddam responded that in [the] past Iraq had not been very interested in the Jordanian pipeline possibility because of the threat that Israel would disrupt it. Now that U.S. companies and the USG [U.S. government] were interested, Iraq would re-examine it." Later that month, Eagleburger personally urged the U.S. Export-Import Bank to begin extending loans to Iraq to "signal our belief in the future viability of the Iraqi economy and secure a U.S. foothold in a potentially large export market." He noted that Iraq "has plans well advanced for an additional 50 percent increase in its oil exports by the end of 1984," a reference to Bechtel's pipeline. A week later, Eagleburger reported that Hussein's government had approved the project.

In 1984, there were at least seven Iraq oil-and-gas related projects involving U.S. companies under consideration by the U.S. Export-Import Bank. The Bank was highly dubious of Iraq's ability to repay these credits but eventually succumbed to the strong pressure of the State Department.

While the deliberations over Bechtel's pipeline and the various other U.S. business interests in Iraq proceeded, Hussein stepped up his military efforts against Iran and the Kurds. From 1983 to 1988, the Iraqi air force dropped between 13,000 and 19,500 chemical bombs on Iran and on the Iraqi Kurdish city of Halabja. On March 5, 1984, the U.S. State Department issued a public statement condemning Iraq's use of chemical weapons in the war against Iran. In private, however, the Reagan administration was eager to ensure that Hussein knew the

U.S. government still supported his regime. Just four days after the public condemnation, the State Department told the Export-Import Bank that it should start granting short-term loans to Iraq "for foreign relations purposes." In addition, briefing notes prepared by the State Department for Rumsfeld's visit to Baghdad that same month explain that "bilateral relations were sharply set back by our March 5 condemnation of Iraq for CW [chemical weapons] use, despite our repeated warning that this issue would emerge sooner or later."[33] The memo places particular attention on Bechtel's Aqaba pipeline deal and notes that Lawrence Eagleburger was working hard to assure Iraq that Export-Import Bank funding for the project was still secure.

Hussein rejected the deal in 1985, citing continued concerns over the pipeline's security as it passed through Israel. As late as 1987, however, Bechtel was still lobbying the U.S. government to continue its support for the project, hoping to work out a deal to which Hussein would eventually agree. Hussein did not budge, which raised serious doubts among some U.S. corporate and government policy-makers as to just how far Hussein could be expected to go in support of U.S. interests.

Halliburton had a more steady and successful relationship working with Saddam Hussein. As detailed by Robert Bryce, in 1981, Brown & Root signed a $2.4 million contract with the Iraq State Oil Company to develop a plan for Iraq's tankers to load oil while repairs were made to the Mina al-Bakr terminal. In 1983, the company was hired to work on two major Iraqi oil pipelines. From 1988 to 1990, it conducted repairs on the Mina al-Bakr terminal. In 1990, Halliburton signed a contract with the Iraqi State Oil Exploration Company to provide equipment and training. In fact, at the moment when Hussein invaded Kuwait, forty Halliburton employees were at work in Iraq.[34]

Arming Hussein

The Reagan administration and, to an even greater extent, the Bush Sr. administration, spent nearly a decade secretly arming Iraq through direct and indirect sales. The direct sales were of "dual-use" materials, which are goods ostensibly made for civilian purposes but have military applications as well. From 1985 to 1990, U.S. corporations provided $782 million in dual-use goods to Iraq.[35] The Reagan and Bush Sr. administrations allowed the sales over the objections of the Pentagon, which believed these products would inevitably be used for military purposes. One government official explained that in March 1985, high-technology export licenses, which previously had not been approved by the U.S. government to Iraq, "started to go through as if someone had suddenly turned a switch."[36] The indirect method involved sales of conventional and chemical weapons to third parties, generally friendly governments, who then sold the weapons to Iraq.[37]

U.S. arms dealers made out handsomely, as did dozens of U.S. multinational corporations, including Bechtel, AT&T, Hewlett Packard, General Motors, and Philip Morris, who supplied Hussein with everything from rice to computers to helicopters.[38] For example, the Lockheed L-100 transport plane was among the dual-use goods sold to Iraq. By 1984, the U.S. Agriculture Department had extended $513 million in credits to Hussein with which to purchase U.S. farm products.[39] However, many of these "agriculture" funds were diverted for military purchases. As described by Alan Friedman, the volume of U.S. farm products supposedly being shipped to Iraq was so inflated that, in some cases, there were single contracts for seeds that exceeded the dietary needs of the entire country.[40]

Unable to sign a contract with Hussein for the Aqaba pipeline, Bechtel instead went to work on Iraq's petrochemical complex. In the summer of 1988, Bechtel signed a contract with Hussein to manage the PC-2 petrochemical complex south of Baghdad—just four months after Hussein used mustard gas against the Iraqi Kurdish city of Halabja. The Bechtel design involved dual-use technology. According to *Middle East Defense News*, "A key feature of the PC-2 project was the plan to manufacture ethylene oxide, a precursor chemical that is easily converted to thyodiglycol, which is used in one step to make mustard gas." According to Friedman, "The U.S. Embassy in Baghdad was pleased as was the Commerce Department which encouraged Bechtel to go ahead." When UN weapons inspectors arrived in 1991, they declared that the PC-2 industrial complex was a major part of the "smoking gun" that proved Iraq was pursuing a weapons-of-mass- destruction program.[41]

In 1988, the State Department was again forced to release a public condemnation of the Iraqi government after Hussein used chemical weapons against the people of Halabja. The next day, the U.S. Senate passed a tough trade sanctions bill against Iraq that would have significantly curtailed U.S. corporate dealings with the country. But the Reagan administration, led by Secretary Shultz, lobbied vehemently against the sanctions, and they were not enacted.

A FAILED COURTSHIP: THE FIRST GULF WAR

Ronald Reagan finished his two terms as president of the United States in 1989 and was succeeded by his vice president, George H.W. Bush. From the moment Bush was elected until Iraqi troops crossed into Kuwait on August 1, 1990, his administration led the most aggressive courtship of Saddam Hussein to date.

A prominent leader in this courtship was President Bush's secretary of state, James A. Baker III. Their shared political careers date back to Bush's unsuccessful bid for the U.S. Senate in 1970, the first of many failed electoral campaigns run by Baker. Baker worked for Gerald Ford until he ran Ford's unsuccessful 1976 presidential reelection campaign, which was followed by Baker's own failed bid for attorney general of Texas in 1978. Baker then ran George H.W. Bush's losing campaign for the Republican Party presidential nomination in 1980 but was awarded the post of White House chief of staff and then treasury secretary by the victorious Ronald Reagan. Baker finally ran a successful political campaign as chairman of Bush's 1989 presidential race. Bush and Baker share more than a long and often losing political history. They are both Texans with long histories and continued heavy involvement in the oil sector. In addition to Baker and his family's personal investments in the industry (at the time of the first Gulf War in

1990, Baker and his immediate family reportedly held interests in Amoco, Exxon, and Texaco), the family law firm, Baker Botts, has represented Texaco, Exxon, Halliburton, and Conoco Phillips, among others, in some cases since 1914 and in many cases for decades.[42]

The end of the Iran-Iraq war in 1988 offered the newly elected Bush administration a fresh opportunity to alter U.S. relations with Iraq. Bush's transition team provided the president-elect with a policy memo on Iraq, which explained that "Saddam Hussein will continue to eliminate those he regards as a threat, torture those he believes have secrets to reveal, and rule without any real concessions to democracy."[43] They argued that engagement was the appropriate response. The United States would have "to decide whether to treat Iraq as a distasteful dictatorship to be shunned where possible, or to recognize Iraq's present and potential power in the region and accord it relatively high priority. We strongly urge the latter view." Two reasons offered in the paper were Iraq's "vast oil reserves," which promised "a lucrative market for US goods," and the fact that U.S. oil imports from Iraq were skyrocketing.

Bush and Baker accepted the advice of the transition team and ran with it. On October 2, 1989, President Bush signed National Security Directive 26, which made explicit U.S. support for Iraq: "We should pursue, and seek to facilitate, opportunities for U.S. firms to participate in the reconstruction of the Iraqi economy, particularly in the energy area, where they do not conflict with our nonproliferation and other significant objectives. Also, as a means of developing access to and influence with the Iraqi defense establishment, the U.S. should consider sales of non-lethal forms of military assistance . . ." President Bush then began discussions of a $1 billion loan guarantee for Iraq one week before Baker met with Tariq Aziz at the State Department.

The British, for their part, were wringing their hands with greed and anticipation. According to Friedman, Iraq was known among British Foreign Secretary John Major's staff as "the big prize." Britain's senior Foreign Office officer wrote in October 1989, "I doubt if there is any future market on such a scale where the UK is potentially so well placed. . . . We must not allow it to go to the French, German, Japanese, etc. The priority of Iraq should be very high."

Meanwhile, the U.S. Congress had imposed U.S. Export-Import Bank financing restrictions on Iraq because of the Halabja massacre. On January 17, 1990, Bush voided the prohibition with a stroke of a pen, stating that it was "not in the national interest of the U.S."[44] Baker then described trade as the "central factor in the U.S-Iraq relationship." During the Bush-Baker tenure, the United States became Iraq's largest supplier of nonmilitary goods, and Iraq became the United States second biggest trading partner in the Middle East. In January 1990, the United States imported eight times more Iraqi oil than it had in 1987. By July 1990, U.S. imports had increased to 1.1 million barrels per day, more than a quarter of Iraq's total oil exports. As Baker would comment, U.S. policy toward Iraq was "not immune from domestic economic considerations."[45]

At the same time, in the early months of 1990, Saddam Hussein turned up his anti-U.S. and anti-Israel rhetoric and actions, which included the public hanging of a British journalist of Iranian descent in Baghdad. In response, several Iraq sanctions bills were submitted in Congress, but every one was successfully fought off by Bush and Baker. Just two months before Hussein invaded Kuwait, on April 21, 1990, a U.S. delegation led by Senator Bob Dole was sent to Iraq in an attempt to placate Hussein. Baker personally sent a cable to the U.S embassy in Baghdad, instructing U.S. Ambassador to Iraq April Glaspie to meet with Hussein and to make it very clear that, "As concerned as we are about Iraq's chemical, nuclear, and missile programs, we are not in any sense preparing the way for a preemptive military unilateral effort to eliminate these programs." In May 1990, the Bush administration was still sharing military intelligence with Hussein. In July, the White House continued the push for the $1 billion in loans to Iraq, while rebuffing efforts by the defense and commerce departments to restrict the export of dual-use technology.[46]

However, a collision course was being driven by Iraq's enormous and growing debt. During the Iran-Iraq war, Hussein built the fourth largest military in the world with financial assistance from many nations, including Saudi Arabia. Hussein thought the money was aid in support for the war effort, but Saudi Arabia thought it was a loan and wanted its money back at the end of the war. Kuwait's oil was looking increasingly attractive, and its actions in OPEC, specifically overproduction that drove down the price of oil for all, were inciting Hussein's anger. In mid-July, Hussein amassed Iraqi troops on the Kuwait border and publicly threatened the use of military force. On July 25, U.S. Ambassador Glaspie was once again dispatched to meet with Hussein. This time, the message she delivered to him would become infamous: "I have direct instruction from the President to seek better relations with Iraq" and "we have no opinion on the Arab-Arab conflicts like your border disagreement with Kuwait."[47]

Less well known is the personal cable sent a few days later from President Bush to Hussein, just five days before the invasion of Kuwait, in which Bush expressed concern but added, "Let me reassure you that my administration continues to desire better relations with Iraq."[48]

At 9:00 P.M. on August 1, 1990, Iraqi forces crossed into Kuwait. On August 15, President Bush declared, "Our jobs, our way of life, our own freedom and the freedom of friendly countries around the world would all suffer if control of the world's great oil reserves fell into the hands of Saddam Hussein."[49] To this Secretary Baker added, "The economic lifeline of the industrial world runs from the Gulf and we cannot permit a dictator such as this to sit astride that economic lifeline."[50] On January 15, 1991, the day before the United States launched attacks against Iraq, President Bush signed National Security Directive 54. The first line states, "Access to Persian Gulf oil and the security of key friendly states in the area are vital to U.S. national security. . . . The U.S. remains committed to defending its vital interests in the region, if necessary through use of military force, against any power with interests inimical to our own."[51] By March 3, the first Gulf War was over. The Iraqi army was successfully expelled from Kuwait, while Saddam Hussein remained in power in Iraq.

There are two main schools of thought as to why the United States did not stop the invasion of Kuwait before it began. The first is that Bush truly was caught off guard. Bush believed he and Hussein were working together, but Hussein had to make a show of aggression to impress both those in and outside of Iraq of his seriousness. Bush did not actually believe that Hussein would invade Kuwait in defiance of U.S. interests (even if those interests had not been stated) and those of most of Iraq's neighbors, and Hussein did not actually believe that Bush would stop him if he did invade. But once Hussein invaded Kuwait, the Bush administration could not allow him to control both his own and Kuwait's oil and threaten Saudi Arabia, particularly since he had demonstrated that he was no longer to be trusted in serving U.S. interests. Hussein had to be removed.

The other school of thought, argued most forcefully by Middle Eastern scholars, is that Bush *allowed* Hussein to invade Kuwait because it provided an excuse to remove Hussein from power, and the war with Iraq, in turn, provided the necessary excuse to bring a significantly increased U.S. military presence into the region—including five hundred thousand U.S. troops in Saudi Arabia. The U.S. military presence not only remained but also spread to more countries and grew dramatically with the second U.S. war against Iraq.

A combination of the two arguments is also possible. Bush did not think Hussein would invade, but once he did, Bush used it as an excuse to wage war and establish an increasingly large U.S. military presence in the region. To date, the writings of and interviews with the key players directly involved do not yield any more concrete conclusions. What is clear is that Hussein had outlived his usefulness to the United States, and a decision was made to stop him. This of course begs the next question: Why did Bush "leave" Hussein in power rather than remove him during the first Gulf War?

Bush administration officials have repeatedly said that they firmly believed Saddam Hussein would be overthrown by internal pressure after losing the war. The White House did not want to perform this task itself, so it ordered the CIA to plan a covert operation to destabilize Hussein's regime, strangle the Iraqi economy, support anti-Saddam resistance groups both inside and outside of Iraq, look for alternative leaders, and ultimately remove Hussein. The United States did not want to do this overtly because it feared that its regional partners, particularly Saudi Arabia, would not approve. Henry Rowen, a Wolfowitz adviser at the time, recalled, "As the war was coming to an end, I went to Cheney and said, 'You know, we could change the government and put in a democracy. The answer he gave was that the Saudis wouldn't like it."[52] Another and probably more important reason was that the Bush White House simply did not want to take over Iraq. Rather, it wanted a more pliable and reliable leader at its helm and worked aggressively to achieve this end.

Dick Cheney, Bush Sr.'s defense secretary, explained the administration's reasoning in April 1991: "Once you've got Baghdad, it's not clear what you do with it. It's not clear what kind of government you would put in place of the one that's currently there. . . . How much credibility is that government going to have if its set up by the U.S. military when its there? . . . I think to have American military engaged in a civil war inside Iraq would fit the definition of a quagmire, and we have absolutely no desire to get bogged down in that fashion."[53]

Paul Wolfowitz, Bush Sr.'s undersecretary of defense policy, expressed similar views in 1997: "A new regime [in Iraq] would have become the United States' responsibility. Conceivably, this could have led the United States into a more or less permanent occupation of a country that could not govern itself, but where the rules of foreign occupier would be increasingly resented."[54] In an earlier article, he wrote that, after the end of the war, "Nothing could have insured Saddam Hussein's removal from power short of a full-scale occupation of Iraq. . . . Even if easy initially, it is unclear how or when it would have ended."[55]

SANCTIONS

The attempt to oust Hussein took the form of full economic sanctions against Iraq, imposed by the United States and the UN Security Council. Implemented just five days after the invasion of Kuwait, the sanctions amounted to an almost complete economic embargo of the nation, with deadly effects. The World Health Organization found that compared to two years prior to the war, infant mortality rates from 1989 to 1994 had doubled, and the mortality rate for children under five increased by six times. UNICEF reported that from 1991 to 1998, some half a million children under the age of five died in "excess" of the number expected to die without sanctions. The child mortality rates were just the most vivid depiction of the desperation faced by the entire nation. In light of catastrophic and well-documented Iraqi suffering under the sanctions regime, on April 14, 1995, the UN Security Council adopted Resolution 986, establishing the Oil-for-Food Program. The program allowed Iraq to sell some of its oil in order to purchase certain limited humanitarian goods. According to Resolution 986, the program was intended to be a "temporary measure to provide for the humanitarian needs of the Iraqi people." It became the country's sole access to such goods until 2003, when the program was officially terminated. The first Iraqi oil exported under the program left Iraq in December of 1996. The first shipments of food arrived in March of 1997.

In a report to the U.S. Congress in October 2004, Charles Duelfer, the U.S. arms investigator in Iraq at the time of the 2003 invasion, reported that Hussein had established a worldwide network of companies and countries, most of them U.S. allies, which secretly helped Iraq generate about $11 billion in illegal income from oil sales under the Oil-for-Food Program. The United Nations initiated an Independent Inquiry into the alleged abuses headed by Paul Volcker, a former chairman of the U.S. Federal Reserve System, which supported the earlier conclusions. Many members of the U.S. Congress jumped on the report as evidence that the United Nations itself could not be trusted. Few addressed the fact that the United States was part of the UN committee that oversaw every contract signed under the Oil-for-Food Program or that billions of dollars were earned from these contracts by U.S. companies, including Halliburton, Chevron, and ExxonMobil.

Between the first half of 1997 and the summer of 2000, while Cheney was Halliburton's CEO, Halliburton subsidiaries Dresser-Rand and Ingersoll Dresser Pump Co. sold water and sewage treatment pumps, spare parts for oil facilities, and pipeline equipment to Iraq through French affiliates. Halliburton ultimately sold more than $73 million in goods and services to Saddam Hussein's regime.[56] In fact, the *Financial Times* reported that Halliburton sold more oil industry products to Hussein than any other U.S. company.[57] Chevron and ExxonMobil both received contracts through the UN Oil-for-Food Program to market Iraqi oil.

THE CLINTON YEARS

In 1993, President Herbert Walker Bush lost his presidency, while Saddam Hussein continued to hold on to his. Bill Clinton began eight years in the White House, during which time he steadfastly maintained economic sanctions against Iraq while conducting sporadic military incursions. Instead of spending the next ten years exclusively plotting ways to oust Hussein, most conservatives were more immediately focused on ousting Bill Clinton. Meanwhile, for many U.S. corporations, the presidency of Bill Clinton provided many new "free trade" opportunities, including the 1994 North American Free Trade Agreement, the 1995 World Trade Organization, ongoing negotiations for a Free Trade Area of the Americas, expanded trade with China, and the proposed Multilateral Agreement on Investment. Clinton was backed by the Republican-led Congress, which whole-heartedly endorsed each new free trade agreement. And with the collapse of the Soviet Union, Americans felt free to look internally. Clinton won his election with a message that focused on domestic U.S. policy: "It's the economy stupid."

Hussein needed money, so he began aggressively courting oil companies to sign lucrative deals for some of Iraq's largest and most important oil fields. The problem, from a U.S. perspective, was that U.S. companies were excluded from the bidding. The contracts went to Russia's Lukoil, France's Total, China National, and other non-U.S. companies. However, as long as the sanctions remained in effect, the contracts were essentially meaningless. As global public pressure mounted for ending the sanctions for purely humanitarian reasons, many in the U.S. business community and the Bush White House-in-waiting saw the writing on the wall. They increasingly began to argue that the only way to ensure U.S. economic advantage in Iraq was to oust Saddam Hussein finally and totally and remake the country in America's image.

In 1997, a group of conservative thinkers, including a preponderance of former Reagan and Bush administration officials, founded the Project for the New American Century (PNAC). Its first letter, written on January 26, 1998, called on President Clinton to use military force to overthrow the regime of Saddam Hussein. The letter states that it "hardly needs to be added that if Saddam does acquire the capability to deliver weapons of mass destruction, as he is almost certain to do if we continue along the present course, the safety of American troops in the region, of our friends and allies like Israel and the moderate Arab states, and a significant portion of the world's supply of oil will all be put at hazard." It was signed by Elliott Abrams, Richard L. Armitage, William J. Bennett, Jeffrey Bergner, John Bolton, Paula Dobriansky, Francis Fukuyama, Robert Kagan, Zalmay Khalilzad, William Kristol, Richard Perle, Peter W. Rodman, Donald Rumsfeld, William Schneider Jr., Vin Weber, Paul Wolfowitz, R. James Woolsey, and Robert B. Zoellick.

A few weeks later, another *Pax Americana* organization, the Center for Peace and Security in the Gulf chaired by Richard Perle, which "promotes international peace through American strength," submitted a similar letter to President Clinton demanding the overthrow of the Hussein regime. In addition to many of the same signatories to the PNAC letter such as Perle, Wolfowitz, Bolton, Abrams, and Khalilzad, the CPSG letter was signed by Donald Rumsfeld and Douglas Feith. Feith would later become a lead architect of the George W. Bush invasion of Iraq as the undersecretary of defense for policy at the Pentagon.

In 1997 corporate America launched USA-Engage, a fiercely anti-sanctions lobbying group formed as an offshoot to the National Foreign Trade Council, whose most vocal supporter, Dick Cheney, had become CEO of Halliburton in 1995. USA-Engage does not make its membership lists publicly available. However, its documents make clear that at least in the year 2005, Halliburton, Chevron, Lockheed Martin, and Bechtel were all members. USA-Engage had a very specific message: It opposed *unilateral* sanctions wherever they occurred, particularly those against oil-rich nations such as Nigeria, Libya, Iran, and Burma. However, it did not oppose *multilateral* sanctions, such as those imposed on Iraq, and in the words of many USA-Engage members, such sanctions were desirable.

In a speech typical of this position, which could easily have appeared on USA-Engage stationery, then Chevron CEO and current

Halliburton board member Kenneth Derr told a San Francisco audience in 1998, "I think the evidence clearly shows that unilateral trade sanctions just don't work. . . . Once the free market genie is out of the bottle, it's hard to keep the hunger for political freedom bottled up." Derr then said, "Iraq possesses huge reserves of oil and gas—reserves I'd love Chevron to have access to." However, Derr was quick to add, "I fully agree with the sanctions we have imposed on Iraq. Why? Because Iraq's behavior has been especially egregious—so much so that other countries have been willing to join the United States by adding sanctions of their own." In his speech, Derr specifically compares Iraq to countries that have been subject to some of the most brutal governments in history, including Algeria, Angola, China, Russia, Burma, Iran, Libya, Nigeria, and Indonesia. For these governments, Derr argues, sanctions are not appropriate, but for Iraq they are.

Derr's argument and similar ones made by Cheney and the others at USA-Engage were spurious because the path they were setting for Iraq was different. Even if sanctions had been lifted, the government of Iraq was not going to sign lucrative trade deals with U.S. companies. This is why it was necessary to overthrow Hussein. Henry Kissinger weighed in on March 23, 1998, in a syndicated Op-Ed declaring U.S. policy on Saddam to be a "shilly-shally strategy," and "we cannot negotiate with Saddam Hussein, we must try to weaken or, if possible, overthrow him."[58]

On October 31, 1998, under intense lobbying by PNAC, CPSG, and others, Congress and President Clinton enacted the Iraq Liberation Act, which states, "It should be the policy of the United States to support efforts to remove the regime headed by Saddam Hussein from power in Iraq and to promote the emergence of a democratic government to replace that regime." It authorized the president to provide "the Iraqi democratic opposition" with $2 million in support of radio and television broadcasting and $97 million in military and other defense support. The money went to the Iraqi National Congress, which had been formed seven years earlier under the guidance and funding of the CIA to unite Iraq's leading Shiite, Kurdish, and secular organizations in their opposition to Hussein. Ahmed Chalabi, a secular Iraqi

Shiite and mathematician who had served as chairman of the Petra Bank of Jordan until convicted in absentia for embezzlement and fraud, was the most recognized leader of the group.

CHENEY'S ENERGY TASK FORCE

In 2000, George W. Bush was named president of the United States by the U.S. Supreme Court in one of the most contested elections in U.S. history. Oil interests and U.S. corporations found a new home inside of the White House. On his tenth day as vice president, Dick Cheney established the National Energy Policy Development Group, widely referred to as "Cheney's Energy Task Force," to draft a proposal for a new energy strategy for the United States. The Task Force met throughout 2001 in closed meetings; its discussions and materials were not made available to Congress, the public, or the media. In other words, the meetings were secret. Even the names of participants in the meetings were not made public until the U.S. Supreme Court, the U.S. General Accounting Office, and several lawsuits filed by environmental organizations finally forced Cheney's hand.

The Task Force included heads of key government agencies and lobbyists and other representatives of the petroleum, coal, nuclear, natural gas, construction, and electricity industries—including Bechtel, Chevron, and Halliburton. Notably absent from the meetings was the head of the Environmental Protection Agency, Christine Todd Whitman. In 2003, she quit and then publicly criticized the administration for its "divisive" policies. At least one meeting included environmentalists and other advocates of clean energy, but their access to the Task Force was so limited in comparison to that of the business interests that they initiated the legal battles, which culminated in the public disclosure of the Task Force's work.

A draft of the Task Force's recommendations was shown to the press in April 2001. The first recommendation—listed under the heading "Strengthening Global Alliances" and followed by a graph showing Iraq oil output to the United States in 2000—was to "make energy security a priority of our trade and foreign policy." The second recommendation was for the United States to "support initiatives by [Mid East] suppliers to open up areas of their energy sectors to foreign investment."

Two years after they were drawn up, legal proceedings forced the Bush White House to reveal a series of lists and maps prepared by the Task Force that outlined Iraq's entire oil productive capacity and the foreign countries and companies lined up for contracts. Compiled in March 2001, the documents include detailed descriptions of Iraq's "super giant oilfields," oil pipelines, refineries, and tanker terminals. Two lists, entitled "Foreign Suitors for Iraqi Oilfield Contracts as of 5 March 2001," list more than sixty companies from some thirty countries with contracts in various stages of discussion for oil and gas projects across Iraq.

Apart from the minimal access provided under the Oil-for-Food Program, the companies within the Task Force were closed out of Iraq's oil market and were watching from the sidelines as the country's oil was divvied up to everyone but them.

With the tragedy of September 11, 2001, a series of paths that had been in development for at least a decade would finally join and find an open door for a march into Iraq. As former Joint Chief of Staff Admiral William J. Crowe once told Colin Powell, "First, to be a great President you have to have a war. All the great Presidents have had their wars. Two, you have to find a war where you are attacked."[59]

THE DRUMBEAT FOR WAR

The U.S. corporations beating the loudest drums for war against Iraq could be heard from the Defense Policy Board (DPB) within the Bush administration and the Committee for the Liberation of Iraq (CLI) from without.

While the Defense Policy Board has been in existence since 1985, it gained its greatest public recognition under President George W. Bush. The DPB is a federal advisory committee to the U.S. Department of Defense. Members are elected by the undersecretary of defense for policy and approved by the secretary of defense. Richard Perle began serving on the DPB in 1987. He then chaired the DPB under Wolfowitz, Rumsfeld, and Feith from 2001 until 2003, when he was forced to step down due to severe conflicts of interest. Seymour Hersh described the DPB thus: "Under Perle's leadership, the policy board has become increasingly influential. He has used it as a bully pulpit, from which to advocate the overthrow of Saddam Hussein and the use of preemptive military action to combat terrorism. Perle had many allies for this approach, such as Paul Wolfowitz, the Deputy Secretary of Defense." Nine members of the Defense Policy Board were connected to companies that received $76 billion dollars in defense contracts during 2001 and 2002. Several companies won contracts in Iraq, including Bechtel, which was represented on the DPB by Jack Sheehan.

The Committee for the Liberation of Iraq (CLI), less well known than the DPB, was founded in 2002 and can be viewed as an extension of the Project for the New American Century. Among the many people who overlapped between the two organizations are Robert Kagan and William Kristol, the cofounders and codirectors of PNAC, and Gary Schmitt, PNAC's executive director. Bruce Jackson left his vice presidency at Lockheed Martin to launch the CLI. The CLI's first president, Randy Scheunemann, is credited with writing the Iraq Liberation Act while working for Senator Trent Lott. Lockheed Martin is a client of Scheunemann's Washington lobbying firm, Orion Strategies. George Shultz, while maintaining his position as director and counsel for Bechtel, served as chairman of the CLI. Richard Perle was also a member.

The Committee's statement of purpose specifically announces that it was formed to replace "the Saddam Hussein regime with a democratic government." It claimed that Hussein had "acquired weapons of mass destruction" and that his government "poses a clear and present danger to its neighbors, to the United States, and to free peoples throughout the world." The Committee existed to mobilize U.S. and international support to end Saddam's regime and to work "beyond the liberation of Iraq to the reconstruction of its economy and the establishment of political pluralism, democratic institutions, and the rule of law." This last sentence provides the CLI proposal for what would differentiate the current Bush administration from the last. Whereas the former President Bush was seeking merely to replace Hussein with a more U.S.-friendly Iraqi leader, George W. Bush would replace Hussein, his economy, his government, and his laws.

More than fifty people ultimately served on the CLI's advisory boards, including many individuals whose companies—such as Bechtel and Lockheed Martin—would receive some of the most lucrative contracts in Iraq. In addition to Jackson and Scheunemann, Lockheed's representatives included Charles Kupperman, Lockheed Martin's vice president for space and strategic missiles, and Douglas Graham, Lockheed's director of defense systems. Retired General William Downing, who also served on the Committee, is on the Board of Directors of SAIC, a company that received seven contracts in Iraq. James Woolsey was a founding member, and his wife, Suzanne Woolsey, serves on Fluor's board, the third largest recipient of contracts in Iraq.

Committee members did more than profit from the war. They played a lead role in carrying the United States into war in the first place. They influenced public opinion through articles in the nation's leading newspapers, appearances on television and radio talk shows, and public speeches. Perhaps the clearest public articulation of the Committee's opinion, and ultimately that of the administration, was expressed by Ronald Reagan's former secretary of state, George Shultz. In an essay published in the *Washington Post* on September 7, 2002, under the headline, "ACT NOW. The danger is immediate. Saddam Hussein must be removed," Shultz laid out the most influential arguments for war, which have since been proven to have the least basis in fact. According to Shultz, the most compelling argument for war was the catastrophic and immediate threat posed to the United States by Hussein's weapons of mass destruction and his links to terror. Shultz writes:

> Self-defense is a valid basis for preemptive action. The evidence is clear that Hussein continues to amass weapons of mass destruction. He has also demonstrated a willingness to use them against internal as well as external targets. By now, the risks of inaction clearly outweigh the risks of action. If there is a rattlesnake in the yard, you don't wait

> for it to strike before you take action in self-defense. The danger is immediate. The making of weapons of mass destruction grows increasingly difficult to counter with each passing day. When the risk is not hundreds of people killed in a conventional attack but tens or hundreds of thousands killed by chemical, biological or nuclear attack, the time factor is even more compelling.

Shultz then argued that Iraq is "a major source of and support for terror and instability," and that by taking Hussein out, "a model can emerge that other Arab societies may look to and emulate for their own transformation and that of the entire region."

Few people reading these words, written by a former U.S. secretary of state and respected scholar, published in one of the world's most prominent newspapers, would feel anything less than terrified.

What Shultz fails to mention in this essay is his long-standing positions at Bechtel and the company's role in providing Hussein with the means to produce chemical weapons. He also fails to mention his personal role throughout the 1980s as a courtier to Hussein. In describing Hussein, Shultz writes, "No other dictator today matches his record of war, oppression, use of weapons of mass destruction and continuing contemptuous violation of international law." And yet, for more than a decade, Shultz, Bush the first, Reagan, Baker, Cheney, Wolfowitz, Rumsfeld, Kissinger, Eagleburger, and other U.S. corporate and government CEOs were content to supply this very same dictator with weapons, money, goods, and services.

The Committee for the Liberation of Iraq disbanded with the launch of the second Gulf War.

~

The corporations, the neoconservatives, and the George W. Bush administration are three interlocking groups with fluid demarcations. The players rotate through each group, generally wearing two to three of the monikers at once, making policy decisions based on the interlocking interests of each group. Iraq represents several things to these players: oil, wealth, regional power, and global power. Iraq presents them with the first opportunity for a truly imperial invasion in recent memory. Not only would the George W. Bush administration invade Iraq, it would fundamentally remake its internal government and policy-making structure such that the new country would serve U.S. interests. The corporations would gain access to the world's second largest oil supply and all of the wealth it generates, and the politicians would have their "new Iran"—a regional ally from which to protect Israel and guarantee U.S. access and hegemony over the entire region as well as much of the world.

SIX

THE ECONOMIC INVASION OF IRAQ

> The U.S. invasion of Iraq was not preemption; it was—like our war on Mexico in 1846—an avaricious, premeditated, unprovoked war against a foe who posed no immediate threat but whose defeat did offer economic advantages.
>
> —*Michael Scheuer, former senior CIA al-Qaeda expert and author,* Imperial Hubris[1]

Imagine that the United States is currently under a foreign military occupation. The foreign occupiers have thrown out the Constitution, the Amendments to the Constitution, and the Bill of Rights. They have actively and publicly participated in the writing of a new constitution to replace all three. There will be a popular vote on the new constitution, although its contents have not been made public and our elected officials have not seen the final version. Some of the contents have been leaked in the press, but the leaks are contradictory. Three days before the vote, a small handful of leaders—not all elected and at least one who represents the foreign occupier—meet and make changes to the constitution. That same day, a paper copy written several weeks earlier is made available to the public. However, there are only enough copies for less than one-third of eligible voters. Few legal experts have read or analyzed it, so those voters lucky enough to get a copy must decipher its meaning on their own. Political and religious leaders who feel some confidence about the larger points of the constitution advise their followers how to vote based on their limited knowledge. You know it is a critical vote that will impact every aspect of your life, so you go to the polls and cast your ballot.

This was the situation faced by the Iraqi public when, on October 15, 2005, 9.8 million people voted in a national referendum on a new constitution for Iraq. Around the world, the image used to capture the event was of Iraqi women, heads covered with the traditional Muslim Hejabb, smiling broadly and raising their freshly dyed purple fingers in a now iconic sign of success. When asked, most Iraqis said they were casting their votes in a spirit of hope, for a chance at change—any sort of change. Few, however, were voting on the actual content of the constitution. As described earlier, three days before the vote, U.S. Ambassador, and long-time Bush Agenda stalwart, Zalmay Khalilzad and a select handful of Iraqi leaders rewrote key provisions of the constitution. That same day, five million paper copies of an earlier draft were distributed among Iraq's 15.5 million eligible voters. Without access to adequate information, discussion, debate, or analysis, Iraqis voted on a document that they knew little about.

Therefore, few Iraqis could have known that the constitution they endorsed locked in the most crucial aspects of the Bush Agenda in Iraq: the military occupation, the economic invasion, and increased U.S. access to Iraq's oil.

The U.S. military invasion of Iraq, which began on March 17, 2003, was completed in six weeks. On May 1, President Bush hopped out of a fighter jet and onto the deck of the carrier ship, the USS *Abraham Lincoln.* Before the ship's assembled crew, he delivered what has since become one of the most infamous speeches of his career—not for its words, which few people likely remember, but for the banner hanging behind the president that read "Mission Accomplished." We know now, if we were unaware then, that the military engagement in Iraq was far from over; thirteen times more American troops have lost their lives in

Iraq since the end of "active engagement" than during it, and an estimated 23,000 Iraqi civilians have died since the president's declaration. While the mission was far from being accomplished, May 1, 2003, did mark an important transition in the U.S. war in Iraq: The military invasion was over, and the U.S. occupation of Iraq was set to begin.

Five days later, President Bush named L. Paul Bremer III presidential envoy to Iraq and the administrator of the U.S.-led occupation government, the Coalition Provisional Authority (CPA). During the next fourteen months, considered the period of formal U.S. occupation, Bremer enacted new laws with full legal force over Iraqis, all but a handful of which remain in effect today. These laws have gone virtually unnoticed and unreported in the United States, yet they go to the heart of the Bush Agenda. They lock in sweeping advantages to U.S. corporations, ensuring long-term U.S. economic gain while guaranteeing few, if any, benefits to the Iraqi people. In fact, they have set in place conditions for the ongoing inadequate provision of basic services, unemployment, underdevelopment, economic inequality, and violence for the foreseeable future.

THE U.S. OCCUPATION GOVERNMENT IN IRAQ: THE COALITION PROVISIONAL AUTHORITY

The United States, on behalf of the Coalition forces, laid out the framework for the occupation of Iraq, including the role of the CPA, in a letter submitted to the UN Security Council on May 8, 2003. The letter stipulated that the CPA was to "exercise powers of government temporarily, and as necessary" in Iraq. As the occupation government, it would provide security, allow the delivery of humanitarian aid, and eliminate weapons of mass destruction. In addition, it would provide "for the responsible administration of the Iraqi financial sector, for humanitarian relief, for economic reconstruction, for the transparent operation and repair of Iraq's infrastructure and natural resources, and for the progressive transfer of administrative responsibilities to such representative institutions of government."

The letter was part of the Bush administration's efforts to secure passage of UN Security Council Resolution 1438, by which the Council would formally acknowledge the occupation of Iraq by the Coalition and identify the CPA as the occupation government. The administration was eager to restore its relationship with the Council after the latter's refusal to endorse the invasion of Iraq, and to receive the United Nations' assistance in the occupation and reconstruction. Some members of the Council saw a humanitarian role for the United Nations now that the invasion was over, while others simply wanted to ensure that the United States would be held to its international obligations as an occupier in the next stage of the war. On May 22, the Council passed the resolution, adding its own specifications for the CPA, including the promotion of the welfare of the Iraqi people through the effective administration of the country, and in particular the restoration of security and stability along with the conditions necessary to ensure that the Iraqi people can "freely determine their own political future." The resolution also calls on the Coalition to "comply fully with their obligations under international law including in particular the Geneva Conventions of 1949 and the Hague Regulations of 1907."

Article 43 of the Hague Regulations (ratified by the United States) requires that an occupying power "take all the measures in his power to restore, and ensure, as far as possible, public order and safety, *while respecting, unless absolutely prevented, the laws in force in the country*" (emphasis added). Provision #363 of the U.S. Army's Law of Land Warfare repeats Article 43 word for word. The legal interpretation of these and related provisions is that an occupier is required to ensure that the lights are on, the water is flowing, the streets are safe, and the basic necessities of life are provided. However, the occupier is not permitted to make changes beyond those necessary to meet these obligations.

Not only would the Bush administration fail to meet the conditions of Article 43 and Provision 363 in Iraq, it would directly contradict them by radically altering Iraq's laws in order to restructure its entire economy. In June 2005, two years after the occupation of Iraq began, the transitional Iraqi government pleaded to the world from the opening page of its National Development Strategy that "we urgently need emergency/humanitarian interventions to provide basic services such as water, electricity, hospitals and schools."[2] The Bush administration failed to provide these necessities because it did not focus its efforts on the immediate provision of needs, but rather on the opening of Iraq to U.S. corporations and the advance of the Bush Agenda.

Both the CPA and its administrator, however, were replacements. The CPA replaced the Office of Reconstruction and Humanitarian Assistance (ORHA) and Paul Bremer replaced Jay Garner, a retired army lieutenant general. The reason for the replacements exposes a good deal about the Bush Agenda in Iraq. On January 20, 2003, Garner was named head of the U.S. Defense Department's ORHA. The staff of ORHA designed and began implementing the administration of postwar Iraq. The ORHA plan called for a U.S.-led occupation government of Iraq to last no more than three months. Two weeks after the "fall of Baghdad," on April 21, 2003, Garner and his staff arrived in Iraq. The very same night, Garner received a phone call from Defense Secretary Donald Rumsfeld informing him that in less than one month he would be replaced by Paul Bremer.[3] There has been a great deal of speculation about the reason for this change, often focusing on turf wars between the U.S. Defense and State Departments (Bremer spent twenty-three years working for the latter). Interdepartmental rivalry did exist, particularly between Donald Rumsfeld and Secretary of State Colin Powell. According to Garner, however, the reason for the change had more to do with differences over how aggressively the United States should be seen to dominate the political and economic process on the ground in Iraq than over who dominated the political process back in Washington.

Garner told BBC reporter Greg Palast, "My preference was to put the Iraqis in charge as soon as we can, and do it with some form of elections. . . . I just thought it was necessary to rapidly get the Iraqis in charge of their destiny."[4] As a retired army lieutenant general who had overseen assistance to the Kurds in Northern Iraq after the first Gulf War, Garner knew that the longer the United States appeared to be in charge, the more antipathy toward the United States would grow and expose U.S. troops to greater danger. Garner also disagreed with many of the Bush administration's more radical economic proposals, such as full privatization of Iraq's 192 state-owned enterprises, arguing that the Iraqis should be in charge of determining their own economic fate. Apparently, this attitude met with disapproval among his superiors in the administration, and Garner was summarily fired. As Garner put it, "The announcement . . . was somewhat abrupt."

Unlike General Jay Garner, corporate consultant Paul Bremer had no qualms about remaking Iraq and its economy in accordance with the Bush Agenda. After replacing Garner, Bremer wasted little time laying the foundation for U.S. political and economic dominance in Iraq. It would not be long before Bremer was christened the "Dictator of Iraq."

THE "DICTATOR OF IRAQ": L. PAUL BREMER III

> The transition from dictatorship to democracy will take time, but it is worth every effort.
>
> —*President George W. Bush, May 1, 2003*[5]

> Bremer is the dictator of Iraq. He has the money. He has the signature. Nothing happens without his agreement in this country.
>
> —*Lakhdar Brahimi, UN special adviser to Iraq, June 2, 2004*[6]

L. Paul Bremer III, "Jerry" to his friends, was administrator of the CPA from May 6, 2003, until June 28, 2004. During this time, the CPA had full executive, legislative, and judicial authority over Iraq and its people, and Paul Bremer had full authority over the CPA.[7] Paul Bremer is an attractive man. He looks younger than his sixty-one years and appears to be in good shape. He seems as comfortable traversing the Iraqi landscape in suit, tie, and army combat boots as he does appearing with MTV talk-show host Gideon Yago, discussing "the vision of most Iraqi youth." According to *USA Today,* while Bremer may not have any specific background in Iraq per se, one reason he was selected to be its de facto prime minister is "his charm."[8]

Bremer does in fact have other qualifications. Chief among them is his deep political and corporate connections dating back nearly four decades. At age 25, Bremer's first job out of graduate school was serving as a U.S. State Department official in Kabul, Afghanistan. This was the beginning of a twenty-three-year-long career at the U.S. State Department during which time Bremer came to know and work with Donald Rumsfeld and serve directly for Secretaries Henry Kissinger and George Shultz, two of the most powerful voices on behalf of U.S. corporations in Iraq and U.S. government engagement with Saddam Hussein. They helped blur the line in U.S.-Iraq relations between business and government interest—a line that Bremer would virtually erase.

After leaving the U.S. State Department in 1989, Bremer became the managing director of Kissinger Associates, where he provided advice and political risk assessments to the largest multinational corporations. In 1999, Bremer served as Chairman of the U.S. government's National Commission on Terrorism. After eleven years with Kissinger, Bremer became chairman and CEO of insurance company Marsh and McLennan's Crisis Consulting Practice, specializing in risk assessments and services for multinational corporations, particularly those operating in "crisis environments." Companies hired Bremer to lay out the various risks associated with operating in foreign countries, such as terrorist attacks or tsunamis, and then purchased insurance against those risks from him.

In a November 2001 paper entitled "New Risks in International Business," Bremer outlined the risks to multinational corporations associated with the implementation of corporate globalization policies.[9] Every policy Bremer describes in this paper was among those he himself implemented in Iraq a year and half later. Bremer walks through the devastating impacts of each policy on the local population—the same impacts that his policies would inflict on Iraq. Bremer warns companies that "the painful consequences of globalization are felt long before its benefits are clear." Bremer cites several specific globalization policies, such as privatization of state enterprises, deregulation of controlled industries, and reductions of tariffs and nontariff barriers to open up trade in goods and services. In the paper, Bremer explains that "privatization of basic services, for example, almost always leads to price increases for those services, which in turn often lead to protests or even physical violence against the operator." As for economic equality, Bremer says, "the process of globalization has a disparate impact on incomes," which in turn causes "political and social tensions." The harmful impact of the elimination of trade protections such as tariffs and quotas on local producers causes "enormous pressure on traditional retailers and trade monopolies" when "opening markets to foreign trade."

In conclusion, Bremer draws attention to a survey of American business executives in which 68 percent stated their belief that globalization increases their risks. "They are surely correct," Bremer asserts, "especially as many companies now have a much larger portion of their balance sheets exposed in emerging markets. Insurance underwriters have responded to the evolution of political risk by developing products that prudent managers and corporate boards would be wise to consider."

Bremer was therefore well aware that his policies would, at a minimum, reduce access to basic services and support for local businesses in favor of foreign businesses. He also knew the policies would increase inequality and political and social tension. However, he believed that he knew how to protect U.S. multinationals from the impact of these policies and therefore the policies went forward, ever clear on who the intended beneficiaries were.

Bremer was aided in Iraq by a staff comprised of people largely like himself—mostly American, mostly male, on loan from United States and a few British government agencies and banks, or institutions such as the IMF. Typical of Bremer's staff was Chris Foote, an economist with the Federal Reserve Bank of Boston who spent a little over two months in Iraq in the summer of 2003. What makes Foote unique is that, after his return, he wrote an article describing his "summer stint as an economist in Baghdad" for the bank's magazine, *Regional Review*.[10] In the photograph accompanying his article, Foote is shown standing with a safari hat on his head and clean, white tennis shoes on his feet, towering over the seven Iraqi men posing at his side. His arms are tightly wrapped around the two men closest to him while he smiles broadly for the camera from an airport runway in Mosul during a stopover on his way to Baghdad.

Foote describes the basic premise of his work in the article: "In many ways, the job is similar to the one I held at the Council of Economic Advisers before coming to Iraq. . . . The questions are all over the map, so I draw more from my experience teaching macroeconomics to undergraduates, and less from my own specialized research." In fact, while in Iraq, Foote rarely left the American-controlled "Green Zone"—moving mainly between Hussein's former Republican Palace (which housed the CPA) and his hotel.

He does tell one story, however, of riding in an armored SUV with an American military escort to determine the condition of a bank outside of the Zone. While on the road, he took a photograph of two young Iraqi boys pressing their faces up against his car window. Foote writes, "I often wish I could hop out of my fish bowl and learn more about Iraq's economy by talking with people in the streets." Foote's expertise, or apparent lack thereof, with respect to Iraq and its economy, did not impact his perceived ability to do his job. To Foote and those who hired him, it did not matter what existed in Iraq before or what the people in the street thought, just what would come next.

BREMER'S BLUEPRINT BY BEARING POINT, INC.

> We don't often get a chance in our lifetimes to see a country with such tremendous oil wealth and virtually no civilian commercial infrastructure get a whole new blueprint.
>
> —*Stephen Thomas, in Baghdad hoping to interest Bechtel in his British company specializing in food-service and telecommunications*[11]

It has been said so often that it is now repeated as gospel that the Bush administration had no plan for post-conflict Iraq. But the gospel is not correct. There was at least one clear plan—an economic plan—the blueprint for which was ready and in Bush administration hands at least two months prior to the invasion.

The 107-page three-year contract between the Bush administration and Bearing Point, Inc. of McLean, Virginia, lays out the president's economic agenda in Iraq. In return for $250 million, Bearing Point provides "technical assistance" to the U.S. Agency for International Development on the restructuring of the Iraqi economy to meet Bush administration goals.[12] In September 2004, Bearing Point's contract was extended for an additional three years. While you have probably never heard of Bearing Point, you are likely to recognize its former name, KPMG Consulting (the consulting division of accounting firm KPMG LLP). In October 2002, KPMG was forced to change its name due to the crisis that engulfed the accounting profession in the wake of the Enron/Arthur Andersen corruption scandals. After the name change, Bearing Point acquired most of Arthur Andersen's worldwide consulting operations. Bearing Point's lawyers and economists provide consulting services to more than two thousand business and government clients worldwide, including all U.S. federal government agencies.

With staff operating in the United States and Iraq, Bearing Point wrote the framework to restructure Iraq from a state-controlled economy to one that guarantees "free markets, free trade and private property"—among other goals—in just three short years. The contract, entitled "Economic Recovery, Reform and Sustained Growth in Iraq," calls on Bearing Point to "begin to reform, revise, extract or otherwise advise on changes to the policies, laws and regulations that impact the [Iraqi] economy." For example, Bearing Point is to recommend changes to laws "that impede private sector development, trade and investment." Bearing Point is performing these tasks under the supervision of the U.S. Agency for International Development and, originally, "under the overall policy direction of the Coalition Provisional Authority." For example, under the contract heading, "Privatization: Assessment and Support," Bearing Point provides analysis and assessment for undertaking a "mass privatization" of Iraq's state-owned industries.

While it may seem odd for the Bush administration to hire an outside company to turn its economic agenda into a specific set of implementable policies in Iraq, privatization of former government services is a trend that predates the Bush administration but which it has taken to new heights. Today, the U.S. federal government spends some $200 billion a year buying everything from interrogation services to trash collection to satellites from the private sector.[13] This, in turn, has led to an increasingly enriching "revolving door" between the public and private sectors, as former government employees enter the private sector and cash in on their intimate knowledge of the contracting process and access to those who make contracting decisions. In fact, earlier chapters have pointed to numerous cases of former and current members of the Bush administration who have participated in this lucrative revolving door.

The U.S. Agency for International Development and Bearing Point signed the company's first Iraq contract on July 24, 2003. However, Bearing Point's plan for Iraq was ready and in Bush administration hands at least five months earlier. Bearing Point's Draft Statement of Work, "Stimulating Economic Recovery, Reform and Sustained Growth in Iraq,"[14] was completed on February 21, 2003. While it was not available to the public, I was made aware of the document in May 2003, and I acquired a copy later that year. The final contract has since been made public through a Freedom of Information Act request and posted on the website of the Center for Public Integrity. The final contract is virtually unchanged from the February draft.

The extent to which the Bearing Point contract sets out to transform the Iraqi economy is astonishing. The company specifies changes in every sector of the Iraqi economy—from trade rules to banking and financial services, to public services, agriculture, housing, media, elections, and the structure of the government itself. It even specifies propaganda tools to sell these policies to the Iraqi public.

The contract asserts, "It should be clearly understood that the efforts undertaken will be designed to establish the basic legal framework for a functioning market economy; taking appropriate advantage of the unique opportunity for rapid progress in this area presented by the current configuration of political circumstances. . . . Reforms are envisioned in the areas of fiscal reform, financial sector reform, trade, legal and regulatory, and privatization."

From May 16, 2003, to June 28, 2004, the day the CPA was dissolved and Bremer hopped on a plane bound for the United States, he enacted exactly 100 Orders in Iraq. The Orders reflect Bearing Point's plan to a tee. Implemented at a fairly steady rate of a few every several days over the course of Bremer's fourteen-month reign, the Orders established new laws that continue to govern Iraq's political and economic structures.

HOW BREMER MADE THE LAW

Bremer's first act of business in Iraq was to define his powers as administrator of the CPA with the enactment of Coalition Provisional Authority Regulation Number 1 on May 16, 2003. Regulations and Orders were his primary tools. Regulations "define the institutions and authorities of the CPA" and Orders are "binding instructions or directives to the Iraqi people that create penal consequences or have a direct bearing on the way Iraqis are regulated, including changes to Iraqi law."[15] Both remain in force until repealed by the administrator or superseded by legislation issued by "democratic institutions of Iraq." Both take precedence over existing Iraqi laws, although all Iraqi laws that do not conflict with the Orders remain in place. The enactment of the Regulations and Orders requires only the approval of the administrator, Paul Bremer.

While one would assume that Bremer's Orders are akin to U.S. bills in that they create new laws, they are instead most similar to U.S. Presidential Executive Orders. Presidential Executive Orders are generally instructions by the president on how *existing* laws should be implemented, not *new laws* created exclusively by the executive branch. Executive Orders, like Bremer Orders, come into force simply by the will of the executive. To enact a bill in the United States, however, is a far more extensive process. The distinction is the difference between democracy and dictatorship.

First, a bill must be introduced by either a member of the House or the Senate and then referred to the appropriate committee. Committees can hold public hearings, get expert advice, seek clarification, write amendments, and so on. If the bill passes out of the committee, it goes back to the House or Senate, where it is debated, potentially amended, and voted on. If it receives a majority of votes, it moves on to repeat the identical process in the other chamber. If any changes are made as it passes out of that chamber, it must then go back to the original chamber for reconsideration until both the House and the Senate pass an identical version of the bill, or else the bill dies. If it succeeds in passing both chambers, it moves on to the executive branch, where it either receives the president's approval and signature to become a law or is vetoed. If the bill is vetoed, it returns back to the Congress, where a three-fourths majority vote can override the veto. If the veto is overturned, the bill becomes law; otherwise, the bill dies. Finally, if the people disagree with the new law, they can turn to the Judiciary to determine if it should be altered or even eliminated.

This long and arduous process is at the heart of American democracy. It is an attempt to ensure (as much as possible) that the nation's laws accurately reflect the will and the needs of those whose lives they will govern. In Iraq, the democratic process was thrown out with the Ba'ath water with the entry of the CPA and Paul Bremer. Bremer wrote the laws. Bremer enacted the laws. And the U.S. military continues to enforce the laws. This may be why UN Special Envoy Lakhdar Brahimi named Bremer the new "Dictator of Iraq."

If Bremer were only enacting limited Orders that sought to fulfill the requirements of an occupying nation as described in the Hague Regulations and the U.S. Army's Law of Land Warfare—such as the rehabilitation of water, electricity, and health care—the rejection of the democratic process might have been legitimate. The Bremer Orders not only failed to meet these obligations but also directly contradicted them by transforming the very foundation of the Iraqi economy. Bremer altered Iraqi law in order to implement an economic model preferred by the Bush administration—a model that has put the potential for Iraqi development fundamentally at risk, while ensuring increased access and economic benefit to U.S. corporations. With the authority vested in him by the Coalition and the United Nations, Bremer proceeded to use his Orders to implement the radical corporate globalization agenda of the Bush administration in Iraq.

FROM CAR HORNS TO CAPITALISM

> "The CPA stands for the Condescending and Patronizing Americans," a Baghdad diplomat told a *Newsweek* reporter. "So there they are, sitting in their palace; 800 people, 17 of whom speak Arabic, one is an expert on Iraq. Living in this cocoon. Writing pages. It's absurd," says one dissident Pentagon official.
>
> —*John Barry and Evan Thomas,* Newsweek, *October 6, 2003*[16]

One gets the impression when reading the Bremer Orders (not to mention the words of President Bush and others in his administration) that prior to the invasion, Iraq was a country without a government structure (other than the dictatorial ravings of Saddam Hussein), and that it lacked public services, businesses, and a skilled and educated citizenry. Nothing could be further from the truth.

Constitutional government in Iraq dates back to 1922. The constitution in place when the United States invaded Iraq was written in 1970, nine years before Saddam Hussein came to power. Although it placed significantly more power in the hands of the president and was clearly badly abused by Hussein, the Iraq constitution resembled the U.S. Constitution in many ways. For example, it included freedom of religion, expression, and association and it guaranteed equality before the law for all citizens without discrimination "because of sex, blood, language, social origin or religion." The Iraq Constitution also guaranteed a free education through the university level and access to maternal, child, and medical care for all. Not only were these services guaranteed, they were delivered. Prior to the first Gulf War in 1991 and even after eight years of war with Iran, Iraq was ranked 15 out of 130 countries on the 1990 United Nations Human Development Index. This index measures national achievements in health, education, and per capita GDP. Iraq was close to the top of the "medium human development" category, "a reflection of the Government's continued investment in basic social services."[17]

Prior to 1990, Iraq had the highest percentage of college-educated citizens in the Middle East and above average overall literacy rates. According to the World Health Organization, prior to 1991 health care reached approximately 97 percent of the urban population and 78 percent of rural residents, while the infant mortality rate was well below average for developing countries. Before 1991, the southern and central regions of Iraq had well-developed water and sanitation systems, and the World Health Organization estimates that 90 percent of the population had access to an abundant quantity of safe drinking water.[18] In fact, after the first Gulf War, when the U.S. military specifically targeted electricity and water systems for attack, Iraqi engineers rebuilt the electricity system in just three months.

In other words, while Iraq was a nation ravaged by a brutal dictator, war, and twelve years of economic sanctions, it was also a country of law, public services, education, and health care that was able to succeed *in spite of* its ruler because of a government and economic structure made functional by a knowledgeable and dedicated citizenry.

It is difficult to overstate how far the Bremer Orders go to overturn the existing Iraqi economic structure. The Orders cover virtually every aspect of Iraqi life, from the use of car horns to the implementation of a market economy. For example, Order #86, implemented on May 20, 2004, established a new traffic code for Iraq. Apparently, the existing Iraqi Traffic Law Number 48 of 1971, which Order 86 specifically replaced, was inadequate at such things as defining "roads," "lanes," and "U-turns," or regulating the appropriate terms by which a vehicle owner can change the color of his or her car, or guaranteeing the level of attention that Iraqi citizens gave to traffic safety. For example, Order #86 requires that "the Traffic Division in each Governorate shall reserve a week every year to try to focus the public's attention on traffic safety, which reduces accidents."[19]

While this Order and others like it are clearly intrusive, all in all, they are essentially benign. On the other end of the spectrum, however, is Order #39 on foreign investment, which does no less than "transition [Iraq] from a . . . centrally planned economy to a market economy" virtually overnight and by U.S. fiat. Similar sentences appear in several of the Orders discussed later in this section. Compare these with Article 1 of the Iraqi constitution that was in place when the United States invaded: "Iraq is a Sovereign People's Democratic Republic. Its basic objective is the realization of one Arab State and the build-up of the socialist system." And Article 12: "the State assumes the responsibility for planning, directing and steering the national economy for the purpose of (a) establishing the socialist system . . ."[20] In short, the Bush administration used the Orders to turn the Iraqi economy on its head.

There were, however, Hussein-era laws that fit into the Bush Agenda and therefore remained untouched—for example, Hussein's 1971 law barring Iraq's half a million public sector workers and those employed by public enterprises from joining or being represented by a union.

Order #39 is the most crucial Order defining Iraq's economy, present and future, and is therefore the focus of much of this chapter. First, however, I will discuss sixteen other key Orders that have transformed Iraq's economic infrastructure by rewriting such basic laws as those on taxation, banking, intellectual property rights, and trade. These Orders determine which Iraqis can hold certain jobs, who controls the government ministries, what is "acceptable" media behavior, how the media will be run and by whom, how elections will be run, and who is eligible to hold public office. Virtually every Orders remains in effect to this day.

THE BREMER ORDERS

Order #1[21] (May 16, 2003) provided for the "De-Ba'athification of Iraqi Society." All Ba'ath party members holding any position "in the top three layers of management in every national government ministry, affiliated corporations and other government institutions (e.g., universities and hospitals)" were removed from their jobs. With this Order, 120,000 of Iraq's most experienced and highest ranking civil servants, including engineers, scientists, university professors, doctors, skilled laborers, and government administrators from every ministry, were fired. The apparent goal of the Order was to eliminate any remnants of Hussein's regime, using Ba'ath party membership as an indicator of loyalty to Hussein and participation, or at least complicity, in his crimes. Under Hussein, however, Ba'ath party membership was a prerequisite for employment in the civil service. For many Iraqis, membership was simply the only route to a good job in the field of their choosing. It was in no way a direct indicator of support for the regime or criminal activity. These were the Iraqis with the most knowledge of the country's water, electricity, sewage, transportation, finance, health care, and education services, among others. And, in the first days of the U.S. occupation of Iraq, they were no longer allowed to work.

David Phillips, author of *Losing Iraq: Inside the Postwar Reconstruction Fiasco* and a former senior U.S. State Department Iraq expert who worked on post-invasion reconstruction planning for almost two years, argued that Order #1 served to remove the "opponents to the liberalization of Iraq's national economy." By eliminating those few Iraqis who would be in a position to know about the Orders, understand their impact, and interfere with their implementation, Bremer locked in the economic fate of the nation.

More than five months later, after government services had slid to a state of total disarray and the failure of Order #1 had grown painfully apparent, Bremer was forced to amend it with CPA Memorandum #8, authorizing a case-by-case review of individuals seeking to return to work. While this amendment was welcome, much of the damage had already been done, and the review process itself was slow and tarnished by political favoritism.

Order #2[22] (May 23, 2003) dissolved Iraqi "entities," including the Iraqi army and intelligence services. This order threw the entire Iraqi army—half a million men—out of work at a time when unemployment in Iraq was estimated at between 50 and 70 percent. With no jobs waiting for them and no way to provide for their families, many were believed to have taken their arms and joined the ranks of the insurgency.

Tragically, this was not the original plan. As early as 2002, U.S. military planners spoke of removing the nine thousand military officers and members of Hussein's various Special Forces, while retaining the four hundred thousand rank-and-file soldiers, the vast majority of whom were originally drafted under fear of death, unless they were charged with a crime and found guilty in public hearings. The soldiers were to provide police and rebuilding services. In addition, all would continue to receive their pay, whether or not they were put to work.

Retired Colonel Scott Feil told the U.S. Senate Foreign Relations Committee in 2002 that the rank-and-file soldiers would be "essential to preserving order" in post-invasion Iraq. For example, seventy thousand soldiers would form a national police force and a special Frontier Guard for Iraq's borders. W. Patrick Lang, a retired chief analyst at the Defense Intelligence Agency, said that carefully screened Iraqi units under U. S. control "would do much better against this enemy than we can."[23]

Instead, Bremer disbanded the military and refused to continue to pay their salaries. He handed security and reconstruction work to private U.S. contractors and the U.S. military. He even eliminated benefits to war widows and disabled veterans who were senior party members.

Like so many other aspects of the Bush invasion and occupation of Iraq, however, this decision was also handed over to a private U.S. company. On March 14, 2003, three days before the invasion, Ronco Consulting Corporation of Washington DC was awarded a $419,000 U.S. Defense Department contract to develop a plan to "disarm, demobilize and reintegrate the Iraqi armed forces."[24] Ronco was an odd choice, given that it is most recognized for its expertise at clearing land mines. In fact, it received demining contracts in both Afghanistan and Iraq. Ronco does have a long government history. For the first ten years or so of its existence, 98 percent of its contracts were with the U.S. Agency for International Development (USAID), and at least two of its chief executives are former USAID officials. However, Ronco probably got the contract because its other area of expertise is privatization consulting, which involves advising governments on how to shift services from the public to the private sector.

Faisal al-Istrabadi, deputy Iraqi ambassador to the United Nations, told David Phillips that disbanding the army "was an atrocious decision. . . . I don't understand why you took 400,000 men who are highly armed and trained, and turn them into your enemies." Phillips estimated that when one includes the families of the fired soldiers, Order #2 turned some 2.4 million Iraqis, roughly ten percent of the entire population, against the United States in the first month of the occupation.

Public outcry was so great and the immediate results so negative that Bremer was once again forced to reverse an Order, and on August 7, 2003, he implemented Order #22, establishing a "New Iraqi Army." All former members of the old army were allowed to sign up again, even those who held former leadership roles in the Ba'ath party as long as they first received CPA permission. Those who joined faced an initial service period of twenty-six brutal months, largely spent fighting fellow Iraqis. They also increasingly became the targets of deadly violent acts. From June 2003 to January 2006, an estimated 750 Iraqis were killed by insurgents while standing in lines waiting to sign up for service, while 3,800 Iraqi military and police officers have died.[25] Out of the original half a million, by the end of 2005, only 105,700 men were counted as part of either the Iraqi National Guard or the Iraqi Armed Forces, and only one Iraqi Army battalion was considered capable of effective combat independent of U.S. forces.

The lack of a viable Iraqi military, in turn, has been the primary justification offered by the Bush administration for the continued presence of U.S. troops in Iraq.

Order #12[26] (June 7, 2003; replaced with **Order #54,**[27] February 24, 2004) outlined the "Trade Liberalization Policy" for Iraq. Among other things, it suspended "all tariffs, customs duties, import taxes, licensing fees and similar surcharges for goods entering or leaving Iraq." Where existing Iraqi law sought to protect the local economy from foreign competition, the trade liberalization law eliminated all protective barriers in one fell swoop—leaving the market suddenly fully exposed. This led to an immediate inflow of cheap foreign consumer products, which, in turn, devastated local producers and sellers who were not prepared to meet the challenge of their global competitors. As discussed in earlier chapters, policies identical to the Trade Liberalization Policy have been implemented at the behest of the WTO, World Bank, and IMF with devastating and even deadly effects in countries such as South Korea, India, and Zambia.

When Iraq's market was opened, some of the first companies through the door were U.S. wheat exporters. Within the first year of the invasion, total U.S. wheat exports to Iraq went from zero to $190 million. Prior to the first Gulf War, U.S. exporters shipped nearly a million tons of wheat each year to Iraq, garnering up to 40 percent of the market in some years. U.S. companies are now aggressively lobbying for even greater access. U.S. Wheat Associates, representing twenty U.S. wheat-producing states and the U.S. Department of Agriculture, have announced that they "are working on activities that will hopefully strengthen the Iraqi Grain Board and provide a better understanding of the U.S. wheat marketing system," including sponsoring trips for members of the Iraqi Grain Board to the United States. John King, vice chairman of the USA Rice Council, told the U.S. House Agriculture Committee in June 2004, "The U.S. rice industry wants to play a major role once again in supplying rice to Iraq . . . renewed Iraqi market access could have a tremendous impact in value-added sales." King added, "The liberation of Iraq in 2003 by coalition forces has brought freedom to the Iraqi people. The resumption of trade has also provided hope for the U.S. rice industry."

Iraq has not been self-sufficient in food production since the 1950s and has always relied on imports, much of these from the United States. The problem emerging today, however, is that without farming and price supports, Iraqis will no longer be able to compete with the imports and contribute their share to the Iraqi farming sector. In addition, Iraqis may no longer be able to afford the import of American products.

Before the invasion, the Iraqi government heavily subsidized the farming sector. Farming "inputs" such as seeds, fertilizer, pesticides, sprinklers, and tractors were subsidized often at a third or even a fourth of the market price. The government leased land for one cent per donam, about six-tenths of an acre, a year. It bought the country's main crops, wheat and barley, at a fixed price, whether they were usable or not. And it ground up the grain and provided it free as flour to the people each month as part of the guaranteed food program in which every family received a basket of flour, sugar, tea, and other necessities.[28]

Bremer began changing these policies shortly after the occupation began. Trevor Flugge, the CPA's senior civil administrator for agriculture, described the CPA's changes. Invoking a WTO's mantra, Flugge explained that subsidizing farming supplies is "all wrong"; instead, the new government would provide assistance in the form of technology and education and "the market will take care of the rest."[29] These same "market measures" implemented under the WTO's Agriculture Agreement led South Korean Lee Kyung Hae and tens of thousands of farmers like him the world over, to take their own lives in a desperate response to the sudden economic dislocation brought upon them, their families, and their communities.

Abu Ahmed Al-Hadithi, an Iraqi vegetable seller at the Al-Adhamiyah market, described the impacts already being experienced in Iraq to Dahr Jamail, one of the only independent American journalists who remained in Iraq throughout the invasion and occupation: "The economic situation is so bad now. The costs of gas and food are going up so high; so even if we make more now, everything is costing more. . . . In Saddam's days we grew all our own vegetables to sell . . . but now so many are coming from outside of Iraq and it is causing us to sell them for less. So I make less profit now, and I have nine people to take care of, and it has made my life very difficult."[30]

Order #14[31] (June 10, 2003) defined "prohibited media activity" as that which, among other things, "incites violence against Coalition Forces or CPA personnel," "advocates alternatives in Iraq's borders by violent means," or "advocates the return to power of the Iraqi Ba'ath party or makes statements that purport to be on behalf of the Iraqi Ba'ath party."

In testimony provided to the U.S. Senate Democratic Policy Committee in 2005, Don North, a CPA contractor who worked on Iraq's media network, described how the CPA enforced its authority over Iraq's media. For example, if the U.S. Army or the CPA found news stories offensive, "they visited the offices of offending newspapers and often left them padlocked and in ruins. There was no mediation, no appeal." North wrote that the CPA's "Code of Conduct, which bans 'intemperate speech that could incite violence,' is selective democracy, similar in spirit if not in effect to censorship by Saddam Hussein." He added, "If the *Washington Post* reported terrorist threats or bin Laden statements in Baghdad today, it would probably be closed down."

Order #14 was the basis for the closure of *al-Arabiya's* office in Baghdad as well as the *al-Hawza* newspaper, published by Shi'a leader Muqtada al-Sadr. A Coalition spokesperson said that the paper was closed because it incited violence against the CPA. Al-Sadr's followers protested the closure by demonstrating in the thousands in several cities across Iraq. Declaring that peaceful protests had become useless, on April 5, 2004, al-Sadr then urged his followers to "terrorize" their enemy.[32]

Order #17[33] (revised on June 27, 2004) granted full immunity from Iraqi laws and the Iraqi legal system to Coalition military forces and all foreign contractors, including private security firms. Non-Iraqi members of the military, corporations, corporate subcontractors and their employees cannot be held subject to Iraq's laws to this day. Thus, if in the course of his or her duties, a soldier or contractor commits murder, torture, rape, dumps toxic chemicals, poisons drinking water, starts an oil spill, rips off an Iraqi subcontractor, abuses an Iraqi employee, or the like, the injured Iraqi has no legal recourse other than to try to bring charges in foreign courts under foreign laws.

As Adam Price, a member of the British Parliament, commented, "How is anyone in Iraq expected to bring a case in the British courts? It is taking the idea of diplomatic immunity and applying it to 130,000 troops. There is a danger that you are actually going from immunity to being able to act with impunity."[34]

For example, neither the U.S. soldiers nor the private security guards found guilty of committing illegal acts against Iraqi prisoners at Abu Ghraib prison can be charged or forced to stand trial in Iraqi courts for their crimes. The Iraqis and their families directly affected by the acts have no legal recourse other than to have their cases heard in American courts. Furthermore, none of the millions of Iraqis denied water and electricity due to the failure of U.S. corporations to deliver these services have legal standing in Iraqi courts.

The Order also gives foreign contractors freedom from all income taxes, corporate taxes, and sales taxes, and denies Iraqis the ability to inspect contractor vehicles or require any sort of licensing or registration fees. Contractors do not have to pay tolls and are granted "freedom of movement without delay throughout Iraq." The U.S. government maintains control of Iraqi civil airspace, so Coalition planes are off-limits to Iraqi inspectors, and Coalition goods can enter the country without import fees or inspections.

It is unusual to have such blanket immunity for nonmilitary actors. However, it is relatively unheard of to have that immunity continue for both the military and nonmilitary *after* the formal occupation period ended and authority was handed over to the Iraqis on June 28, 2004. Rather than eliminate or renegotiate the Coalition's terms of immunity as most of the world anticipated, Bremer explicitly excluded Order #17 from the transfer of authority to interim Prime Minister Iyad Allawi. Instead, Order #17 remains under U.S. authority and will be in effect until the last soldier of the last Coalition military unit departs Iraq.

Order #37[35] (September 19, 2003; amended with **Order #49**,[36] February 19, 2004) replaced Iraq's progressive tax strategy (by which the more you earned, the more you paid in taxes) with a flat tax—that long-desired but never-achieved dream of the American right wing. The law dropped the existing tax rate on corporations from a high of 40 percent to the flat rate of 15 percent, which is now in effect for both individuals and corporations. As the *Washington Post* reported, "It took L. Paul Bremer, the U.S. administrator in Baghdad, no more than a stroke of the pen Sept. 15 to accomplish what eluded the likes of publisher Steve Forbes, Reps. Jack Kemp (R-N.Y.) and Richard K. Armey (R-Tex.), and Sen. Phil Gramm (R-Tex.) over the course of a decade and two presidential campaigns."[37] While the poorest Iraqis are exempt from paying taxes, everyone else—whether they barely make a living or are among the wealthiest in society—pays the same tax rate.

Order #40[38] (September 19, 2003; replaced with **Order #94**,[39] June 6, 2004), the "Bank Law," opens the Iraqi banking sector to foreign ownership. Under Order #40, foreign banks were allowed to enter this previously closed sector and purchase up to 50 percent of an Iraqi bank. The total number of licenses for banks controlled by foreign companies was limited to six through December 31, 2008. One year later, Bremer expanded the Bank Law with Order #94, allowing foreign banks to purchase 100 percent of Iraqi banks and to open subsidiaries and branches without restriction. In addition, banks owned by Iraqis are not to be granted any legal preferences over foreign-owned banks (although the opposite is possible). Therefore, the Iraqi government can not require, for example, that a certain percentage or a minimum number of banks in Iraq be owned by Iraqis, or that tax breaks or subsidies be given to small, community owned and run banks committed to providing services for groups or locations that a foreign bank might consider unprofitable.

The HSBC Bank of London was one of the first foreign banks authorized to operate in Iraq and to purchase majority ownership (70 percent) of a private Iraqi bank, the Dar Elsalam Investment Bank, with fourteen branches across Iraq.

JPMorgan Chase received an early contract to run the Trade Bank of Iraq, a consortium of thirteen banks. Chase is the second-largest bank in the United States and it counts Condoleezza Rice as a former board member. Chase was charged with knowingly manipulating Enron's shareholders. On June 14, 2005, Chase agreed to a $2.2 billion settlement with the University of California, the lead plaintiff in the Enron shareholders' class action suit. A University press release explained that JPMorgan Chase allegedly "set up false investments in clandestinely controlled Enron partnerships, used offshore companies to disguise loans and facilitated phony sales of phantom Enron assets. As a result, Enron executives were able to deceive investors by reporting increased cash flow from operations and by moving billions of dollars of debt off Enron's balance sheet, thereby artificially inflating securities prices."

When the banking sector of New Zealand was abruptly opened to foreign private investment in 2001, every one of the nation's banks, including the bank of New Zealand, came under foreign control. Affordable financial services and low-cost loans quickly dried up—so much so that the government of New Zealand has proposed setting up a new bank, the People's Bank, to be owned and operated by the government itself in order to redress the inequities of the foreign-owned banks.[40] Here it is interesting to note the U.S. banking laws that Bremer did not bring to Iraq. For example, basic consumer protections such as the Community Reinvestment Act, which requires banks to make credit available in low-income neighborhoods, or the Truth in Lending Act, which requires full disclosure to consumers of the cost of loans.

Order #62[41] (February 26, 2004) enabled Bremer to determine which Iraqis could run for or hold public offices. "When determined necessary for security and public order within Iraq, the Administrator of the CPA may disqualify an individual from participating in an election as a candidate, and for accepting a nomination to, or holding public office, at any level," if that individual has, among other things, "publicly espoused political philosophies or legal doctrines contrary to the democratic order and rule of law being established in Iraq." In other words, if Bremer decided that someone opposed the occupation, the CPA, or its laws, he could simply keep them from serving in local, regional, or national government—or in any of the government ministries. This is one of the small handful of Orders that was actually rescinded on June 28 with the handover of authority to the Iraqis. Apparently, Bremer did not wish Allawi to possess this power.

Order #65[42] (March 20, 2004) established an Iraqi Communications and Media Commission, and Bremer appointed its members. The commission has sole responsibility for licensing and regulating telecommunications, broadcasting, information services, and other media in Iraq. The order stipulates that it will establish a regulatory regime that provides for "the competitive provision" of these services. Furthermore, it has the authority to shut down news agencies, extract written apologies from newspapers, and seize publishing and broadcast equipment.

Order #57[43] and **Order #77** place American representatives in key decision-making positions within each government ministry for terms that last five years—well after the permanent elected government of Iraq takes office in 2006. These two Orders in particular bear a striking resemblance to laws imposed on Iraq by Britain after World War I. Britain's occupation of Baghdad began in March 1917. Under a mandate of the League of Nations in 1921, Britain was to begin preparing Iraq for self-government under British tutelage. In order to ensure its indirect rule of Iraq, the British required that its officials be appointed to specified posts in eighteen departments of the new Iraqi government "to act as advisers and inspectors."[44] **Order #57** (February 5, 2004) established an Inspector General—handpicked by Bremer—with five-year terms within every

Iraqi Ministry. The Inspector Generals can, among other things, perform audits and investigations, promulgate policies and procedures, and have full access to all offices, employees, contracts, and all other materials of the Ministries. **Order #77**[45] (April 18, 2004) established the Board of Supreme Audit. Bremer appointed the board president and his two deputies, who are to serve five-year terms. The auditors can be removed only with a two-thirds vote of Iraq's parliament. To date, no such vote has occurred. This Board oversees inspectors in every Ministry, with wide-ranging authority to review government contracts, audit classified programs, and prescribe regulations and procedures.

Order #80[46] (April 26, 2004), **Order #81**[47] (April 26, 2004), and **Order #83**[48] (May 1, 2004) rewrote Iraq's patent, trademark, and copyright laws, just two months before the handover, to ensure guaranteed access and protections to foreign products and producers. All three Orders have identical preambles that explain the need to "transition [Iraq] from a non-transparent centrally planned economy to a free market economy," which recognizes that "companies, lenders and entrepreneurs require a fair, efficient, and predictable environment for protection of their intellectual property." In order to achieve this end, they change Iraq's laws so that they come into compliance with highly controversial WTO requirements on intellectual property rights.

The World Health Organization found that the imposition of the same intellectual property rights requirement would increase the price of medicines for HIV, AIDS, and malaria by some 200 percent in Andean nations. As enforced by the WTO, these rules have made the practice of sharing and saving seeds by farmers illegal as agribusiness companies increasingly patent the rights to seeds. Pharmaceutical companies have used these rules to patent indigenous knowledge and make traditional medicines inaccessible to those who have used them free of charge for generations. Such regulations make it increasingly difficult for governments to regulate in the public interest at the expense of private industry.

Order #97[49] (June 7, 2004), the "Political Parties and Entities Law," may go the furthest to demonstrate the farce of the June 28, 2004, handover of authority. With Order #97, Bremer established a seven-member commission that has the power to disqualify political parties and any of the candidates they support. Political parties or candidates can be disqualified for several reasons, including if they do not follow a code of conduct established by the commission or if they are deemed by the commission to be "associated with armed forces, militias, use of terrorism, or incitement to violence, or if they have been directly or indirectly financed by such forces." In this way, the Bush administration retained a firm stamp of approval over any party or individual that was even allowed to participate in Iraq's elections.

Order #100[50] (June 28, 2004) handed over authority for all remaining Orders to the newly appointed Iraqi interim prime minister, Dr. Iyad Allawi. The Orders were then upheld as Iraqi law with the passage of the October 15, 2005, constitution.

ORDER #39: FOREIGN INVESTMENT

Foreign investment rules have been the cherry on top of the corporate globalization pie for decades. They have long been widely sought after but rarely achieved. It was the adamant opposition of developing country governments to these same laws that contributed significantly to the collapse of WTO talks in Seattle in 1999 and again in Cancun in 2003. To avoid another collapse, the provisions were removed altogether from the agenda of the WTO's 2005 Hong Kong ministerial. Foreign investment provisions continue to be one of the most controversial elements of the North American Free Trade Agreement (NAFTA) between the United States, Mexico, and Canada. Global opposition to these laws led to the defeat of the Multilateral Agreement on Investment. In Iraq, however, there was no opposition to overcome. Bremer simply put the foreign investment provisions into force with the stroke of a pen.

Order #39 (September 19, 2003) is the foreign investment Order.[51] It includes the following provisions: (1) privatization of Iraq's state-owned enterprises; (2) 100 percent foreign ownership of Iraqi businesses; (3) "national treatment"—which means no preferences for local over foreign businesses, which has allowed for a U.S. corporate invasion of Iraq; (4) unrestricted, tax-free remittance of all profits and other funds; (5) forty-year ownership licenses; and (6) the right to take legal disputes out of Iraq's courts and into international tribunals. More than any other Order, it fully embodies the Bush Agenda in Iraq: the creation of a U.S. corporate haven that will act as a model and jumping-off point for the rest of the region.

Provision #1: Privatization

Under the heading, "Treatment of Foreign Investors," Order #39 states, "The amount of foreign participation in newly formed or existing business entities in Iraq shall not be limited, unless otherwise expressly provided herein." Business entities are defined as including, but not limited to, "state-owned enterprises." Under the heading, "Areas of Foreign Investment," Order #39 states, "Foreign investment may take place with respect to all economic sectors in Iraq, except . . . the natural resources sector involving primary extraction and initial processing remains prohibited. In addition, this Order does not apply to banks and insurance companies." This exception means that the extraction and initial processing of oil in Iraq is excluded from the provisions of the Order. (This exclusion and the impact of the Bremer Orders and the Bush Agenda on Iraq's oil prize are the topic of the final section of this chapter. The banking and insurance sectors are the subject of several other Orders, including Order #40 discussed earlier.)

Order #39 therefore allows for the privatization of all of Iraq's 192 government owned industries (except oil extraction), including everything from water to electricity, schools to hospitals, factories to airlines, newspapers to television stations, food to housing programs. The Order overturns all or part of Articles 13 and 16 of the pre-invasion constitution. These articles specifically *prohibit* private ownership of "natural resources and the basic means of production" and explain that "ownership is a social function, to be exercised within the objectives of the Society and the plans of the State." Furthermore, in Article 11, the State ensures "maternal and child care"; Article 27 guarantees free education in "primary, secondary, and university stages, for all citizens"; and Article 33 guarantees "free medical services" in both cities and rural areas. Privatization of these sectors under Bremer's Order #39 would certainly lead to the violation of Articles 11, 27, and 33, which is explicitly illegal under the Hague Regulations.

"Privatization" occurs when a public entity, such as a drinking and sewage water system—owned, operated, and/or managed (usually all three) by a government—is turned over to a private entity such as a corporation. The United States, for example, has an almost fully state-run water system. Governments at the federal, state, and local level own, operate, manage, and *pay for* most of the nation's water systems. Water is a resource provided as a service of the government, and therefore the fees collected are intended to support the system rather than to generate extra revenue or profits. In the United States, water is also a resource that is heavily subsidized by the government to ensure that all Americans have access to water and sewage services. Thus, while many Americans may question the quality of their water, virtually everyone can afford it. However, when water is provided by a private entity, it becomes a resource that must generate revenue. To change the system from one that supports itself to one that also generates revenue, the company must either reduce costs or increase fees. Usually, a combination of the two is chosen, often resulting in increased prices, reduced services, and reductions in numbers of workers and/or worker wages or benefits.

As Paul Bremer himself noted in November 2001, "Privatization of basic services, for example, almost always leads to price increases for those services, which in turn often lead to protests or even physical violence against the operator."

The CPA began conducting in-depth assessments of Iraq's largest state-owned enterprises (SOEs), considered to have the greatest potential for privatization in the first months of the occupation. These assessments are posted on the U.S. Commerce Department website and include numbers of employees, amount of damage suffered during and after the war, the amount of money necessary to bring the facility to prewar levels, whether the company can be considered profitable and commercially viable, and its potential for privatization. Many companies were determined to be "good candidates for privatization."

In October 2003, Thomas Foley, director of private sector development for the CPA, announced a list of the first state enterprises to be privatized. The list included cement and fertilizer plants, phosphate and sulfur mines, pharmaceutical factories, and the country's airline.[52] With anywhere from 50 to 70 percent of the Iraqi workforce unemployed at the time, additional layoffs were unacceptable and the list was met with immediate resistance by Iraq's workers.

Again, in Bremer's own words, "Restructuring inefficient state enterprises requires laying off workers."[53] Or, to quote Bearing Point's contract, "The need for new job creation will only increase when workers are laid off from these [state-owned] firms as they are restructured for privatization, or liquidation." Even those workers who still had jobs in Iraq at the time (remember that approximately 620,000 had lost their jobs under Orders #1 and #2) only received "emergency pay," which was mandated by the CPA and amounted to about half of what they made before the war. At the same time, prices skyrocketed due to inflation rates of as high as 36 percent and the social safety net was being decimated, as one government program after another was terminated or rolled back. In response to the organized resistance and negative publicity that followed, Bremer was forced to halt immediate privatization.

The process continued, however, just on a slower quieter track.

In April 2004, Iraq's minister of public works said that she was considering privatizing Iraq's water sector to "fund essential works."[54] As an occupier, the U.S. government was obligated to ensure that water was provided to the people of Iraq. It is therefore telling that the discussion turned not to the failings of the U.S. government, nor of Bechtel (which has the contract to rebuild the water system), but rather to privatization of the service.

The Bush administration's commitment to Iraq's privatization effort was reaffirmed in September 2004 when Bearing Point's contract was extended for another three years with a renewed emphasis on privatization of the electricity and telecommunication sectors as part of its effort to develop "a policy-enabling environment for private sector-led growth in Iraq." Bearing Point is now aided by the Louis Berger

Group of New Jersey, which received a contract in October 2004 to provide (among other services) assistance in restructuring and privatization of state-owned enterprises.

In February 2004, the U.S. Commerce Department held a "Doing Business in Iraq" conference in Washington DC, attended by some five hundred U.S. companies, including Bechtel, Boeing, Caterpillar, Microsoft, Motorola, and Fluor. This conference took place immediately following vocal criticism by the Iraqi Governing Council's top representative in Washington that the United States was passing over Iraqi firms in awarding reconstruction contracts. In December 2004, the U.S. Government's Overseas Private Investment Corporation and the World Bank's International Finance Corporation were among the hosts of the "Iraqi-Jordanian Investment Symposium" in Amman, Jordan. The symposium "exposed the multitude of business opportunities in Iraq" for, among others, "privatization specialists." Interested companies can frequent websites such as www.export.gov/Iraq or www.iraqprocurment.com to learn more about similar meetings taking place monthly in their home countries and neighborhoods.

The impact of Bremer Order #39 and the influence of the Bush administration crystallized in May 2005, when Mohammad Abdullah, the interim minister of industry, announced plans for Iraq's Ministry of Industry and Minerals to "partially privatize most of its 46 state-owned companies as part of the government's plan to establish a liberal, free-market economy."[55] Abdullah said, "We have plans to develop and pave the way for domestic and foreign investment in these sectors."

A few months later, the U.S. Department of Commerce posted lists of all Iraqi state-owned enterprises, including descriptions of their estimated value, on the department's www.Export.gov/Iraq website under the heading "Pursue Iraqi Tenders and State-Owned Enterprise (SOE) Opportunities." According to one fact sheet, "The textile industry in Iraq has the potential to be turned around through private sector investment and leasing strategies." In addition to textiles, the website provides lists in construction materials, engineering industries, food and drugs, industrial services, and chemical and petrochemical sectors.

In June 2005, the transitional Iraqi government announced that its national development strategy for 2005 to 2007 included "laying the groundwork for eventual privatization of State-owned-enterprises . . ."[56] Manufacturing, electricity, telecommunications, and transportation topped the government's privatization list.

Then, in November 2005, the U.S. Commerce Department arranged a three-city American tour for the new Interim Iraqi Minister of Industry and Minerals, Osama al-Najafi, to discuss investment opportunities in his ministry, especially in the areas of cement production, mining, and chemical production. More than 170 representatives of U.S. corporations, banks, financial institutions, and investment funds attended the minister's events in Washington, DC, New York City, and Detroit. According to the U.S. Commerce Department, "At all three of the minister's stops he emphasized the desire for more foreign participation in Iraq's economy."

The transitional Iraqi government hosted a similar conference in London a few months earlier to present the international business community with the opportunity to invest in Iraq's newly partially privatized telecommunications sector. Hundreds of companies attended the conference, where they heard papers by U.S. telecommunications giants Nortel and Motorola. As described by Iraq's Minister of Communications, Jowan Masum, the government hoped to signal "to the investor that the telecommunications sector is not wholly state owned and is not monopolised."[57]

Such conferences and meetings among global corporations to discuss privatization and investment in Iraq have taken place virtually unabated since before the invasion. There are many examples to draw from in addition to those already listed, such as "Iraq Procurement 2005—Meet the Buyers," at which Iraqi ministers from industries such as agriculture, construction and housing, education, industry, technology, science and technology, telecommunications, and water, met with U.S. and other global corporations in Amman, Jordan, to "further their business relations with the rest of the world." Corporate sponsors of the event included Shell, DHL, Raytheon, Midas, DynCorp, Hertz,

Microsoft, and Pfizer, among others. The sponsors of the 2004 conference also included ChevronTexaco and ExxonMobil.

Provision #2: 100 Percent Foreign Ownership of Iraqi Businesses

Order #39 allows for 100 percent ownership of Iraq's businesses by non-Iraqis. It contradicts and overrides Iraqi commercial laws that prohibited "investment in, and establishment of, companies in Iraq by foreigners who are not resident citizens of Arab countries."[58]

Bans on foreign ownership and requirements that foreign companies partner with local businesses are common tools used by developed and developing countries alike, particularly at the local level, to ensure local economic development. Such requirements help guarantee that money earned in a country or community stays there and services local growth and employment.

To date, the lack of security in Iraq and the ever-increasing hostility toward the occupation has kept most foreign corporations from purchasing Iraqi state-owned or private businesses, or at least from publicly acknowledging that they have or intend to do so. However, the ongoing violence has not stopped U.S. corporations from taking full advantage of Order #39's other provisions.

Provision #3: National Treatment

Order #39 states, "A foreign investor shall be entitled to make foreign investments in Iraq on terms no less favorable than those applicable to an Iraqi investor, unless otherwise provided herein." "National Treatment" is a standard element in trade and investment agreements, which restricts governments from preferencing domestic businesses or workers over foreign businesses or workers. National Treatment made it unnecessary for Bremer, and impossible for the Iraqi government, to require that Iraqis be given preference in the reconstruction effort over Americans. Foreign companies operating in Iraq were therefore granted the right to perform all of the reconstruction work without having to hire Iraqi workers, use Iraqi companies, or even use Iraqi products.

The result is a U.S. corporate invasion of Iraq. According to the U.S. General Services Administration's Federal Procurement Data System and figures collected by the Center for Public Integrity, a Washington, DC-based think tank, by mid-2005, more than 150 U.S. companies had been awarded contracts totaling more than $50 billion—more than twice Iraq's entire GDP—to work on the Iraq War and reconstruction. The largest single recipient, Halliburton, has contracts worth more than $11 billion. Excluding Halliburton, the seven largest contracts, totaling nearly $23 billion, are for companies working almost exclusively on some form of construction—be it of buildings, highways, bridges, electricity, water, oil, or sanitation systems. These companies are: Parsons Corporation of Pasadena, CA ($5.3 billion); Fluor Corporation of Aliso Viejo, CA ($3.75 billion); Washington Group International of Boise, ID ($3.1 billion); Shaw Group of Baton Rouge, LA ($3 billion); Bechtel Corporation of San Francisco, CA ($2.8 billion), Perini Corporation of Framingham, MA ($2.5 billion); and Contrack International, Inc. of Arlington, VA ($2.3 billion). All of the reconstruction work for which these companies have been contracted can and should be done by Iraqis themselves.

National Treatment is also a powerful tool used by corporations to circumvent domestic regulations on the environment, public health, and worker and consumer safety. Virtually every challenge brought to such laws under similar foreign investment rules in the NAFTA includes claims that the government violated National Treatment. The fact that NAFTA's National Treatment has allowed multinational corporations to overturn laws in Canada, the United States, and Mexico bodes poorly for Iraqi environmental, health, and public interest laws—or those that any new government may wish to enact.

For example, a U.S. corporation used NAFTA's National Treatment and Foreign Investment provisions to force the Canadian government to reverse a ban on the gasoline additive MMT, a known carcinogen. MMT is banned in Europe and California. In 1997, the Canadian parliament passed its own ban of MMT (which is made in the United

States and sold in Canada) because it was proven to pollute ground water and was, in the words of Canadian Prime Minister Jean Chretien, an "insidious neurotoxin." Virginia's Ethyl Corporation, the producer of Canada's MMT, first tried to stop the legislation in the Canadian parliament. When it was unsuccessful, it turned to the NAFTA, where it launched a $250 million suit. Ethyl argued that it should be reimbursed for future profits it would have earned from the sale of MMT and that the mere *discussion* of the ban in the Canadian Parliament damaged Ethyl's "good name" and therefore its profitability. Ethyl demanded compensation or the elimination of the law. Because the Canadian government believed it would lose the case at a NAFTA tribunal, it reversed the ban, paid Ethyl $13 million, and wrote the company a formal letter of apology.

Provision #4: Unrestricted Repatriation of Profits

This provision may be the purest act of imperial hubris put in place by Bremer. Under "Implementing Foreign Investment," Order #39 authorizes foreign investors to "transfer abroad without delay all funds associated with [their] investment, including: shares or profits and dividends; proceeds from the sale or other disposition of its foreign investment or a portion thereof; interest, royalty payments, management fees, other fees and payments made under a contract; and other transfers approved by the Ministry of Trade."

Foreign investors can put their money wherever they like and take it out whenever they want to, "without delay." Nothing need be reinvested locally to service the floundering Iraqi economy. Nothing need be targeted to help specifically damaged areas, communities, or services. U.S. corporations are therefore invited to enter the Iraqi economy, exploit a nation at its most vulnerable point, with no obligation to reinvest in the country at a time when rebuilding Iraq is professed to be the Bush administration's most vital assignment.

U.S. corporations have reaped staggering revenues from their Iraqi operations. However, because of Order #39, they are not required to reinvest a cent of this money in Iraq. Chevron, Bechtel, and

Halliburton have each experienced skyrocketing returns to their Iraqi endeavors. As described by the *Financial Times,* for example, Halliburton received steep "profits from their Iraq operations."[59] Both Bechtel and Halliburton have cost-plus contracts in Iraq, which means they are paid for all costs associated with their work plus a guaranteed additional fixed fee above the costs.

While not reinvested in Iraq, much of this money was "invested" in the Republican Party's 2004 election-year coffers—helping to ensure the continuation of the Bush Agenda both in the White House and in the Congress. According to the nonpartisan Center for Responsive Politics, each of these companies was among the leaders in its industry in 2003–2004 election-cycle contributions, with most of the donations going to Republicans. Halliburton gave 85 percent of its political contributions to Republicans, Chevron gave 83 percent, and Bechtel, the most egalitarian of the three, gave 53 percent.[60]

Chapter 3 details how the IMF has, for decades, required the same repatriation of capital rules as those provided by Order #39. These rules are cited by the world's leading economists as primary contributors to financial collapse the world over, including the East Asian financial crisis of 1998–1999. Nations that had once been characterized as the "East Asian Tigers" because of their thriving economies suddenly crashed when the IMF set restrictions on the ability of their governments to regulate which sectors of their economies received foreign investment and how long or in what quantities investment had to stay in a country. When foreign investors started playing with these nations' currencies as if they were in a global casino, the governments were powerless to act. Foreign investors shot up the value of the currencies with their investments and then pulled them out quickly to make a profit. The economies started to crumble, whereupon all the investors pulled out at the same time—sending the economies into total collapse. Had investment restrictions been allowed to remain in place, the financial crises could have been avoided and the lives consequently lost to unemployment and poverty saved.

The stage has now been set for a similar crisis in Iraq.

Provision #5: Forty-Year Leases

Provision #5 allows Iraqis to maintain a semblance of domestic control over foreign investors by denying them the right to own private real property (land or things permanently attached to it, such as buildings) outright. It does allow, however, for the next best thing: Foreign investors or companies are granted forty-year licenses with unlimited renewal options for the lease of Iraqi real estate. According to the dispute settlement procedure for Order #39, if such contracts are broken, the foreign company can turn to international courts rather than Iraq's domestic laws for arbitration.

Provision #6: Dispute Settlement

If a legal dispute arises over any of the rights granted under Order #39, foreign companies and investors can reject Iraq's domestic courts and turn to international tribunals instead. Any international trade or investment agreement that both countries have signed is available. If the Bush administration is successful in implementing its trade goals for Iraq, Iraq will soon be a World Trade Organization member and the United States will have a Bilateral Investment Treaty with Iraq. Both the WTO and the World Bank's International Centre for the Settlement of Investment Disputes (ICSID), which adjudicates most disputes over Bilateral Investment Treaties, are closed door courts that have demonstrated consistent biases in favor of corporations in their rulings over issues of serious public concern. In fact, since it was created, the WTO has ruled that every environmental, health, or public safety policy it has reviewed, except one, is illegal and must be eliminated or changed. While developing country laws have been hit the hardest, U.S. laws designed to protect health and environmental standards have also fallen by the wayside. Both the U.S. Clean Air Act and the U.S. Endangered Species Act have been weakened as a result of WTO rulings.

At both the WTO and ICSID, only the parties directly involved in the cases can participate in the legal proceedings, which are confidential, with limited access to either the press or the public. There are no conflict of interest restrictions on the arbitrators and qualifications are limited to their knowledge of trade and investment law. The ICSID, in particular, was originally designed to handle the most technical contractual disputes between companies and governments. As trade and investment laws have increasingly moved into environmental, social, labor, human rights, and other areas of public interest, the ICSID and the WTO are ruling on issues far beyond their structural mandate.

THE ON-THE-GROUND IMPACT OF BREMER'S ORDERS AND THE BUSH AGENDA

Iraqi Companies Excluded

Iraq has companies, private and public, that can perform much if not all of the work that has thus far been provided to U.S. companies. After the first Gulf War, while Iraq was cut off from the world by economic sanctions, the Iraqis rebuilt bridges, roads, buildings, electricity, water, and sanitation systems, and performed all other postwar recovery work themselves. The reconstruction was not perfect and suffered greatly from a lack of resources, but the skill and the desire were certainly there.

A young college-educated Iraqi woman who goes by the name "Riverbend" has kept a uniquely insightful Web log of her life in Baghdad since the war began. In August 2003, she captured the sentiments of millions of Iraqis when she wrote, "Instead of bringing in thousands of foreign companies that are going to want billions of dollars, why aren't the Iraqi engineers, electricians, and laborers being taken advantage of? Thousands of people who have no work would love to be able to rebuild Iraq. . . . No one is being given a chance."[61]

Among Iraq's state-owned enterprises are companies that perform construction, engineering, architecture, information technology, and other high-tech services, as well as factories that produce everything from concrete to ship containers, sulfur to industrial ceiling fans, phosphates to pharmaceuticals, and trucks to swimming trunks. In fact, construction is one of Iraq's largest sectors, with forty-five factories designed to produce essential building materials. The companies are divided among the various government ministries, with the majority in the Ministry of Industry. These companies have been essentially idle since the war, primarily from a lack of financial and material resources.

A list of forty-three companies assessed by the CPA as excellent candidates for privatization, due to the high quality of their work and conditions of operations, includes: the State Company for Industrial Design and Construction (SIDAC), which rebuilt eighty bridges after the 1991 Gulf War and, according to the CPA, is "staffed with highly qualified engineers . . . with forty years experience on large local construction commissioning projects"; the State Industrial Design and Consultation Company, which provides engineering, construction, and project management; the State Company for Information Systems, which provides consultation, design, and support for information systems; Al-Nasr Al-Adheem State Company, which makes all kinds of heavy equipment for oil, petrochemical, electricity, and water industries; the State Companies for Cement, which manufactures all varieties of cement in seventeen different factories across Iraq; and the General Systems Company, which provides site work, engineering, project management, and software programming. Each of these companies, among dozens of others, could perform work that has been contracted to U.S. companies.

An extremely useful public list of private Iraqi companies is the *PortAl Iraq* website—"The first stop for doing business in Iraq." On this site (www.portaliraq.com), you will find pop up ads for "Bullet Proof VESTS," "Ballistic Blankets!" and "Armored Cars, both new AND pre-owned." It also provides a list of hundreds of private Iraqi businesses. A review of the companies reveals dozens with over thirty years of experience in all of the work areas for which American companies have been hired. Experienced construction companies include the Al-Iraqi Group, Al-Fadhaa Co. Ltd., Al-Fijaj for Construction Co. Ltd., Al-Ghodwa, Al-Maur Construction Company, Al-Qabas Group, Al-Manhal Construction, Botan Group (with particular expertise in the building of highways, dams, and industrial buildings), and the

Edward Allose Company. Experienced Information Technology companies include Alathar Co. Limited Communication Services; Alqedwa Group Co. LTD.; Altam Commercial Agencies; Iraq SatNet; and Nahj Tech Co., which created Iraq's National Internet ISP. There is also Dewan Al-Emara Architects and Engineers, with over thirty years of experience, and Al–Makhzoumi, which specializes in oil field services.

Both public and private Iraq companies have been hired by the U.S. government in the reconstruction, and increasingly so. However, they have been hired almost exclusively as subcontractors to the American companies, and/or for small short-term projects, and have received a pittance in work compared to their American counterparts.

Hiring Iraqi companies in the place of American companies would mean not only more money for Iraq but also increased savings for the American taxpayer. In a September 30, 2003, letter to U.S. Office of Management and Budget Director Joshua Bolton, Congressman Henry Waxman of California explained that members of the Iraqi Governing Council (IGC) estimated that the costs to American taxpayers of many reconstruction projects could be reduced by 90 percent if the projects were awarded to local Iraqi companies. In one example, IGC member Judge Abdul Latif said that non-Iraqi contractors charged $25 million to refurbish twenty police stations in Basra while an Iraqi company could have done the same work for $5 million and then used the rest of the money to restore every government building in Basra. IGC member and civil engineer Songul Chapouk said that the CPA renovated ten houses in Baghdad for members of the Council for $700,000, while Iraqi firms could have built ten houses from scratch at that price and employed more Iraqis in the process. The U.S. general in charge of northern Iraq, Major General David Petraeus, told a congressional delegation including Waxman's staff that U.S. engineers estimated it would cost $15 million to bring a cement plant in northern Iraq back to Western production standards. When the project was offered to local Iraqis instead, they were able to get the cement plant running again for just $80,000, although not necessarily to Western standards.

The testimony of former senior CPA official Franklin Willis to the U.S. Senate Democratic Policy Committee in 2005 reflects statements common among Iraqis and critics of war profiteering in Iraq but rarely voiced from within the CPA itself. Willis explained that U.S. reconstruction funds were misused by the CPA and the Bush administration because large private U.S. contractors were favored over local Iraqi companies:

> I would urge the Committee to examine the contracts being let under the $18 billion Supplemental Authorization approved by the Congress 14 months ago. While a few big infrastructure projects are required, many, many small projects are essential. They can be done by Iraqi companies, they are visible to the populace and the money gets into Iraqi hands and into the Iraqi economy.
>
> For example, Washington Group International has received a $40 million contract to clean, repair and upgrade a portion of the Sadr City wastewater system. Their administrative, management, and security costs I am advised have eaten up $25 million, with minimum salaries for their personnel in place in Baghdad at $250,000 per year. $15 million makes it to the ground for labor and materials. That's 63 percent overhead. . . . The Army overhead in place is 0 percent; the Army Corps of Engineers 6.5 percent.
>
> By contrast, companies like Bechtel, Parsons, Washington Group International, Black and Veatch, etc. have ID/IQ (indefinite delivery/indefinite quantity) open-ended contracts for millions and millions of dollars. How much is eaten up in overhead, what reaches Iraq on the ground?

Iraqi Workers Excluded

Bremer rejected domestic skill in favor of U.S. corporations, promising that the U.S. companies would hire Iraqis. At the time of the U.S. handover of authority from the CPA to the Iraqi interim government in June 2004, while two million Iraqis were unemployed, only some 25,000 had jobs in the reconstruction effort. The reason was twofold. First, very few of the promised reconstruction efforts were even under way. In mid-2004, less than 140 of 2,300 promised construction projects had begun, and there were widespread reports of waste, fraud, and abuse in the projects that had started.[62] Second, as U.S. contractors are not required to hire Iraqis, many have chosen instead to import workers from the United States and around the world. The reason most often cited by U.S. contractors is that they do not trust the Iraqis and have hired workers from countries where they have contracted work before, such as the United States, Nepal, India, and Indonesia.

For example, Halliburton subsidiary KBR hired the Tamimi Company to provide food services for sixty thousand U.S. soldiers in Iraq. Tamimi does not use Iraqi workers because "Iraqis are a security threat," according to a company manager. Instead, the firm brought in 1,800 workers from Pakistan, India, Nepal, and Bangladesh to do the work.[63] Halliburton also shipped about 500 American workers a week to Iraq, paying four times the amount the same workers could earn in Texas, offering up to $8,000 a month to drive oil trucks.[64] Tragically, the horrific murders of many of these workers have become all too well known around the world. In total, an estimated three hundred civilian contractors have been killed in Iraq, while over 260 foreign nationals have been kidnapped.[65]

KBR is performing two categories of work in Iraq: The first is the provision of military support services, which, until about a decade ago, was performed by the military itself. The second is the rebuilding of Iraq's oil infrastructure. KBR received both contracts without competition. According to the U.S. Army's classified contingency plan for repairing Iraq's infrastructure, KBR was the only company with the skills, resources, and security clearances to do the job on short notice. KBR also happened to be the author of the U.S. Army's contingency plan.[66]

In 1992, then Defense Secretary Dick Cheney authorized a classified study to determine whether private companies should handle the military's civil logistics under the Logistics Civil Augmentation Program (LOGCAP). The logistics under consideration included preparing and serving meals, washing laundry, driving and repairing nonmilitary vehicles (trucks versus tanks), building and cleaning bases, and delivering fuel and water. Cheney's Pentagon paid KBR $9 million to conduct the study, and lo and behold, KBR concluded that the services should be privatized. The following year, Cheney's Pentagon awarded KBR the first privatized LOGCAP and, three years after that, Cheney was hired as the CEO of Halliburton. In 1997, under President Clinton, Halliburton lost this contract due to fraudulent billing practices in Bosnia. But in 2001, under Vice President Cheney, KBR got the contract back just in time for the invasions of Afghanistan and Iraq.

As the occupation continued into its second year, more companies, including KBR, began hiring Iraqis. Unfortunately, the insurgency had grown as well, and Iraqis working for American companies were increasingly viewed as collaborators of a hated occupation.

Dahr Jamail, an independent American reporter in Iraq, interviewed Ahlam Abt Al-Hassan on May 19, 2005. Exactly one year earlier, Al-Hassan, then twenty-four years old, had been shot on her way to work at KBR at the U.S. military base in Diwaniyah.[67] She told Jamail that, as she waited for a taxi, she was shot twice in the head by a member of the Mehdi army. Before the incident, threats against KBR employees had been growing, so much so that Ahlam was told by KBR to stay home from work for two weeks. "After this, I went back to work because my bosses told me the security was better," she told Jamail. "My bosses had told me it was secure now." They were wrong. After being shot, Ahlam was taken to a hospital in Hilla, where she told the attendants that she worked for KBR. KBR was contacted, and a spokesperson said that they would come see her. No one came.

After almost two months in hospitals and following several surgeries, Ahlam's life was saved, but her eyesight was lost. Ahlam explained that while she was in the hospital and unable to speak, a phone call came from "Mr. Jeff," Ahlam's only identification for her KBR employer. She has not heard from KBR since, although, according to Jamail, she has tried desperately to contact them for months. She needs further medical attention but has neither the money nor the access. Jamail writes that Ahlam's friends, family, and colleagues are trying to hold Halliburton accountable. However, Bremer Order #17 granted U.S. companies in Iraq full immunity from Iraq's laws. Therefore, Ahlam's only possible recourse is to gain standing in American courts and attempt to sue Halliburton in the United States.

The Failure to Restore Electricity

The following excerpts from Riverbend's Web log are typical of the experiences of millions of Iraqis across the nation:

> August 18, 2003: Normal day today. We were up at early morning, did the usual "around the house things," you know—check if the water tank is full, try to determine when the electricity will be off, checked if there was enough cooking gas.

> December 16, 2003: The electricity only returned a couple of hours ago. We've been without electricity for almost 72 hours—other areas have it worse. Today we heard the electricity won't be back to pre-war levels until the middle of next year. We heard about Saddam's capture the day before yesterday, around noon. There was no electricity, so we couldn't watch TV.

> January 27, 2005: I have to make this fast. We have about two hours of electricity—hopefully. . . . Unfortunately, the electricity situation has deteriorated. We're getting about four hours for every twenty hours in our area—I'm not quite sure what's going on in the other areas. It feels like we're almost cut off from each other.

As an occupier, the Bush administration was required to ensure that basic necessities were provided to the Iraqi people. It failed in this mission because it did not focus its efforts on the immediate provision of needs, but rather on the opening of Iraq to private foreign corporations. Throughout the years of the occupation, Iraqis have continually pointed to the lack of electricity as a primary source of unrest. In March 2005, Iraqis listed "inadequate electricity" as the issue they most wanted the government to deal with, before crime, security, terrorists, or even the drafting of the constitution.[68]

Except for a few brief "luxurious" months in the summer of 2004, when nationwide electricity averages just barely surpassed prewar levels, electricity in Iraq has remained far below prewar levels and significantly below U.S. stated goals. Before the war, Iraq received about 4,400 megawatts of service a day. The U.S. government was supposed to have nationwide electricity up to 6,000 megawatts by July 2004. Instead, in December 2005, electricity was still only at 3,750 megawatts.

For Baghdad, the goals were simpler—just to reach the prewar level of 2,500 megawatts. The highest level ever reached in Baghdad during the occupation was 1,485 megawatts in September 2004. In April 2005 (the latest date for which data is available), service was a dismal 854 megawatts.[69] The U.S. State Department reports that in the last week of December 2005, Baghdad's residents had electricity for just over five hours a day, while the national average was 11.8 hours a day.

Riverbend's Web entries therefore typify the state of the nation: on-again, off-again service, generally in three- to four-hour blocks. But even that schedule is usually punctuated by frequent blackouts. Back in June 2004, UN special envoy Lakhdar Brahimi told the United States that, after security, the lack of reliable electricity was the number one problem facing Iraq and would continue to destabilize the nation if not adequately addressed. U.S. Air Force Colonel Sam Gardiner, author of a U.S. government study of the likely effect that U.S. bombardment would have on Iraq's power system, summed up the feelings of many when he said, "Frankly, if we had just given the Iraqis some baling wire and a little bit of space to keep things running, it would have been better. But instead we've let big U.S. companies go in with plans for major overhauls."[70]

Mohsen Hassan, technical director for power generation at the Iraqi ministry of electricity said, "We, the Iraqi engineers, can repair anything. But we need money and spare parts and so far Bechtel has provided us with neither. The only thing that the company has given us so far is promises."[71] It seems fair to conclude that a more successful approach would have been to provide money and parts to the Iraqis and let them rebuild the system.

Bechtel's "Lost Summer"

The Bechtel Corporation was one of a select handful of U.S. companies that received a quiet "request for proposals" from the Bush administration more than a month before the invasion of Iraq.[72] On April 17, 2003, Bechtel was awarded a $680 million contract for work in Iraq.[73] In September 2003, it received an additional $350 million, and then, on January 6, 2004, it received a second contract—bringing Bechtel's combined total to more than $2.8 billion.

In Bechtel's first contract, "Iraq Infrastructure I," the company is to "provide the successful design, rehabilitation, upgrading, reconstruction and construction in Iraq of one port, five airports, electric power systems, road networks and rail systems, municipal water and sanitation services, school and health facilities, select government buildings, and irrigation systems as well as institutional capacity building for operation and maintenance of roadmaps for future longer term needs and investments in support of the Iraq Infrastructure Reconstruction Program." The second contract, "Iraq Infrastructure II," is for more of the same. Bechtel is not the only company working on these projects, to be sure, but it was the first—and for a considerable amount of time, the largest—to receive such contracts for reconstruction in Iraq. These contracts have, in turn, led Bechtel's non-U.S.-generated revenues to increase by 158 percent in 2003. The company had record revenues yet again in 2004—$17.4 billion overall, topping the company's 2003 take by 6.4 percent.

Bechtel is the largest engineering company in the world. It has built electricity systems throughout the Middle East and other regions worldwide. It was certainly qualified to build a shining new electricity system in Iraq. Unfortunately for Bechtel, Iraq did not have a system needing to be "built" so much as "rebuilt" during a time of war in an increasingly hostile environment. Depending on whom you ask, either Bechtel or the Bush administration decided that, instead of getting Iraq's electricity system up and running as quickly as possible, a countrywide assessment of all systems was necessary before any reconstruction could begin. The assessment took five long months.[74] These happened to be summer months in a country where temperatures regularly top 125 degrees Fahrenheit. No electricity meant no fans, no ice, no cold drinks, no air conditioners, and a lack of clean water and reliable sewage treatment. It is difficult to exaggerate the extent to which Iraqis suffered during those five months. Thus, the summer following the March invasion was a particularly crucial period in which Iraqi goodwill all but evaporated; I call it "Bechtel's lost summer."

"If they think we're used to this, they're wrong," Mohammad Kasim Hamady of Basra told David Baker of the *San Francisco Chronicle* in September 2003. Hamady said that he was tired of sleeping on the roof and enduring the rashes that Basra's 125-degree air triggers on his skin. He was tired of seeing his children wander listlessly through the house in their underwear when the blackouts hit. Voicing an opinion shared by many Iraqis, he warned, if the Americans and the British can't fix Iraq's infrastructure, resistance to their presence will grow. "We didn't ask them to come," Hamady said. "We need all the basic needs. If they don't like that, they can go home."

It certainly made sense to assess the situation before building, but much of this assessment could have been done prior to the invasion as part of the post-invasion planning. After the invasion, short of turning the reconstruction over to the Iraqis, all of the assessment should have been done in direct discussions and partnerships with Iraqi engineers, such as Mohsen Hassan, who had run the systems for decades, as well as in conjunction with the international humanitarian organizations already on the ground, such as CARE and the World Health Organization—organizations that the Iraqis were used to working with and trusted. What the Bechtel employees discovered after five long months was that two wars and twelve years of economic sanctions had taken their toll. The systems were in far greater need and more difficult to repair than they had assumed. The water, electricity, and sewage systems, for example, were more intimately interlinked than anticipated, and parts were less accessible than expected—a problem greatly exaggerated when the Bush administration banned countries that had not supported the invasion from profiting from the occupation. Thus, Iraq's electricity and water systems, built in Russia, Germany, and France, were unable to receive replacement parts from these nations. Of course, this was all information that expert Iraqis could have easily conveyed to Bechtel from the start, had Bremer not fired the vast majority of them and had Bechtel asked.

"These systems, the repairs they are not all on some blueprint somewhere," Gazwan Muktar, a retired electrical engineer, told author Christian Parenti in Baghdad. "You need to have the people who spent twenty years running these irrigation canals or power plants to be there. They know the tricks; they know the quirks. But the foreign contracts ignore Iraqis, and as a result they get nowhere!"[75]

The critical time spent and lost on the assessment, only to learn that the work was more difficult than anticipated, bred increasing hostility toward the invasion. The lack of water, electricity, and sewage services led to increasing acts of sabotage against all foreign contractors, including Bechtel. It seemed that every time Bechtel built a water tank, overnight it would disappear. Tragically, Bechtel employees also became targets, facing acts of violence, kidnapping, and murder. These factors all combined to reduce Bechtel's ability to fulfill its contracts with the U.S. government and its obligations to the Iraqi people. They did not, however, reduce Bechtel's financial rewards. In April 2004, the Bush administration reduced the expectations for Bechtel's contract, but not its dollar figure. Good for Bechtel, but too bad for the people of Iraq.

The Failure to Provide Water and Sewage

> E. was the first to hear it. We were sitting in the living room and he suddenly jumped up, alert, "Do you hear that?" He asked. I strained my ears for either the sound of a plane or helicopter or gun shots. Nothing . . . except, wait . . . something . . . like a small stream of . . . water? Could it be? Was it back? We both ran into the bathroom where we had the faucets turned on for the last eight days in anticipation of water. Sure enough, there it was—a little stream of water that kept coming and going as if undecided. E. and I did a little victory dance in front of the sink with some celebratory hoots and clapping.
>
> I almost didn't sleep last night. I kept worrying the water would be cut off again. I actually crept downstairs at 4 A.M. to see if it was still there and found E. standing in the bathroom doorway doing the same. My mother is calling the syndrome "water anxiety."
>
> —Riverbend's Web log, January 27, 2005

Electricity in Iraq controls water supply and sewage systems. Without the one, you cannot have the others. The World Health Organization reports that before the first Gulf War in 1991, virtually all Iraqis (a full 90 percent) had access to an abundant quantity of safe drinking water and that the south and center of Iraq had a well-developed water and sanitation system.[76] Both the water and sanitation systems suffered greatly during the sanctions period, and comparable figures do not exist preceding the second Gulf War. What is known is that Bechtel's contract—like that of the California-based Fluor Corporation, one of the largest construction companies in the world and the current recipient of $3.75 billion in Iraq reconstruction contracts—included the rebuilding of Iraq's water and sewage systems. Bechtel was also put in charge of managing this reconstruction.

Bechtel's first contract, for which it has been paid $680 million, states, "The contractor will commence repairs of water infrastructure in ten urban areas within the first month. Within the first six months the contractor will repair or rehabilitate critical water treatment, pumping and distribution systems in fifteen urban areas. Within twelve months potable water supply will be restored in all urban centers, by the end of the program approximately 45 urban water systems will be repaired and put in good operational condition, and an environmentally sound solid waste disposal will be established."

Although Bechtel failed to meet these requirements, the U.S. government neither canceled Bechtel's contract, demanded its (our) money back, nor looked for Iraqis to replace the company. Rather, it revised the contract, moving dates back and changing commitments. According to the U.S. Agency for International Development (USAID), one year and a half after Bechtel entered Iraq, "water meant for consumption is pumped through the system largely untreated while raw waste flows untreated directly into city streets, rivers or marshlands. Many rural communities are not connected to main water or sewer lines, have no access to potable water and suffer from health problems related to poor sewage disposal." Furthermore, "Baghdad's three sewage treatment plants, which together comprise three-quarters of the nation's sewage treatment capacity, are inoperable, allowing the waste from 3.8 million people to flow untreated directly into the Tigris River. . . . Water that is pumped through the system is largely untreated, especially in the South."

A May 2004 UN survey found that 80 percent of families in rural areas used unsafe water. This includes many in the south, who, cut off from adequate drinking and sewage systems, turn to the Tigris and Euphrates for their water supply.

Entesar Hadi lives in southern Baghdad. She has two children, both of whom suffer from intestinal infections. In early 2005, she said that her life had been "completely paralyzed" by the lack of water. "I have collected rain in a pot to wash the toilet," she said.[77] In December 2004, Sadr City, home to more than two million Iraqis, was described thus: "Lakes of sewage-blotched stagnant water and piles of rotting garbage still dot the streets of Sadr City . . . 'Just a few days ago, you couldn't walk this street because the sewers were overflowing. Now they've taken care of it,' said Abu Mustafa, a mattress merchant. . . . Children step over a stream of sewage that runs through their classroom in a school on Moqtar Street, one of 25 areas with major sewage backups."[78] Such conditions remain are all too common in post-invasion Iraq.

The UN survey also found that a 51 percent majority of Iraqi families living in urban areas in the south can literally look out of their window and see raw sewage in the streets. The same was true for 40 percent of all families living in all urban areas. Only 37 percent of all households are even connected to sewage networks, but this nationwide average covers up a staggering urban-rural difference: 47 percent in the urban areas and only *3 percent* in the rural areas are connected.

Hassan Mehdi Mohammed lives in a small village with his wife and eight children, about an hour drive south of Baghdad. He told an interviewer that he would welcome Bechtel into his village to work on water service: "We like to hear that companies are coming here and we can work for them. . . . And where are these companies? They have done nothing to help."[79]

Bechtel's contract also included the restoration of Iraq's schools. Reporters with *Southern Exposure* magazine visited four Baghdad schools, all listed as renovated by Bechtel. They found rain leaking through ceilings, power shortages, new paint peeling, and floors that had not been completely repaired. New brass taps and doors had been painted, but toilets and sinks had not been touched. At Hawa School, for example, the headmistress pointed out toilets for which a new water system—pipes, taps, and a motor to pump the water—had been installed. However, the motor did not work, so the toilets were filled with unflushed sewage.[80]

Al-Ani is a Ph.D. civil engineer with forty years of experience and is one of the top experts in water treatment in Iraq. He is an employee of the General Co. for Water Projects, a state-owned company bypassed by Bremer during the reconstruction. Bechtel was hired to manage a project that General Co. used to run, the expansion of the Sharkh Dijlah, or East Tigris River, Water Treatment Plant. In the case of the Sharkh Dijlah, Bechtel did not overlook Iraqi expertise; rather, Bechtel appropriated it. According to city officials, Bechtel spent four months studying the General Co. plans for expansion, concluded they were adequate, modified them slightly, and reissued orders for parts from the same supplier. General Co.'s 187 workers, including Al-Ani, continued to collect their government salaries, but they were not put to work. Instead, they spent their days playing video games, reading books, and discussing religion and politics. Firas Abdul Hadi, a twenty-year-old civil engineer from Hilla, told a reporter, "It is beyond boring. It is painful."[81]

There has been some success. Bechtel has built power stations, dredged rivers, rebuilt ports, roads, and bridges, and restored water systems. However, the success has fallen far short of expectation, demand, and Bechtel's original contractual obligations; it has happened far too slowly and in shocking disconnect from those who actually live with the results of the work. One such example is Basra, cited by Bechtel and the Bush administration as a shining example of its work—literally a glistening modern water facility. A new central water pumping and purification station and fourteen decrepit substations were refurbished with glossy blue machines that can move along more and cleaner water than before. The problem? The facility was not adequately connected by pipes to the homes in the area, and the electricity is still intermittent. So the water tower sparkles while the pipes in peoples' homes continue to run dry.[82]

Nobody at Bechtel or in the U.S. government denies that the water and electricity reconstruction has failed. According to the U.S. State Department, of 249 water and sewage projects originally planned, only 64 have been completed. However, Bechtel and some Bush administration officials point the finger at the Iraqis. According to Bechtel, of more than forty water plants it has built that are now being run by the Iraqis, "not one is being operated properly." U.S. officials say the same: "None of the 19 electrical facilities that has undergone U.S.-funded repair work is being run correctly."[83] They blame a poor Iraqi work ethic and a lack of knowledge and skill in running the plants.

Iraqis may be unable to run the systems built by Bechtel in Iraq, but a poor work ethic and lack of knowledge are not to blame. Recall that Bremer fired the upper echelons of Iraqi management, sidestepped skilled engineers and workers, hired Bechtel to build state-of-the-art facilities foreign to these workers, and then handed the systems over as a fait accompli, whether or not they were even connected to the homes they were intended to serve. Baghdad's Mayor Alaa Tamimi, an engineer who returned to Iraq after years of exile to help rebuild the country, said that U.S. officials "made a lot of decisions themselves, and the decisions were wrong. This is our country. It's our city. They didn't accept that."[84]

The other problem is money. Iraqis simply do not have enough of it to run the expensive new facilities that they have been handed. The money has gone to U.S. contractors to (largely fail to) build Iraq's systems, rather than to the Iraqis to run the systems after they have been rebuilt. Recall that in April 2004, Iraq's public works minister said that, due to a lack of resources, she was considering privatizing the water system to "fund essential works."[85] Lack of money and skill in running public sectors has always been used as a reason for privatization. Bechtel may well position itself as the only company with the ability to run the facilities that it has built, opening the door for its entrance as a privatizer.

Why would Bechtel be interested in privatizing Iraq's water? After its oil, water is Iraq's second most valuable resource. Iraq is home to the most extensive river system in the Middle East, including the Tigris and Euphrates rivers and the Greater and Lesser Zab Rivers. It also has a sophisticated system of dams and river control projects. Stephen C. Pelletiere, a former CIA senior political analyst on Iraq during the Iran-Iraq war, has written that "America could alter the destiny of the Middle East in a way that probably could not be challenged for decades—not solely by controlling Iraq's oil, but by controlling its water. Even if America didn't occupy the county, once Mr. Hussein's Ba'ath Party is driven from power, many lucrative opportunities would open up for American companies."[86] Bechtel has already become one such company.

In 1995, World Bank Vice President Ismail Serageldin declared, "Many of the wars of this century were about oil, but the wars of the next century will be about water."[87] To which a *Fortune* magazine article added five years later, "water promises to be to the 21st century what oil was to the 20th century: a precious commodity that determines the wealth of nations." Thus, the article concluded, "If you're looking for a safe harbor in stocks, a place that promises steady, consistent returns well into the next century, try the ultimate un-Internet play: water."[88] Investments in water systems therefore have the potential to be both highly lucrative and politically advantageous.

After learning that Bechtel was managing Iraq's water reconstruction and could be in a position to privatize the nation's water, a group of citizens from Cochabamba, Bolivia wrote an "open letter to the people of Iraq" warning them of what they could expect from Bechtel in the wake of their own disastrous privatization experience with the company five years earlier: "We write you now because we fear that you might be made victims of additional suffering at the hands of a multinational corporation—Bechtel." For its part, Bechtel does not plan to be expelled from Iraq as it was from Bolivia. As Cliff Mumm, head of Bechtel's Iraq operation has said, Iraq "has two rivers, it's fertile, it's sitting on an ocean of oil. Iraq ought to be a major player in the world. And we want to be working for them long term."[89]

Following the Money

Who is paying Bechtel and the other U.S. corporations for their failure in Iraq and why are the remaining funds insufficient to pay for Iraqis to run those systems that have been built? Billions of dollars of U.S. money committed to reconstruction have gone unspent in Iraq, been wasted, or are simply unaccounted for. The U.S. government's General Accounting Office reported in June 2004 that the CPA had spent virtually all of Iraq's money during the occupation but relatively little of its own.[90] As the occupation began, there were two primary pots of money earmarked for reconstruction. The largest was the approximately $24 billion of U.S. taxpayer money, appropriated by Congress in October 2003. The second was the Development Fund for Iraq, worth $18.4 billion. The Fund was primarily money from Iraq's oil revenues and was controlled by the CPA until authority for the Fund was transferred to the Interim Iraqi Government on June 28, 2004. The CPA spent approximately $13 billion from the Fund but only about $8.2 billion of the U.S. appropriation. There were significantly more stringent reporting requirements (although, I would argue, not stringent enough) on the U.S. appropriation than on the Fund, for which there was virtually no accounting. To this day, a full $8.8 billion from the Fund remains completely unaccounted for while audits of U.S. taxpayer funds have found contract files "unavailable, incomplete, inconsistent and unreliable."[91]

Public scrutiny of the U.S. congressional appropriation has been intense. There have been dozens of protests at the headquarters of companies operating in Iraq, particularly at Halliburton and Bechtel. In San Francisco, protestors by the hundreds have sat down in front of Bechtel's corporate headquarters time and again in the weeks leading up to and years following the invasion. Members of Congress, led by Henry Waxman of California, have demanded accounting and public watchdog groups have tried to conduct their own.

Congressman Waxman has done more than any single individual to expose Halliburton's misuse of Iraqi and U.S. taxpayer monies. Halliburton has been found guilty and is under investigation for over $1.5 billion in overcharges for its Iraq services by the U.S. Defense Department's Contract Audit Agency. The agency found that Halliburton overcharged upwards of $212 million for gasoline in Iraq. In one instance, Halliburton requested $27.5 million to ship an amount of heating and cooking fuel worth $82,100 from Kuwait to Iraq. The agency reported an additional $219 million in overcharges for Halliburton's Restore Iraq's Oil contract, $813 million under Halliburton's LOGCAP contract, and additional evidence of "unsupported costs" totaling $442 million—for a grand total of $1.5 billion.

Halliburton was also found to have colluded with the U.S. Defense Department to keep these charges out of public purview for more than five months. A report prepared by the U.S. Senate Democratic Policy Committee in June 2005 found, "At Halliburton's request, and despite the urging of Army officials for a 'sanity check,' the Defense Department concealed the magnitude of Halliburton's questioned and unsupported costs . . . the Defense Department redacted every mention of every questioned and unsupported charge from every audit turned over to the international auditors. In total, references to excessive charges were blacked out over 460 times."

U.S. soldiers stationed in Iraq who, after the Iraqis themselves, have borne the brunt of the administration's "free market" miscalculations, have made some of the loudest demands for change in the reconstruction effort. As Brigadier General Mark O'Neill, assistant commander of the Third Infantry Division, Baghdad, said in early 2005, "It's a form of ammunition. If you can get trash picked up, get water running, get electricity flowing, the sewers working so people's quality of life is improving, then you have to fire fewer rounds of the other type of ammunition." These prophetic words were eventually heard by the Bush administration.

In April 2005, the U.S. State Department announced that it was shifting the spending priorities of the remaining U.S. reconstruction appropriation. It would focus on Iraqi subcontractors, because they "are somewhat less susceptible to insurgency attacks and are not burdened by the same heavy overhead expenses of foreign firms." Also there would be a shift away from large infrastructure projects to smaller immediate-term contracts. The shift did have some success. Over the second half of 2005, the numbers of Iraqis employed under U.S. government administered projects averaged around 120,000 per week (although they have fallen off again in 2006).[92] Overall unemployment, however, remains at between 25 and 40 percent in January 2006, while the rate for young men is estimated at a disastrous 37 percent.[93] Iraqis welcome the U.S. government jobs, but complain that the length of employment remains limited, often lasting just a few days or weeks, and that they continue to work far too regularly under American rather than Iraqi companies, making them targets of the insurgency.

The most glaring problem with the U.S. State Department's plan, however, was that it included a massive shift of funds, $3.4 billion, away from water and electricity projects toward the training and equipping of the Iraqi army and police forces. Five long-range water projects were specifically canceled to pay for a new "private sector development" program. Because of the rising violence and growing insurgency, contractors use an average of 25 percent of the funds allocated to them for security provisions rather than the delivery of service, further draining the available reconstruction funds. The insurgency also continues to restrict the ability of contractors to perform their work.

The full failure of the reconstruction was revealed in a January 2006 U.S. government audit. Although more than 93 percent of the U.S. appropriation has been spent or committed to specific companies and projects, as much as 60 percent of all water and sewer projects will not be completed. Projects related to drinking water that were expected to benefit about eight million people will now benefit only about 2.75 million and only two of the ten planned sewerage projects will be completed. More than 125 of a planned 425 electricity projects will be left undone. Plans for four gas-powered generating plants and a diesel plant were canceled.[94]

One could argue that just as "free trade" is a misnomer for trade policies that are heavily regulated to favor multinational corporations, so too is the "free-market" established by the Bremer Orders a misnomer for a heavily state-controlled economy that directs the economy to favor foreign corporations and investors rather than national interests.

THE HANDOVERS THAT WEREN'T

> You set up these things and they begin to develop a certain life and momentum on their own—and it's harder to reverse course.
>
> —*Paul Bremer, on the Bremer Orders, June 27, 2004*[95]

The Bremer Orders are useless to the Bush administration and its corporate allies if they are not followed by the Iraqi government. Thus, the administration crafted a process for the transition of power from the CPA to a permanent elected Iraqi government to guarantee (as much as possible) that the Orders would be upheld and that a new government would emerge. This new government, while not necessarily perfect, would at least be friendlier to U.S. interests as defined by the Bush administration than the post-1990 regime of Saddam Hussein was.

On June 28, 2004, in a secret ceremony in Baghdad, the invasion and occupation of Iraq by the United States was officially brought to an end. Held two days ahead of schedule to confound insurgents who threatened attacks, the handover of authority from the Americans to the Iraqis took place without a hitch. Photographs released after the fact show Iyad Allawi, the newly appointed interim prime minister of Iraq, shaking hands and smiling broadly as he accepted a blue leather-bound folio from the departing de facto prime minister, Paul Bremer. This was the last day of the CPA and of Bremer's term in Iraq. While the ceremony was beautiful, with each man dressed in formal attire—Allawi in traditional Arab robes and Bremer in an impeccable suit and tie, the pomp far exceeded the substance of the event. It was a handover in name only, not in deed. For, not only did 140,000 troops remain on the streets of Iraq under U.S. control, but virtually all of Bremer's 100 Orders remained in effect.

The guidelines for the handover were established four months earlier when, on March 8, 2004, the U.S.-appointed Iraqi Governing Council (IGC) adopted a new interim constitution for Iraq—the Law of Administration for the State of Iraq for the Transitional Period, referred to as the Transitional Administration Law, or TAL. The TAL was written by the IGC and Bremer to replace the existing Iraqi constitution.

The TAL established a three-stage process to transition Iraq from a country governed by a foreign occupation force to one ruled by a permanent elected government. The first stage began when the CPA and the IGC were disbanded and replaced by the Iraqi interim government. The interim government stayed in power until stage two was complete with the January 30, 2005, National Assembly elections, which ushered in the transitional Iraqi government. The TAL remained the constitution of Iraq until, on October 15, 2005, Iraqis accepted the new permanent constitution, followed by the elections for a permanent Iraqi government and the completion of stage three on December 15, 2005. The new Iraqi government was to take office no later than December 31, although it had not yet done so in early 2006 as this book went to publication.

The Iraqi Interim Government

The Iraqi interim government was supposed to be chosen by the U.S.-appointed IGC and Bremer with advice from others in and outside of Iraq, including the United Nations. In reality, Bremer and the Bush administration controlled the process. While public attention focused on the conflict between the Americans and the Iraqis over who would serve in the largely ceremonial post of president, the Bush administration quietly and aggressively fought for and won its handpicked candidate for prime minister, Dr. Iyad Allawi.

Allawi "was the Americans' choice," according to Lakhdar Brahimi, UN special envoy to Iraq.[96] Brahimi described how both he and the IGC were only informed after the fact of the Bush administration's selection of Allawi as prime minister. Allawi was an obvious choice for the Americans. Having left Iraq in 1971 for London at the age of twenty-five, by 1979 he had organized a leading anti-Hussein network that became the Iraqi National Accord in 1990 and later became a direct financial beneficiary of the CIA. Allawi worked with the CIA on a failed coup attempt against Hussein in 1996.[97] He returned to Iraq after the 2003 invasion and was appointed to the IGC where he chaired the Security Committee. After Allawi became prime minister, authority for Bremer's Orders was transferred to him with Order #100.

The TAL made it either impossible or extremely difficult for the interim government to overturn or replace Bremer's Orders. Article 26 of the TAL specifically stipulated that "the laws, regulations, orders and directives issued by the CPA . . . shall remain in force until rescinded or amended by legislation duly enacted and having the force of law." Section 2 of the Annex to the TAL[98] conceivably allowed for the overturning of existing Orders or issuing of new ones, but only with agreement of the president, the two vice presidents, a majority of the one hundred-member National Council, the Judicial Authority, and the prime minister. Finally, Section 1 of the Annex to the TAL denied the interim government the ability to take "any actions affecting Iraq's destiny beyond the limited interim period," which ended with the election of an Iraqi government "by 31 December 2004, but no later than January 2005." The identical sentence appeared in UN Security Council Resolution 1546, which outlined Iraq's transition to sovereignty. Thus, while the interim government could technically have overturned less far-reaching Orders, it was beyond its authority to make any fundamental changes.

Not only did the Allawi government leave the Bremer Orders in place, it dutifully enforced them as aggressively as Bremer himself under the firm guidance of interim Finance Minister Adel Abdul Mahdi. In fact, Mahdi emerged as one of the most aggressive proponents of the Bush economic agenda in Iraq. An economist trained in France, Mahdi served on the IGC. Although considered a religious moderate, he is a senior member of the Supreme Council of the Islamic Revolution, Iraq's largest Shiite political party. In an October 2004 speech at the conservative American Enterprise Institute in Washington, DC, Mahdi assured his listeners that Iraq was making "a good start on a broad range of structural and legal reforms that are critical to making the transition from the centralized economy to an economy based on private ownership, open markets, transparency, and the rule of law." While in Washington, he met with President Bush and Vice President Cheney and became the administration's second choice for prime minister in the January 30, 2005 elections, after Allawi. Following the elections, he was appointed as one of two vice presidents, a position from which he has yielded extensive influence over Iraq and its laws. Mahdi has been a particularly strong advocate on behalf of U.S. corporate access to Iraq's oil.

The Election That Wasn't: January 30, 2005

While President Bush touts the January 30, 2005, Iraqi election as a milestone for democracy in the Middle East, by all international election standards, the election was illegitimate. First and foremost, the elections contravened the 1907 Hague Convention prohibition on occupying powers creating any permanent changes in the government of the occupied territory. The Iraqi elections were arranged under an electoral law and by an electoral commission installed and backed by Paul Bremer and the U.S. government. For example, Bremer's seven-member commission was able to reject any potential candidate or party running in the elections. Election guidelines also allowed members of the IGC and the Iraqi interim government to participate in the elections, both of which were appointed by the U.S. government—directly (in the case of the former) and indirectly (in the case of the latter).

The U.S.-based Carter Center, acknowledged as the world's authoritative agency on election monitoring, stated the day before elections took place that none of its key criteria for determining the legitimacy of elections had been met in Iraq. Those criteria include the ability of voters to vote in a free and secure environment and without fear or intimidation, the ability of candidates to have access to voters for campaigning, and a freely chosen and independent election commission.

In Iraq, elections took place under a declared state of martial law. Heavily armed occupation troops were visible throughout the country. The troops enforced shoot-to-kill curfews in many areas, a prohibition on the use of cars or trucks, closure of the airport and borders, and closure of roads. Even with the military lockdown, nine suicide bombings took place on the day of elections, killing at least forty-four people. The primary targets were polling places. Three thousand of the candidates running for office did not make their candidacies public until the day of elections out of fear of assassination. Their fear was well placed: Eight candidates were assassinated. There were no international monitors in the country. (Election monitors analyze election laws, assess voter registration processes, voter education efforts, and the openness of campaigns. Their presence during elections helps ensure that voters can safely and secretly cast their ballots and deters interference or fraud in the voting process.) Unlike elections in Afghanistan (with 122 monitors) and Palestine (with 800 monitors), whose elections were also held under difficult circumstances, including occupation, Iraq was deemed too dangerous for international election monitors.

Most Sunni political parties boycotted the elections, including the Iraqi Islamic Party, the leading Sunni Arab political group. The boycott, combined with the intimidation in their neighborhoods, led Sunni Arabs to be all but absent from the elections. In Al-Anbar province, the Sunni heartland, only two percent of eligible voters cast ballots. As Rosil Shayeb, a twenty-four-year-old Sunni Arab told the *Los Angeles Times,* "If the elections were legitimate and honest, I would vote. But how can we vote for a government under the control of occupation forces?" Out of a deep belief in hope and conviction for change, however, millions of Iraqis braved all of these threats and came out to vote. Out of a population of over 25 million people, it is estimated that 8.5 million Iraqis cast ballots, a turnout rate of approximately 34 percent of all Iraqis and 58 percent of registered voters.[99]

As anticipated, the United Iraqi Alliance (UIA), the leading Shiite political party pulled together by Grand Ayatollah Ali Sistani, received almost half of all votes cast—48 percent. The Kurdistan Alliance received 26 percent. The National Assembly then had the task of appointing the president and the two vice presidents, who then named the prime minister. This process would take another three months. While Mahdi was the original front-runner for prime minister for the UIA, the competing parties struck a deal such that on April 6, Mahdi became deputy president with Ghazi Ajil al-Yawer, a Sunni Muslim and former president of Iraq's interim government. Jalal Talabani of the Patriotic Union of Kurdistan was named president. The prime minister's office went to another U.S.-backed candidate, Ibrahim al-Jaafari, a Shi'a Muslim, former deputy president of the interim government and member of the IGC.

The elections did not deliver the perfect outcome for the Bush administration. Bremer had not wanted elections at all and had initially argued for a system of provincial caucuses, and the administration would have preferred to have Allawi or Mahdi as transitional prime minister. But the most important criteria for the transitional government was met: It was friendlier to the Bush Agenda than the Hussein regime of the previous thirteen years. For example, the transitional government's 2005–2007 National Development Strategy released in June 2005 announced that the Iraqi government's fundamental national vision was one of "restoring Iraq to its rightful place in the world community as a prosperous and market oriented regional economic powerhouse." More important, the Iraqi transitional government proved its willingness to work with the administration in the drafting of Iraq's new constitution—which locked in the Bremer Orders, the economic transformation of Iraq from a state to a market economy, the military occupation, and increased U.S. access to Iraq's oil.

IRAQ'S NEW CONSTITUTION

Due to the three months of political jockeying over the makeup of the Presidency Council, serious work on the constitution did not begin until the end of May 2005. The TAL set the arbitrary deadlines of August 15 for the National Assembly to draft a new constitution and October 15 for a public referendum to reject or adopt it. If the deadlines were not met, the Assembly was to dissolve, new elections would be held, and the process would begin anew. However, the TAL provided one way out: If, by August 1, the Assembly decided that it needed more time, it could request a six-month extension. As work got under way, it became clear that more time was needed, particularly when Arab Sunni representation was added to the constitutional drafting committee (because of their election boycott, few Sunnis held seats in the National Assembly and therefore they were largely excluded from the drafting process). It also became clear that the Bush administration was not going to allow the Iraqis a six-month delay. President Bush was blunt in a White House press conference at the end of June about the August 15 deadline: "That's the timetable. And we're going to stay on that timetable. And it's important for the Iraqi people to know we are."

Both Iraqi and American support for the occupation was dropping precipitously, and Bush had no intention of bringing troops home any time soon. Keeping Iraq to America's "democracy deadline" served to provide an example of success in the war effort and to make clear that success was only possible with the continued presence of the United States. To this end, and to ensure that Bush administration interests were met in the constitution, the administration publicly stepped up its direct involvement in negotiations. On July 27, Donald Rumsfeld was dispatched to Iraq where he told reporters, "We don't want any delays, now's the time to get on with it."[100] U.S. Ambassador Zalmay Khalilzad was an increasingly dominant public force in the negotiations, even taking the highly unusual step of releasing an Op-Ed in the *Wall Street Journal* outlining U.S. requirements for Iraq's constitution on August 7.

The pressure worked and the Iraqis did not request the necessary six-month extension. The August 15 deadline, however, was also missed, but no one held the National Assembly to account. The Iraqi public was overwhelmed with meeting basic needs and simply staying alive. They knew that a constitution was being negotiated, but it was hardly the most pressing issue of the day. The U.S. government wanted the Assembly to keep working to ensure that the October 15 deadline would not be missed, and it certainly did not want to risk holding new elections for a new National Assembly and starting the entire constitutional drafting process again. Thus, the Assembly continued working on the constitution virtually up until the day of the vote, thereby eviscerating one of the truly democratic provisions of the TAL, which required a two-month period after the constitution was drafted, before it was voted on, for public distribution, debate, and analysis. Instead, Khalilzad and a handpicked group of Iraqi government officials made changes to the constitution even after the public document had been printed and distributed. Iraqis voted on a constitution, which they had not even read.

The Bush administration claimed that it was heavily involved in the constitutional drafting process in order to ensure a separation of "mosque and state" and greater protections for women. If these were the administration's actual goals, then it failed miserably. Simply put, Islamic law is granted a much stronger role in the new constitution than in its predecessor, and women are afforded far fewer legal protections. The administration succeeded, however, in ensuring that the constitution locked in the Bremer Orders, continued the economic transformation, allowed for the continued military occupation, and increased U.S. access to Iraq's oil.

Bremer Orders Upheld

The Bremer Orders are untouched by the new constitution. Unlike the TAL, for example, which the constitution specifically repeals (except for a few key provisions relating to the Kurds), the Orders are not mentioned and are thereby accepted and upheld as Iraqi law. As explained by constitutional law scholar Nathan J. Brown, senior associate of the Carnegie Endowment for International Peace, "Implicitly the body of legislation issued by decree by the CPA continues in effect until modified, because it is currently treated as valid Iraqi legislation."[101] A committee has been established in the Ministry of Justice to review the CPA legislation. Thus far, the committee has not repealed a single Bremer Order.

As a result of leaving the Bremer Orders unaltered, specifically the National Treatment provision of Order #39, the constitution allows U.S. corporations to maintain their lock on Iraq's reconstruction. There is no language in the constitution to contradict their continued presence and the U.S. government's authority over them as Bremer Order #17, granting contractors full immunity, is still operable as well.

Furthermore, U.S. Ambassador Zalmay Khalilzad continues to preside over the largest available source of money in Iraq for the foreseeable future (the U.S. $24 billion Congressional appropriation) and the largest embassy in the world. Anyone who wants access to that money will need to continue to play by Bush administration rules.

U.S. Economic Invasion Advanced

Several articles of the constitution commit Iraq to the continued transformation of its economy along the lines of the Bush Agenda including privatization of Iraq's state-owned enterprises. These articles are best understood in the context of the Bremer Orders and in relation to what existed previously in the 1970 constitution. The new constitution reads:

> Article 25: The State guarantees the reform of the Iraqi economy in accordance with modern economic principles to insure the full investment of its resources, diversification of its sources and the encouragement and the development of the private sector.
>
> Article 26: The State guarantees the encouragement of investments in the various sectors. This will be organized by law.

Compare these to the following articles from the 1970 Iraq constitution:

> Article 12: The State assumes the responsibility for planning, directing and steering the national economy for the purposes of (a) Establishing the socialist system on scientific and revolutionary foundations. (b) Realizing the economic Arab unity.
>
> Article 13: National resources and basic means of production are owned by the People. They are directly invested by the Central Authority in the Iraqi Republic, according to exigencies of the general planning of the national economy.

Thus, Iraq is committed to continuing the process of economic reform using "modern economic principles," which, in the context of the transformation already under way, means the corporate globalization/market principles of the Bremer Orders versus the social welfare principles that formerly guided the economy. "Diversification of sources and encouragement and development of the private sector," combined with Article 26, refer to *foreign* private investment in "various sectors." The clause "This will be organized by law" means that the extent of the provision will be determined by legislation. Of course, the legislation already exists in the form of the Bremer Orders, which specifically allow for private foreign investment in all areas of the economy (except primary resource extraction, which is discussed later). New legislation can change the Bremer Orders, but this article protects the constitutionality of the Orders related to foreign private investment.

The second clause of Article 27 allows for the privatization of state property:

> Regulations pertaining to preserving and administrating state property, the conditions set for using it and the cases when giving up any of the property may be allowed shall be regulated by law.

Again, the law already exists in Bremer Order #39, which allows for the privatization of all of Iraq's state-owned enterprises. New legislation could contradict the privatization aspects of Order #39, but this clause secures the constitutionality of privatization.

Attempts by Iraqi parliamentarians to include strong language on economic and social rights similar to those included in the 1970 Iraq constitution did not survive the drafting process. Whereas the 1970 constitution guaranteed free health care, education, and child and maternal care, the only parallel guarantee in the 2005 constitution is for free education and limited guarantees for free health care in specified circumstances, regulated by the law.

Military Occupation Maintained

Iraq's parliamentarians also tried to include specific language in the constitution prohibiting a permanent foreign military presence in Iraq and the building of foreign military bases.[102] Riverbend translated one such provision from an early draft of the constitution from Arabic to English on her website: "It is forbidden for Iraq to be used as a base or corridor for foreign troops. It is forbidden to have foreign military bases in Iraq." This provision and others like it did not make it into the final draft. There were also unsuccessful attempts by parliamentarians to limit specifically the mandate of the presence of the troops. Instead, the final version was moot on the entire topic of a foreign military presence.

Currently, coalition forces total 183,000 in Iraq, 160,000 of which are from the United States, and there are four permanent U.S. military bases under construction. The U.S. military will likely remain in Iraq until U.S. access to oil is solidly and securely in place. For U.S. corporations, this has meant ensuring the installment of a new, legal, and permanent Iraqi government with which they can sign permanent contracts. It also means security for corporate operations, including facilities, pipes, offices, and ships. With the support of Iraq's laws and its constitution, the U.S. military will be the key to securing continued U.S. access to oil.

WINNING IRAQ'S OIL PRIZE

> Although the final decision for inviting foreign investment ultimately rests with a representative Iraqi government, I believe in due course the invitation will come.
>
> —*Peter J. Robertson, Chevron vice chairman, 2003*[103]

Increased access to Iraq's oil sector has always been at the heart of the Bush Agenda in Iraq. Securing this access is an ongoing process. The first stage was getting Iraq's oil infrastructure up and running as quickly as possible after the invasion. The second was the creation of a legal infrastructure opening the country to private foreign ownership and investment. Third was the election of an internationally recognized legitimate Iraqi government with the authority to sign contracts with foreign oil companies. Fourth is the completion and passage of a new national oil law. The final stage will occur when U.S. oil companies have not only signed contracts, but have gotten safely to work.

Oil Production

When U.S. soldiers invaded Iraq, priority was given to the protection of the Oil Ministry, oil facilities, and oil infrastructure. The Bush administration spoke frankly of the need to rebuild Iraq's oil infrastructure so the proceeds could pay for the larger reconstruction effort. Halliburton received an enormous contract toward this end, as did several other U.S. firms. In addition to the 160,000 U.S. and other foreign troops stationed in Iraq, Operation Task Force Shield employed approximately 14,000 security guards deployed along Iraq's oil pipelines in 175 critical installations, including 120 mobile patrols, to provide continual protection against sabotage.[104] The results of these efforts have been mixed.

Oil production, which was halted during the invasion, resumed shortly after the occupation began and reached prewar levels of 2.5 million barrels per day (mbd) in March 2004.[105] With a few significant bumps, production has held remarkably steady at approximately 2.0 mbd ever since. While respectable, this level is far below the Bush administration goal of 3.0 mbd. Exports reached prewar levels in June 2004 with 1.8 mbd, but they have encountered greater fluctuations than have production levels due to regular acts of sabotage and the failure to fully repair and restore pipelines. Nevertheless, exports averaged roughly 1.6 mbd and 1.4 mbd in 2004 and 2005, respectively. Again, these figures are well below the administration's goal of 2.3 mbd.[106]

Of course, the administration not only wants oil to flow, but it wants oil to flow to the United States. In this regard, it has been successful. U.S. oil imports from Iraq over the last thirty years have been wildly erratic and, more often than not, nonexistent. Since the 2003 invasion, however, imports have been far more steady and at consistently sizeable levels. In October 2003, U.S. imports of Iraqi oil reached 734.5 thousand barrels a day, a record surpassed only three times in the previous thirty years (1973 is the earliest date for which data is available from the U.S. Department of Energy). April 2004 imports went higher, and higher still in August 2004 with eight hundred thousand barrels per day—the postwar peak and the second highest import level in thirty years. On average, the United States has imported 542.6 thousand barrels of Iraqi oil per day since the invasion.[107] Only four of the last thirty years have seen higher averages: 1989, 1999, 2000, and 2001—although these year-end averages hide dramatic variations between months. The war has therefore delivered regularity and significant monetary benefits. Thanks to this steady flow of oil and the dramatic increase in oil prices in 2004, between 2003 and 2004 the value of U.S. imports of Iraqi oil increased by 86 percent, and then increased again in the first three quarters of 2005.[108]

Three months after the invasion, Chevron received one of the first contracts to market Iraqi oil. It has since signed subsequent longer-term deals with Iraq's State Oil Marketing Organization, as have ExxonMobil and Marathon, among others.[109] Data on exactly how much Iraqi oil each company is selling is not publicly available. However, in August 2005, Energy Intelligence Research reported that more than 50 percent of all Iraq's oil exports went to the United States that month, the majority of which was delivered by Chevron, ExxonMobil, and Marathon, with the rest delivered by Shell and BP.[110]

Iraq's oil has therefore already contributed to skyrocketing oil company profits. So, too, it seems, has the myth of a dramatically reduced oil supply from the Middle East due to the Iraq War.

Iraq's New Petroleum Law

Like Iraq's economy generally, its oil sector is experiencing a radical transformation at the hands of the Bush administration and its allies. U.S. oil company executives were brought in to advise the Bush administration on Iraq oil policy six months prior to the invasion. Philip Carroll, former CEO of both Shell Oil's U.S. division and the Fluor Corporation, was the first adviser. He was succeeded by Rob McKee, a former executive of ConocoPhillips, who is and was at the time of his Iraq service the chairman of Eventure, a subsidiary of Halliburton.[111] The administration was also advised by the U.S. State Department's Future of Iraq Project's Oil and Energy Working Group—of which Ibrahim Bahr al-Uloum was a member. Uloum is a U.S.-educated oil engineer who served as Iraqi Minister of Oil from September 2003 to June 2004, and again from May 2005 to the present.

Meeting four times between December 2002 and April 2003, members of the U.S. State Department's Oil and Energy Working Group agreed that Iraq "should be opened to international oil companies as quickly as possible after the war."[112] Bearing Point's contract reflects this opinion, stating the need for "private-sector involvement in strategic sectors, including privatization, asset sales, concessions, leases and management contracts, especially those in the oil and supporting industries."

Differences appear to have emerged in the Working Group over just how far the "opening" should go. An Energy Intelligence Research report on the proceedings of the Working Group, stated: "U.S. policy makers argue that the oil wealth of the country would need to be broken up and distributed more widely, not concentrated in the hands of the central government. Privatization is seen as one of the best way to do this."[113] There are varying degrees of privatization. At one end of the spectrum would be private companies literally owning Iraq's oil in the ground or the full privatization of Iraq's National Oil Company. Few advisers argued for such changes, as such a wholesale transformation of Iraq's most vital industry would surely incite mass opposition from the Iraqi people. They did argue for a level of privatization that would turn some of the formerly state-controlled activities over to the private sector.

The model that won out was the Production Sharing Agreement (PSA). None of the top oil producers in the Middle East use PSAs because they favor private companies at the expense of the exporting governments. PSAs turn the entire exploration, drilling, and infrastructure-building process over to private companies under contracts with terms of twenty-five to forty years that lock in the laws in effect at the time the contract was signed. This means that, while a future Iraqi government could change Iraq's laws, including the Bremer Orders, its changes would not impact oil contracts signed while the current laws are in effect.

According to the Working Group, PSAs "can induce many billions of dollars of foreign direct investment into Iraq, but only with the right terms, conditions, regulatory framework, laws, oil industry structure and perceived attitude to foreign participation."[114] The "regulatory framework," many of the "laws," and the "perceived attitude to foreign participation" were established by the Bremer Orders. A new national petroleum law was needed, however, to address everything else on the Working Group's list.

When Bremer left Iraq in June 2004, he bequeathed the Bush economic agenda to interim Prime Minister Allawi and interim Finance Minister Mahdi. Just two months later, Allawi submitted guidelines for a new petroleum law to Iraq's Supreme Council for Oil Policy. The guidelines declared "an end to the centrally planned and state-dominated Iraqi economy" and suggested the "Iraqi government to disengage from running the oil sector, including management of the planned Iraq National Oil Company (INOC), and that the INOC be partly privatized in the future." Allawi recommended that the Iraqi field services industry "should be exclusively based in the private sector, that domestic wholesale and retail marketing of petroleum products should be gradually transferred to the private sector, and that major refinery expansions or grassroots refineries should be built by the local and foreign private sectors."[115] Most important, Allawi's guidelines turned all undeveloped oil and gas fields over to private international oil companies.

Just seventeen of Iraq's eighty known oil fields have been developed. According to an Energy Intelligence Research (EIR) report boldly titled, *Iraqi Oil & Gas: A Bonanza-in-Waiting,* only modest development work has been carried out in recent years anywhere in Iraq and no significant exploration has been done in the last twenty years.[116] The largest known fields are evenly distributed between the north and the south of Iraq with eleven fields each. Other fields are located in central Iraq and the EIR reports that East Baghdad could be a giant field once it has been fully appraised. The largest known fields are Kirkuk in the north and Rumaila in the south, but there is potential for vast amounts of oil across virtually all of Iraq, including in the western desert. EIR estimates that while Iraq's proven oil reserves are second only to Saudi Arabia's, "its 200 billion-plus barrels of probable reserves could put it in competition for the top spot." Therefore, the development of untapped fields could yield billions of gallons of oil, and trillions of dollars in revenue.

The plans for Iraq's new petroleum law were made public at a press conference in Washington, DC, hosted by the U.S. government. On December 22, 2004, Mahdi joined U.S. Undersecretary of State Alan Larson at the National Press Club and announced Iraq's plans for a new petroleum law to open the oil sector to private foreign investment. Mahdi explained, "So I think this is very promising to the American investors and to American enterprise, certainly to oil companies."[117]

He described that, under the proposed law, foreign companies would gain access both to "downstream" and "maybe even upstream" oil investment in Iraq. ("Downstream" refers to refining, distribution, and marketing of oil. "Upstream" refers to exploration and production.) A few weeks later, Mahdi was appointed one of Iraq's new deputy presidents. The transitional Iraqi government's June 2005 National Development Strategy Paper states that Iraq's oil sector plan "invites foreign companies to participate in developing the oil and gas fields in shared production contracts."

Iraq's petroleum law is set for implementation in 2006. It reportedly adopts Allawi's recommendation that currently producing oil fields are to be developed by Iraq's National Oil Company, while *all new fields* are opened to private companies using PSAs, giving private companies control of 64 percent of known reserves. If a further 100 billion barrels are found, as is widely predicted, foreign companies could control 81 percent of Iraq's oil—or 87 percent if 200 billion are found, as the Oil Ministry predicts. Officials in the Iraqi Oil Ministry reportedly plan to begin signing long-term contracts with foreign oil companies during the first nine months of 2006.[118]

Before new oil contracts could be signed, the existing contracts had to be erased. This all-important step was taken back in May 2003. As reported in Energy Intelligence Research, the U.S.-appointed senior adviser to the Iraqi Oil Ministry, Thamer al-Ghadban, announced that few, if any, of the dozens of contracts signed with foreign oil companies under the Hussein regime would be honored. In June 2004, after being appointed Iraq's minister of oil, al-Ghadban told Shell Oil Company's in-house magazine that 2005 would be the "year of dialogue" with multinational oil companies.

By May 2005, approximately thirty international oil companies had signed Memoranda of Understanding with Iraq, generally for the training of Iraqi staff, consulting work, and studies.[119] For example, Chevron has been flying Iraqi oil engineers to the United States free of charge for four-week training courses since early 2004.[120] ExxonMobil signed a Memoranda of Understanding with the Oil Ministry in late 2004, laying the groundwork to provide technical assistance and conduct joint studies. In January 2005, BP signed a contract to study the Rumaila oil field near Basra, and Royal Dutch/Shell Group signed an agreement to study the Kirkuk field. Shell is also helping write a master plan for Iraq's natural gas sector for free. The purpose of all of these free services and memoranda of understanding is to keep their foot in the door and be the first companies in line when the real deals finally become available.

Iraq's Constitution, Federalism, and Oil

In spring 1994, Paul Wolfowitz reflected in the conservative political journal, *The National Interest,* on the failure of the first Gulf War to remove Saddam Hussein from power and expressed regret for not putting more trust into Iraq's Shi'a majority: "Some U.S. government officials at the time appeared to believe—despite strong urgings to the contrary from the Saudis—that we had less to fear from Saddam Hussein than we would from a Shi'a government in Baghdad. It was clearly a mistake. . . . In part it was a failure to recognize the enormous differences between the Arab Shi'a of southern Iraq and the Shi'a extremists who governed in Tehran."

The new constitution of Iraq places the greatest authority among the Bush administration's "allies" in Iraq: the Kurds in the north, the Shi'a in the south, and the Iraqi federal government in Baghdad. None are perfect allies, but, as Wolfowitz explains, they are better than Saddam—and that is what ultimately matters.

Control of oil is distributed among the three groups and appears to leave the new petroleum law untouched. The provisions in the constitution only specify their application to current oil fields, preserving the petroleum law's authority over new fields. The constitution specifically endorses the use of "market principles and encouraging investment," which allows for the law's various modes of privatization, including the use of foreign private companies.

While Article 108 states that "Oil and gas are the ownership of all the people of Iraq in all the regions and provinces," it is immediately qualified with Article 109, which reads:

> First: The federal government will administer oil and gas extracted from *current* fields in cooperation with the governments of the producing regions and provinces on condition that the revenues will be distributed fairly in a manner compatible with the demographical distribution all over the country (emphasis added).
>
> Second: The federal government with the producing regional and governorate governments shall together formulate the necessary strategic policies to develop the oil and gas wealth in a way that achieves the highest benefit to the Iraqi people using the most advanced techniques of *market principles and encouraging investment* (emphasis added).

Article 107 gives the federal government exclusive authority to negotiate, sign, and ratify international treaties, agreements, and foreign sovereign economic and trade policy. This provision would appear to apply to agreements signed with foreign oil companies, but it is unclear.

In fact, much remains unclear. As Nathan Brown commented, "Constitutional provisions for oil resources are fraught with ambiguity. . ." However, this ambiguity works to the benefit of the petroleum law, which clarifies the very provisions left vague in the constitution. In other words, the constitution does nothing to contradict the petroleum law, but rather reinforces its core provisions. Thus, if we consider the oil provisions of the constitution, the Bremer Orders, the constitutional commitment to the economic transformation already underway, and the new petroleum law, it appears that expanded private foreign corporate access to Iraq's oil wealth is all but guaranteed.

The December 15, 2005, elections established an internationally recognized permanent Iraqi government with the authority to sign oil contracts. The Bush administration, for its part, spent the month of December reiterating its commitment to "stay the course in Iraq" and would not speak of the withdrawal of U.S. troops in any significant numbers prior to the end of 2006, giving U.S. oil companies enough time to sign contracts and get to work under the full protection of the U.S. and Coalition forces. Thus, amid all the talk of training Iraqi soldiers, heading off a civil war, and protecting Iraq's fledgling democracy, the overriding agenda determining the timetable for the end of the Iraq war may in fact be the "oil timeline."

~

While violence increases daily in Iraq and the resistance grows, the Bush administration can be confident about a few things. First, the economic restructuring is well in place and moving forward. The banking, investment, patent, copyright, foreign ownership, commercial, utilities, taxes, media, and trade laws, among others, have been changed according to plan, and Iraq is on its way to WTO membership. Second, U.S. corporations continue to earn billions of dollars for work in Iraq and have the potential to earn far more. Third, a government is in place that, while not ideal, is certainly preferable to the previous regime in terms of its willingness to advance Bush administration goals. Fourth, and most important to many, the oil sector has been opened to U.S. corporate access and control. Everything may not have gone exactly according to the Bush administration's plans in Iraq, but all things considered, Bush's key political and corporate allies have much to be optimistic about.

As President Bush has repeatedly said, Iraq is only the beginning. In the name of spreading peace and democracy, he has revealed plans to take his administration's model of imperial-style corporate globalization from Iraq to the rest of the Middle East through a new regional free trade area. Having begun in Iraq, U.S. corporations are once again in the lead, eager to expand their own interests elsewhere. Next to come in the president's economic invasion of the world is the proposed U.S.–Middle East Free Trade Area.

SEVEN

EXPORTING "FREE TRADE" IN PLACE OF "FREEDOM" TO THE MIDDLE EAST

THE U.S.–MIDDLE EAST FREE TRADE AREA

For many conservatives, Iraq is now the test case for whether the U.S. can engender American-style free-market capitalism within the Arab world.

—*Neil King Jr.,* Wall Street Journal, *May 1, 2003*

The military intervention in Iraq and the continued talk about the need for change in the Middle East, all this constituted a major change in the international scene and placed immense pressures on governments.

—*Hazim El-Biblawi, Egyptian economist, 2005*

The economic policies implemented in Iraq are the same policies that U.S. corporations and the Bush administration would like to put in place throughout the Middle East and the world. The traditional mechanism for implementing such policies, at least in recent U.S. history, has been the so-called free trade agreement. The

Middle East, however, insulated by oil revenue, has been largely immune to the need to succumb to the sacrifices required under such agreements. Thus, few agreements have been signed and even fewer implemented in the region—denying U.S. corporations the level of access they desire to one of the wealthiest areas of the world.

With the invasion and occupation of Iraq, however, the Bush administration has demonstrated that it will defy the will of its allies and of global public opinion to use military force to implement its agenda. President Bush followed up military action with a demand that all nations unequivocally demonstrate whether they are with or against the United States. Just one month after declaring "mission accomplished" in Iraq, Bush proposed a U.S.–Middle East Free Trade Area to spread the economic invasion of Iraq to the entire region. The result has been a new level of acquiescence among previously recalcitrant nations. With two-thirds of the world's oil standing like a prize at the finish line, these free trade negotiations are moving forward at breakneck speed, as country after country is pulled onto the corporate globalization bandwagon.

Chevron, Bechtel, Halliburton, and Lockheed Martin have longstanding and sizeable interests throughout much, but certainly not all, of the Middle East. They have made enormous investments in the region and know that there is much more money to be made. Reconstruction contracts are one thing, but long-term access to the entire region on the corporations' own terms is far superior. To this end, each company is a participant in the U.S.–Middle East Free Trade Coalition, a business lobby created for the sole purpose of advancing the president's Middle East Free Trade Area.

PRESIDENT BUSH'S COMMENCEMENT SPEECH

> Across the globe, free markets and trade have helped defeat poverty, and taught men and women the habits of liberty. So I propose the establishment of a U.S.–Middle East Free Trade Area within a decade, to bring the Middle East into an expanding circle of opportunity, to provide hope for the people who live in that region.
>
> —*President George W. Bush, May 9, 2003*

President Bush delivered the commencement address to the graduating class of the University of South Carolina on May 9, 2003. It had been just one week since his nine Democratic challengers met in the state for their first debate of the election season. The South Carolina Democratic Party chairman called the president's speech the official kickoff to the 2004 presidential campaign, while the state's Republican Party chairman said it would solidify South Carolina as "Bush country." Bush used the twenty-five-minute address to make his case for expanded corporate globalization in the Middle East. Freedom, the president argued, would be advanced through a regionwide area in the Middle East that would adopt the same radical economic policies implemented in Iraq. Bush did not mention specifics, only that "free trade" in the Middle East would assure its people opportunity, prosperity, liberty, and freedom.

The president told the audience, "The Middle East presents many obstacles to the advance of freedom. And I understand this transformation will be difficult." He expressed his faith in the ability of free trade to overcome these obstacles and explained that the advance of freedom in the Middle East "is in our national interest." In pursuit of that interest, President Bush was sending Secretary of State Colin Powell and U.S. Trade Representative Robert Zoellick to Jordan the following month to open discussions on economic, political, and social progress in the Middle East. "Progress," the president emphasized, "will require increased trade, the engine of economic development."

While President Bush received the audience's applause, Amanda Martin received her diploma. Dressed in her all-black graduation robe, Amanda stood in the 90-plus degree South Carolina heat outside of the hall where the president spoke. Joined by her family and over 150 other students, professors, and community members, Amanda participated in an "alternative graduation ceremony" in front of the building. Rather than receive certificates for her masters' degrees in public health and social work, Amanda and the other graduates received copies of the U.S. Constitution.

When asked by a local reporter why she was protesting her own graduation, Amanda replied, "Fifty-seven textile plants in South Carolina alone have been closed and 55,000 people have lost their jobs because of free-trade agreements. And he's bringing his free-trade stump speech to our graduation ceremony."[1] In a later interview, Amanda told me about her personal connection to these statistics through her job as director of the New Haven Community Center in Greenwood, about ninety miles away. There she worked with undocumented men and women who had been driven to South Carolina, mainly from Mexico, in search of work in local plants. She told me that South Carolina alone had lost more than 70,000 textile jobs since 2000 and that the textile industry had cut more than half the jobs in the state. "NAFTA and other trade deals," she said, "sent those jobs overseas to the lowest bidder, further exploiting workers in developing countries and forcing them to turn around and search for work in the U.S. in an endless race to the bottom." Amanda and the 150 others who protested with her feared more of the same in both the United States and the Middle East if the president had his way with the U.S.–Middle East Free Trade Area (MEFTA).

U.S. TRADE WITH THE MIDDLE EAST

> With one-fourth of the world's proven oil reserves and some of the lowest production costs, Saudi Arabia is a key market for U.S. energy services firms. Saudi Arabia's opening of its energy markets presents major opportunities for U.S. firms.
>
> —*Office of the U.S. Trade Representative, Saudi Arabia's Accession Agreement to the WTO, September 2005, step one of MEFTA*

Many of the reasons to negotiate a corporate globalization agreement with the Middle East overlap with those for invading Iraq, particularly the desire to increase U.S. access to the region's resources—including oil, wealth, and human capital. Oil, of course, attracts oil, gas, and energy service companies, among others; the region's wealth provides a market for U.S. services and products; and the human capital attracts companies seeking to produce goods more cheaply and with fewer restrictions to sell across the region, the world, or even back to the United States.

Today, almost all trade between the United States and the Middle East is in just two interrelated areas: oil and weapons. They sell us oil, and we sell them oil services, technology, and weapons.

According to the U.S. International Trade Commission, more than half of all U.S. imports from the Middle East are oil and gas. However, while almost two-thirds of the world's oil supply is found in the Middle East, and almost half of all oil consumed in the United States comes from abroad, less than 20 percent of U.S. foreign oil is imported from the Middle East—and almost half of that comes from just one country: Saudi Arabia. After Saudi Arabia, the United States imports its largest percentage of Middle Eastern oil from Iraq. U.S. oil companies and the Bush administration expect the MEFTA to expand their access to and control over the region's oil. In fact, it already has done so, as demonstrated by Saudi Arabia's bilateral accession agreement with the United States to the WTO.

The world's top five oil-rich nations—which, combined, account for 54 percent of the world's proven oil reserves—are included in the proposed MEFTA. They are, in order of known reserves: Saudi Arabia, Iraq, Kuwait, Iran, and the United Arab Emirates. Of the five, four are actively involved in the MEFTA negotiating process, while the Bush administration is trying to convince the fifth—Iran—that it is in its best interest to be "with us" rather than "against us." Each of these countries, as well as all of the countries with sizeable quantities of oil that are listed in the MEFTA, have fully or partially nationalized oil sectors with limited arrangements for foreign private companies. U.S. companies do have some contracts to explore for, extract, and sell oil in the Middle East, but only with a few of the countries, and in nearly all cases only in partnership with the state-controlled oil companies and under heavily guided rules of the foreign governments. U.S. corporations hope that the MEFTA will redress this situation through new rules on foreign investment and ownership, the elimination of tariffs and quotas, and privatization.

The MEFTA proposes to cover twenty countries in the Middle East and North Africa. As listed by the U.S. Congressional Research Service, these are: Algeria, Bahrain, Cyprus, Egypt, the Gaza Strip/West Bank (otherwise known as Palestine), Iran, Iraq, Israel, Jordan, Kuwait, Lebanon, Libya, Morocco, Oman, Qatar, Saudi Arabia, Syria, The United Arab Emirates, Tunisia, and Yemen. Notably absent from this list is Afghanistan. One possible explanation is that Afghanistan's primary export is heroin made from poppy grown throughout the country. If Afghanistan were included in international trade negotiations, the heroin trade could come under significantly greater international focus than it has to date. Until the 2003 invasion of Iraq, the United States had trade agreements with just two of the twenty MEFTA countries: Israel, signed in 1985, and Jordan, signed in 2000.

U.S. exports to the Middle East fall largely into two categories: (1) machinery and technology, primarily to support the oil sector, and (2) weapons, including "dual-use technology" such as helicopters and computers, most of which are arguably used to protect oil and/or those who profit from it. In fact, the United States is the world's largest arms exporter and the Middle East is far and away its largest customer. Between 2001 and 2004, the leading Middle East purchasers of U.S. arms spent $16.3 billion between them (Egypt $5.7 billion, Israel $4.4 billion, Saudi Arabia $3.8 billion, Kuwait $1.8 billion, and Oman $960 million). Asia was at a distant second, with its top five purchasers of U.S. arms spending just $9.5 billion.[2]

U.S. trade with the Middle East is so limited because large sectors of Middle Eastern economies are simply closed to foreign or even private companies. Additionally, there are high tariffs on many foreign goods, and the governments exercise a far greater level of control over foreign companies than most U.S. multinationals are accustomed to in the age of corporate globalization. Thus, you can visit the Hard Rock Café, buy a Coke, and go to a Citibank in Beirut, but it is the Lebanese government that calls the shots—at least much more so than Hard Rock Café International, Inc., the Coca-Cola Company, or Citigroup are used to in most countries around the world today.

Furthermore, if you peek at the label on your GAP jeans or your Liz Claiborne sweater, you will not see "Made in Oman." If you go to Lebanon, you will not find a single Wal-Mart, you will be unable to fill your prescription for *Cialis* (Eli Lilly's version of *Viagra*), and you will far more likely find a bottle of *Saha* water than Pepsi's *Aquafina*. If you go to Saudi Arabia, you will need to get cash at the Saudi Cairo Bank because you will not find any Citibank ATMs. And when in Egypt, you will not find a Chevron station, but you will be able to fill your tank at the local *Misr.* U.S. corporations are certainly troubled by the fact that they do not own many stores, factories, or banks in the Middle East, or sell large quantities of pharmaceutical drugs, bottled water, or jeans, or have access to more than a few of the region's oil reserves, but it does not mean that the people of the Middle East are doing without such products and services.

All told, just 1 percent of U.S. foreign direct investment goes to the Middle East, less than 4 percent of all U.S. exports go there, and less than 5 percent of U.S. imports come from the region. U.S. corporations see the MEFTA as the leading route to change.

U.S.–MIDDLE EAST FREE TRADE COALITION

In October 2004, two of the oldest and most influential U.S. corporate organizations focused on international trade joined forces to increase their access to the Middle East. The National Foreign Trade Council and the Business Council for International Understanding teamed up to form the U.S.–Middle East Free Trade Coalition, with the specific purpose of advocating for the MEFTA.

National Foreign Trade Council

Founded in 1914, the National Foreign Trade Council (NFTC) is arguably the most powerful U.S. corporate lobbying association on international trade. It wields enormous influence, has wide access to policy makers, and knows how to use its membership clout. While CEO of Halliburton, Dick Cheney remarked, "From the point of view of Halliburton, one of the most valuable organizations we are a part of is the NFTC." The group has four hundred member companies and its company board of directors includes Bechtel, Chevron, and Halliburton, as well as the largest corporation in the world—Wal-Mart. In March 2005, the NFTC released its "Trade and Investment Agenda." Out of the long list of trade and investment negotiations scheduled to appear before Congress in 2005, just two made the group's priority agenda: the Middle East Free Trade Area and the completion of the WTO negotiations begun in Doha, Qatar, but stymied in Cancun. To advance these trade goals, the NFTC lobbies Congress and meets with senior administration officials and "world leaders and high-level decision-makers."

The Business Council for International Understanding

The Business Council for International Understanding (BCIU) was founded in 1955 as a place for "promoting dialogue and action between the business and government communities for the purpose of expanding international commerce." It hosts meetings between U.S. business, U.S. government officials, and their foreign counterparts. In essence, it lets the U.S. government know what the U.S. business community is looking for and allows the U.S. government to turn to the business community to help advance its goals. The group's members comprise a veritable "who's who" of global corporate power, including Chevron, Bechtel, Halliburton, and Lockheed Martin, as well as every major U.S. oil company; financial behemoths such as Citigroup and JP Morgan Chase; consumer companies such as Coca-Cola, Ford Motor, and Estee Lauder; and our old friend, Henry Kissinger, through Kissinger McLarty Associates—the DC branch of Kissinger Associates.

In 2002, for example, the BCIU hosted a series of events on Middle Eastern trade, including discussion sessions with the chairman of the interim government of Afghanistan, the deputy assistant secretary of commerce for the Middle East and Africa, the U.S. ambassador to Saudi Arabia, the U.S. ambassador to Bahrain, and then-U.S. ambassador to the United Nations, John Negroponte. They met to discuss Iraq in a session moderated by Bechtel, which included the U.S. director of Arabian Peninsula Affairs at the U.S. Department of State. They also held a series of "Middle East Regional Teleconferences" with U.S. ambassadors in Bahrain, United Arab Emirates, Qatar, Yemen, Saudi Arabia, Kuwait, and Oman.

The NFTC and the BCIU joined forces by forming the U.S.–Middle East Free Trade Coalition to advocate for one coordinated approach to the entire region. The 110 members of the coalition include all of the companies cited earlier whose products and businesses you will not find readily available in the Middle East: Chevron, The Gap, Liz Claiborne, Eli Lilly and Co., Citigroup, Wal-Mart Stores, Inc., and PepsiCo Incorporated. The members also include Bechtel and Lockheed Martin, as well as several business associations, including the Coalition of Service Industries, of which Halliburton is a member. Other notable members include ExxonMobil, Pharmaceutical Research and Manufacturers of America, General Motors, General Electric, Motorola, and Hewlett Packard.

Jeffrey Donald, senior vice president of BCIU and a spokesperson for the Coalition, described for me in an interview the member companies' interests in the MEFTA, which he summarized as "gaining greater predictability and flexibility for their Middle East operations." Oil companies, he said, are interested in the privatization of currently nationalized oil sectors. Pharmaceutical companies are interested in higher levels of intellectual property rights protections. The U.S. service sector, the largest sector of the American economy, wants to sell its services to more countries in the Middle East with fewer restrictions. For example, companies that provide water, electricity, or oil services, are interested in privatizing currently state-run services and securing foreign investment protections like those in Bremer Order #39, because Donald explained, "if you have a thirty year time horizon, you care a lot about investment rules." The Coalition is working hand in hand with the Bush administration to see that these interests, and others, are met by the president's 2013 deadline if not sooner.

ENTER ROBERT ZOELLICK

Robert Zoellick, current undersecretary of state and former U.S. trade representative, is the architect of the MEFTA and is widely considered the force driving it toward completion from inside the Bush administration. At age fifty-two, Zoellick looks and acts like a man who should be sitting at the helm of the British East India Company. He wears a tight, well-groomed mustache and parts his hair to the side in a strict unwavering line, exposing a large forehead and a receding hairline. He is never without his small wire-rimmed reading glasses—most often found on the tip of his nose.

Zoellick's career has been spent in service to the Bush family and its closest allies. All along, his work has focused on international trade policy as a form of statecraft. Zoellick is a protégé of a close Bush family friend, James Baker III, whom Zoellick worked for in the Reagan administration. In the administration of Bush Sr., Zoellick was the U.S. State Department's lead negotiator on the development of both the North American Free Trade Agreement and the World Trade Organization. In 1992, he was named White House deputy chief of staff and assistant to President Bush. He then served as foreign policy advisor for governor and then presidential candidate George W. Bush. When candidate Bush needed help in Florida, Zoellick was on the legal team headed by Baker that fought the Florida recount. After their success, Zoellick was appointed U.S. trade representative in 2001 and then undersecretary of state in 2005. Back in 1998, Zoellick also signed the Project for the New American Century letter to President Clinton that called for the invasion of Iraq.

While advising presidential candidate George W. Bush, in January 2000, Zoellick set out several of the core provisions of the Bush Agenda in "Campaign 2000: A Republican Foreign Policy," in *Foreign Affairs.* Zoellick described corporate globalization as a tool of national security strategy and *Pax Americana,* explaining that superior U.S. military might and economic prowess should be harnessed and expanded through corporate globalization agreements—all in the name of greater world peace. He emphasized the importance of the new century as a time to assert the primacy of the United States. In defining the differences between a Republican and Democratic foreign policy, Zoellick looked to the Middle East. He argued that the United States should be willing to replace the brutal regimes of Iraq and North Korea because they threaten U.S. "vital interests, such as maintaining access to oil in the Persian Gulf." He added, "As a new generation of leaders gain authority in the Middle East, possible peace agreements can be buttressed by drawing these societies into information-age economics and integrating their economies into world markets."

Zoellick became the U.S. trade representative at a time when the debate on corporation globalization and its institutions had fundamentally shifted from "How fast?" to "For whom and at what cost?" Zoellick responded to an increasing wave of opposition by setting out a negotiating framework that has since crystallized in the MEFTA process. The collapse of talks at the WTO ministerial meeting in Seattle, Washington, in 1999 was a moment of profound transformation. Not only did fifty thousand people turn out on the streets to oppose the institution, but developing country delegates stood up as a bloc and said they would no longer simply follow the demands of the United States and the European Union. The public displays of opposition to corporate globalization continued to grow in the months and years that followed, mirroring what was happening inside the negotiations: government leaders the world over were rejecting the onrush of corporate globalization.

Five months after talks collapsed in Seattle, thirty thousand people demonstrated against the World Bank and International Monetary Fund in Washington DC. In January 2001, over ten thousand people gathered in the Brazilian city of Porto Alegre for the first annual World Social Forum, an event organized for the sole purpose of discussing meaningful alternatives to corporate globalization. In the years that followed, annual participation at the Forum would top one hundred thousand people. In April 2001, sixty thousand people protested against the Free Trade Area of the Americas in Quebec City, Canada. In July, two hundred thousand people gathered in Genoa, Italy, to voice opposition to corporate globalization, as the Group of Eight industrialized countries (G8) held its annual meeting. The event is remembered, however, for the tragic death of nineteen-year-old Carlo Giuliani, who was shot by police while participating in a protest. Opposition to corporate globalization was even on display in the Middle East, when in November 2001, over one thousand people gathered in Beirut, Lebanon, for the World Forum on Globalization and Global Trade.

Opposition to Corporate Globalization in Beirut

Beirut is a city of extreme contradictions. It has yet to recover from the bombings that it suffered in the fifteen long years of war fought on its soil between Lebanese, Israelis, Syrians, and Palestinians. Although 1991 marked the official end of the wars, large craters still rip through the skeletal remains of buildings throughout the city, exposing concrete, pipes no longer attached to water mains, and bits of carpeting and wallpaper that just refuse to turn to dust. While the buildings are all but shells, people still call them home. The Hard Rock Café, one of the largest you are likely to see anywhere, sits directly in front of one such building. From one vantage point, the entire building is eclipsed by the enormous red electric guitar jutting out from the side of the restaurant.

Beirut, with all its contradictions, provided the setting for a series of events organized by critics of corporate globalization leading up to and during the WTO's November 2001 ministerial meeting held in Doha, Qatar. The critics' events were held in Beirut rather than in Qatar because the strict monarchy that governs the latter had made social and political organizing all but illegal and greatly restricted access to foreigners.

Ziad Abdel Samad, a lead organizer of the Beirut events, is executive director of the Beirut-based Arab NGO Network for Development, which includes nongovernmental organizations (NGOs) from Lebanon, Palestine, Iraq, Jordan, Bahrain, Yemen, Egypt, Sudan, Tunis, Algeria, Morocco, and Mauritania. The Arab NGO Network was a member of a larger coalition, the Lebanese Platform on the WTO, which included organizations representing teachers, women, the disabled, environmental concerns, children's rights, student groups, trade unions, and farmers.

As a representative of the International Forum on Globalization, I attended the Lebanese Platform on the WTO's opening press conference. Although offered in Arabic, the presentations were highly familiar. I have heard them in a dozen languages, spoken on virtually every continent, and in towns and cities across the United States. Samad explained the purpose of the gathering by recalling what happened in Seattle a year and a half earlier. Activists from Lebanon had joined tens of thousands of people to speak out against the WTO and demand changes to the global economic system. Samad described how the main powers behind the WTO ignored the interests and demands of developing countries, pressuring their governments to open markets to foreign trade and investment on terms that greatly benefit multinational corporations but harm their citizens. The people of Lebanon, Samad declared, now demanded that their government reject WTO membership and any similar trade deals.

Waddah Fakky, an elderly, small-framed, and soft-spoken representative of the Union of Farmers from southern Lebanon, spoke next. Fakky described how existing free trade policies implemented without conditions to address local impacts were already harming small farmers across Lebanon. Lebanon grows potatoes, olives, and bananas, but, Fakky stressed, the ability of local farmers to survive on earnings from these crops was being decimated by cheaper imports. The farmers do not oppose imports, explained Fakky, but they must be provided with supports that allow the local farmers to compete. He stressed that these problems would only be exacerbated if Lebanon joined the WTO. Fakky's views were seconded by Bchara Chaaya of the General Confederation of Lebanese Trade Unions, who described how unions across the Middle East were organizing to oppose the WTO due to a predicted increase in the power of corporate monopolies, which were already causing vast unemployment and reduced wages across the region.

Four months later, on November 5, 2001, and while the WTO met in Qatar, citizens from across the Middle East and the world gathered in Beirut for the World Forum on Globalization and Global Trade. After several days of lectures and discussions, the declaration released by the Forum—two years prior to the U.S. invasion of Iraq—was prescient. It was entitled "No to the Militarization of Globalization: Toward a Global Front for Global Peace and Social Justice."

Zoellick's "Competition Strategy"

The opposition to corporate globalization, both among civil society and its elected officials, had undermined the ability of the United States to sign trade deals. The opponents were not demanding an end to trade, but an end to one-sided trade rules that benefit developed over developing countries and multinational corporations over local businesses, workers, and communities. In response, the United States could have opened itself to greater negotiation and flexibility in its trade discussions. Instead, Zoellick and the Bush administration responded with even greater unilateralism, choosing to flex American muscle to force recalcitrant countries into line.

In his 2000 *Foreign Affairs* article, Zoellick set out his own version of "You are either with us, or against us": "If some regions are too slow to open their markets, the United States should move on to others. America should spur a competitive dynamic for openness and transparency. Competition can work wonders: when the United States pursued NAFTA and APEC [Asia Pacific Economic Cooperation], the EU finally felt the pressure to complete the global Uruguay Round trade negotiations. If others hold back in the new WTO round, the United States should repeat this strategy of regionalism with a global goal in order to break the logjam." In other words, pit region against region and country against country to come up with a deal.

For Zoellick, the tragedies of September 11 became *the* indisputable argument for countries to sign his free trade bottom line. On September 14, 2001, literally before the dust from the terrorist attacks had cleared, Zoellick released a statement on the importance of September 11 to the upcoming WTO negotiations in Doha, Qatar, in November 2001. Apparently brushing aside the very real concerns for the safety of those who planned to attend the WTO ministerial, Zoellick asserted that now, more than ever, the world needed to come together to advance free trade:

> America has been attacked by those who want us to retreat from world leadership. Let there be no misunderstanding: the United States will continue to advance the values that define this nation—openness, opportunity, democracy and compassion. Trade reinforces these values, serving as an engine of growth and a source of hope for workers and families in the United States and the world. Trade is particularly vital today for developing nations that are increasingly relying on the international economy to overcome poverty and create opportunity. While

> we will take every possible step to ensure security, it is important that the World Trade Organization meeting in Doha proceed so that the world trading system can continue to promote international growth, development, and openness.[3]

He was successful, not only in getting countries to Doha, but also in kicking off a new round of negotiations, known as the Doha Round of the WTO. In Zoellick's words, he had finally "erased the stain of Seattle."[4]

Zoellick's tactics are less than reassuring for countries now involved in the MEFTA process. Delegates from developing countries entered Doha virtually united in a set of demands, including fundamental changes to existing WTO rules, opposition to the launch of a new round of negotiations, and opposition to any new WTO agreements. Also united were citizens' groups who had signed a WTO "shrink or sink" statement, demanding that the WTO "shrink" its current set of rules or "sink" as an organization altogether. Once in Doha, however, all of their demands were quickly rejected, and the unity of the developing country delegates was systematically destroyed.

Zoellick set the tone before the meeting began by saying, "As much as developing countries may need debt relief and development aid, a prerequisite for their long-term economic growth is full participation with the global economy and trading system. Doha is the best opportunity we will have in the next ten to fifteen years to expedite this integration. It is an opportunity neither we nor the developing world can afford to miss."[5] The not-so-hidden threat was that countries would either play ball in Doha or their debt relief and development aid would be cut off. This threat was also bluntly voiced in the backrooms of Doha. In *Power Politics in the WTO,* Aileen Kwa of Focus on the Global South, a research organization based in Thailand, reported that delegates from Haiti and the Dominican Republic were told by U.S. government officials to support the expansion of the WTO's government procurement agreement or risk cancellation of their preferential trade arrangements. The report also describes how Pakistan, traditionally a stalwart among developing countries against the WTO, was quiet in Doha, after receiving a massive aid package of grants, loans, and debt reduction from the United States. Nigeria, which had issued an official communiqué denouncing the draft WTO declaration before Doha, came out loudly supporting it on November 14—another flip-flop difficult to separate from the Bush administration's promise of a big economic and military aid package in the interim. In response to these and similar actions, members of the European Parliament issued a Joint Motion for Resolution after the ministerial, stating their opposition to the "heavy-handed and divisive tactics" used by the U.S. government, among others.

Zoellick's questionable tactics were not limited to developing country delegates. He also abandoned promises made to senior U.S. Senator John D. Rockefeller, Democrat from West Virginia. According to the senator, Zoellick "promised me, facing me, looking directly in my eyes, in my office, that he would not in any way compromise various fair trade and anti-dumping laws [and] he immediately did so within the first five minutes" of arriving in Doha.[6] Zoellick also ignored a legal requirement from the U.S. Congress that he seek to establish a working group at the WTO focused specifically on addressing labor issues.

Zoellick's tactics worked. However, he knew that developing country resistance had not been overcome in Doha; it had simply been quieted. In order to achieve the expanded access to foreign markets desired by U.S. companies, particularly in the Middle East, bilateral and regional deals would be necessary.

The invasion and occupation of Iraq provided Zoellick's final impetus. Doha, Qatar became the staging ground for U.S. military operations against Iraq. Countries across the Middle East knew that they had to prove that they were either with the United States or against it, and that there would be very real consequence if they were found to be in the latter category. The Bush Agenda makes clear that military might and economic prowess go hand in hand in furthering America's interests, especially in expanding U.S. corporate access in the Middle East. As Charles Dittrich of the National Foreign Trade Council told me in an interview, "The administration looks at the MEFTA as a key component of its Middle East policy . . . [and] things are moving surprisingly fast."

THE U.S.–MIDDLE EAST FREE TRADE AREA

Zoellick describes the MEFTA as "a region-wide commitment to open trade with the United States." Utilizing his competition strategy, the MEFTA is devised as a five-stage negotiating process with individual countries: (1) World Trade Organization membership, which the U.S. government will actively support for those "peaceful" countries that seek it; (2) the United States will apply the Generalized System of Preferences to certain products from countries involved in the MEFTA; (3) signing a Trade and Investment Framework Agreement; (4) signing a Bilateral Investment Treaty; (5) signing a Free Trade Agreement. If all goes according to plan, each of the individual Free Trade Agreements will then be linked together to form one U.S.–Middle East Free Trade Area by 2013.

The MEFTA is completely unique as a process of multiple bilateral negotiations that will culminate in a regionwide trade agreement. In the MEFTA, the United States negotiates one-on-one with each country, which means each country—often an economy half the size of one state in the United States—must try to negotiate a deal that serves its own interests with the most economically and militarily dominant nation in the world. The reality is that there can be no "negotiation" between such thoroughly unequal pairings. This is why developing countries prefer to negotiate internally among small blocs before inviting larger countries into the fold. Short of this process, they prefer at least to negotiate as a group, so that they can come together to form blocs to assert their interests, as was done at the WTO negotiations in Seattle and again in Cancun. It is also through such blocs that the United States has been stymied in its attempts to form a Free Trade Area of the Americas in the Western hemisphere.

In the wake of September 11 and the invasion of Iraq, Zoellick and the Bush administration are using the opportunity to dictate to the

Middle Eastern countries that they will be dealt with individually and that the cumulative effect of the negotiations will be pieced together to form one final MEFTA. Those who do not participate face the now proven threat of regime change, or at least the Bush administration's disfavor in the form of reduced loans, aid, trade in arms, military support, and the like. Thus, the negotiations are moving forward at breakneck speed as countries of the Middle East fight over one another to be the next to negotiate. It is in every country's best interest to get in line quickly because, as the negotiations move forward, the negotiating positions crystallize and the United States demands from country number four the same outcomes that it received from countries number one, two, and three. The agreements are forthcoming, but the question is to what extent countries will be able to set their own terms.

MEFTA Step One: World Trade Organization Membership

Of the MEFTA countries, Palestine, Iran, Iraq, Lebanon, Syria, Yemen, Algeria, and Libya are not members of the WTO. The United States is currently involved in accession negotiations with Iraq, Lebanon, Yemen, Algeria, and Libya. Becoming a WTO member requires that three-fourths of the existing 148 member nations vote in favor of a country's bid. Next, a review committee is established to determine the conditions under which the country can join. If a country is accepted, in most cases the country's home parliament or legislature must then ratify the agreement before membership is complete. Being a member of the WTO locks a country into a variety of commitments to alter its laws to come into compliance with WTO requirements.

IRAQ'S WORLD TRADE ORGANIZATION BID. Bearing Point's contract set a February 2004 target date for Iraq to begin the process of joining the WTO. Like clockwork, on February 11, 2004, even without an elected government and over European complaints that the "demand was irregular and premature,"[7] Iraq obtained WTO observer status: the first step toward full membership. Peter Allgeier, the deputy U.S. trade representative presiding at the ceremony at the WTO in Geneva, explained that Iraq was now on its way to bringing its laws into compliance with WTO rules. "In pursuing this request, Iraq has made clear its commitment to the rule of law and the principles that underlie the WTO," Allgeier told the WTO's General Council. "Of particular note is the importance that they attach to the WTO as part of their ongoing efforts to rebuild their economy. We believe that these efforts hold great promise to the people of Iraq."[8] Less than one year later, the WTO agreed to accept Iraq's bid and begin the review and negotiation process toward full membership. Few doubt that Iraq will ultimately be accepted, as the Bremer Orders have already transformed the country's laws to meet and even surpass WTO rules.

SAUDI ARABIA'S WORLD TRADE ORGANIZATION BID

> Saudi Arabia has made a broad range of positive commitments resulting in the substantial opening of its energy services market. These commitments will allow U.S. energy services firms to compete on a level playing field for energy services projects associated with oil and gas exploration and development.
>
> —*from Saudi Arabia's September 2005 accession agreement with the United States for WTO membership*

When the MEFTA process began in May 2003, Saudi Arabia was not a member of the WTO. Officially, the Saudi's have been trying to join the WTO since its inception. Unofficially, the Saudi government's bid has been largely symbolic, as it has steadfastly refused to relinquish its control over its own laws, particularly those related to foreign corporations and investors. Furthermore, WTO membership means opening trade with Israel and, until November 2005, such trade was unacceptable to the Saudis. Domestic resistance has also been strong for all of the same reasons cited earlier. In September 2005, the Bush administration succeeded where all others had failed by convincing Saudi Arabia to fundamentally transform its laws in full Bremer Order fashion and allow an unprecedented level of foreign private corporate access and control. Two months later, Saudi Arabia became a member of the WTO.

Much of the Bush administration's power lies in its willingness to pursue its own agenda even against and in conflict with the wishes of longtime allies. Thus, the Saudi government was (and likely remains) justifiably concerned about the lengths to which the Bush administration would go to achieve increased corporate access in Saudi Arabia; some even fear that "regime change" could find its way to Saudi soil. Short of overthrowing the Saudi government, however, there are many other ways that the Bush administration could retaliate. An estimated five thousand U.S. troops stationed in Saudi Arabia provide direct military support to the Saudi government. At times of heightened need, the U.S. government has increased its support; for example, it sent some five hundred thousand troops to Saudi Arabia after Saddam Hussein invaded Kuwait.[9] In addition, U.S. arms sales to Saudi Arabia totaled $9.5 billion from 1997 to 2004 alone, not including dual-use technology.[10] For a government that is militarily resisting domestic opposition to its continued rule, these U.S. military services arguably provide the Saudi government its lifeblood. The Saudi government has therefore taken extraordinary and unprecedented steps to transform its economy and demonstrate to the Bush administration and its corporate allies that it is a Bush Agenda adherent.

In May 2005, the Council of Saudi Chambers of Commerce and Industry launched a ten-day, five-city whirlwind tour of the United States. In special forums in New York, Atlanta, Chicago, San Francisco, and Houston, participants were "introduced to $623 billion in new investment opportunities in the Kingdom through 2020; that is in addition to more than $800 billion in privatization opportunities expected in the next ten years," according to the Saudi embassy's press release. Investment opportunities are found in Saudi Arabia's electricity, water, telecommunications, business services, gas, and petrochemical sectors. The forums were in fact notification of "the largest call for foreign investment in decades" made by the Saudi government.[11]

The tour came with a glossy, full-color brochure entitled "Saudi Arabia, Gateway to Opportunities." The cover photograph is of an ancient wooden door with intricate gold woodwork. It is filled with colorful photographs and maps, graphs demonstrating the country's economic strength, enormous dollar amounts, and long lists of areas of investment opportunities. Both the tour and the brochure were sponsored by Saudi Arabia and cosponsored by Chevron, ExxonMobil, the National Association of Manufacturers, the U.S. Department of Commerce, and the U.S. Department of Energy, among others.

The National Association of Manufacturers (NAM) is one of the oldest and most powerful U.S. industry lobbying groups. Founded in 1895 by the likes of Standard Oil, AT&T, DuPont, General Electric, General Motors, and IBM, today it has fourteen thousand members who manufacture everything from pharmaceuticals to weapons. NAM does not publicly release its list of members, but Halliburton, Chevron, Bechtel, and Lockheed Martin are each named in NAM publications and events. The purpose of the group is to open up new areas of access for U.S. manufacturers and reduce government regulation. As the president's brother and the governor of Florida, Jeb Bush is quoted as saying on NAM's publicity material, "I'm a strong supporter of the National Association of Manufacturers because it's a major policy leader in Washington, D.C. and state capitals all across our country to advocate for manufacturing."

Prince Bandar bin Sultan, the Saudi ambassador to the United States and longtime friend and associate of the Bush family, gave the tour his personal endorsement: "Saudi Arabia is at the crossroads of commerce in the Middle East and is home to a vibrant and growing economy, sustained both by its vast oil reserves and by its ever-growing private sector. These forums are a way to promote strong economic ties between our countries to accompany the strong political ties that link Saudi Arabia and the U.S."[12]

In September 2005, the United States got these commitments in writing, so to speak, with the signing of a bilateral agreement for WTO accession with Saudi Arabia. When a new country wants to join the WTO, any existing member can require that it first negotiate a bilateral set of commitments. The agreed on commitments apply to all WTO member nations, although the focus is on U.S. firms here. As quoted earlier, the accession agreement signed with the United States commits Saudi Arabia to open its oil sector to foreign companies, meaning any oil sector project open to private bidding cannot give preference to Saudi companies over foreign companies, including U.S. firms. This is a "national treatment" provision like that included in Bremer Order #39 in Iraq.

The accession agreement even allows U.S. businesses to challenge preferences given to state-owned enterprises. It commits Saudi Arabia to "ensuring that state-owned or controlled enterprises or those with special or exclusive privileges will make purchases and sales of goods and services based on commercial considerations, and firms from WTO members will be allowed to compete for sales to, and purchases from, these Saudi enterprises on non-discriminatory terms."[13] "Non-discriminatory" is trade-speak for "national treatment." Saudi government contracts must therefore be opened to private U.S. companies on equal footing with Saudi government companies and those with "special privileges," which likely refers to the many companies owned by Saudi royalty, particularly Saudi princes. The provision also appears to allow the U.S. government to challenge Saudi Arabia if, for example, the United States believes that Saudi Arabia is limiting its oil exports for "political" rather than "commercial considerations."

The Saudis have committed to opening new markets to U.S. businesses beyond oil and energy services. These include banking and securities, insurance, and telecommunications. For example, Saudi Arabia formerly required that foreign banks partner with Saudi banks in a minority shareholding position. Now, as in Iraq, foreign banks can establish direct branches. Saudi Arabia has also agreed to "liberalize" its "environmental services market." In international trade parlance, this means full or partial privatization of water and sewage systems, among others.

Saudi Arabia also committed "to reduce its domestic support to agriculture" while simultaneously opening more of its market to U.S. agriculture products by lowering tariffs. Similarly, it committed to reducing tariffs on U.S. industrial goods. The disastrous impacts of these combined measures have already been discussed in South Korea, India, Mexico, Zambia, China, and elsewhere around the world.

The accession agreement led the Bush administration to endorse Saudi Arabia's full membership to the WTO, which was granted on November 25, 2005.

MEFTA Step Two: Generalized System of Preferences

The Generalized System of Preferences (GSP) was created in 1976. Today, it provides preferential treatment, in the form of tariff-free entry, for some 4,650 products from 144 countries and territories into the United States. The GSP can be considered as a political sweetener in the MEFTA process. In return for their participation, special access is granted to some of the products entering the United States from countries participating in the MEFTA. For example, as a part of Iraq's GSP benefits granted in June 2005, certain fresh or packed whole dates shipped from Iraq can now enter the U.S. duty-free. Of the MEFTA nations, Jordan, Morocco, Bahrain, Egypt, Lebanon, Algeria, Tunisia, Oman, Yemen, and most recently Iraq, have GSP privileges.

Unique to the MEFTA process has been the requirement that certain countries sign Qualified Industrial Zones (QIZ) trilateral agreements. These are agreements between the United States, Israel, and a third country. The only apparent reason for the agreements is to ensure a direct financial benefit for Israel. Jordan signed the first QIZ, which eliminates tariffs for Jordanian exports to the United States with an 8 percent Israeli component. In other words, in order to receive this preferential tariff treatment, a portion of Jordan's exports must go to Israel, the Israelis must add to them, and then the goods can enter the United States duty-free. For example, a Jordanian textile mill produces a carpet, sends the carpet to Israel where an Israeli textile mill adds fringe, and then the final product is sent duty-free to the United States.

The Jordanian ambassador to the United States, Karim Kawar, told a Washington, DC, audience, "The basic premise behind [the MEFTA] is that if you raise the water in the harbor, then all boats will rise."[14] Unfortunately for Jordan, this has not been the effect of the agreement thus far. The QIZ initially benefited mainly Asian apparel manufacturers seeking to circumvent U.S. quota restrictions. These manufacturers set up factories in Jordan and then, rather than use Jordanian materials, imported most of their fabrics directly from Asia and then, if they used Jordanians at all, the manufacturers employed them as assembly workers prior to shipping the apparel to Israel.[15] Thus, the overall economic benefit to Jordanian industry has been minimal.

EGYPT'S QUALIFIED INDUSTRIAL ZONES. The Egyptian government signed a QIZ in mid-2005 as part of its MEFTA process. Egypt's QIZ requires an 11.8 percent Israeli component to goods prior to duty-free entry into the United States. There was significant public opposition to the QIZ because it meant signing a deal with Israel, but also as a part of the MEFTA.

Egypt's leading exports are clothes and textiles. Egyptians fear that the impact of the various components of the MEFTA will be the same as that described earlier in Zambia: Local manufacturers will be shut out by foreign products and its people will become cheap labor for foreign companies. Such concerns were strongly reinforced when Israeli Vice Premier Ehud Olmert told Egypt's *Daily Star* newspaper that the QIZ would enable Israeli manufacturers to "take advantage of the lower labor costs in Egypt."[16] In fact, when Zoellick, Olmert, and Egyptian Foreign Trade Minister Rachid Mohammed Rachid presented the QIZ at a press conference in Cairo, they were met by thirty people carrying a banner that read, "Egyptian Workers Are Not For Sale!"

While trying to convince Egyptians that free trade policies were necessary in Egypt, Trade Minister Rachid did concede that the loss of industry and jobs was not only likely but also expected: "In pursing a regional market you must accept that some weaker units will have to close. . . . People want a regional market and they want to protect their industries. But it does not work like that. It is like wanting to swim without getting wet."[17] Setting aside public opposition to the process, when asked how the MEFTA negotiations were proceeding, Rachid replied, "We are getting there."

MEFTA Step Three: Trade and Investment Framework Agreements

Trade and Investment Framework Agreements (TIFA) demonstrate a country's commitment to the MEFTA, and they have moved briskly since the invasion of Iraq. To date, the Bush administration has signed a total of fourteen TIFAs—and twelve of those fourteen since the invasion of Iraq. Algeria and Bahrain signed before the war, while agreements with Tunisia, Saudi Arabia, Kuwait, Yemen, the United Arab Emirates, Qatar, Egypt, Morocco, Oman, and Iraq all followed the war. The TIFAs are negotiation platforms that do not lock countries into specific commitments, but rather signify their willingness sit down and start negotiating on terms that meet Bush administration demands.

Each TIFA agreement is virtually identical. In most cases, the name of the country is the most significant change. Each agreement is just three to four pages long and entitled, "Agreement Between the Government of the United States of America and the Government of ____________ Concerning the Development of Trade and Investment Relations." Each country recognizes, for example, the benefits of "increased international trade and investment," "the essential role of private investment"; that "foreign direct investment confers positive benefits"; "the increased importance of services"; and the need to "facilitate greater access to the markets of both countries," among other issues.

The United States uses the TIFA to identify changes that it wants to see in the other country's laws. In most cases, these include privatization of state-owned industries, new investment protections for foreign companies, the elimination of domestic content requirements and technology transfer, elimination of requirements for a certain percentage of local investment by foreign companies, the elimination of tariffs and quotas, the elimination of local price support systems, opening all sectors to foreign investment, the elimination of rules requiring that foreign companies partner with local companies, and the like.

According to Zoellick, the agreements "encourage private sector participation through business councils that drive trade agendas and help us address the specific concerns of business."[18] Zoellick does not mention participation of any other groups, such as labor, human rights, environmental, social justice, or women's groups. In fact, the TIFAs refer only to the need "to seek the advice of the private sector."

MEFTA Step Four: Bilateral Investment Treaties

Of the MEFTA countries, the United States currently has Bilateral Investment Treaties (BITs) with Israel, Jordan, Morocco, Egypt, Tunisia, and Bahrain. The Bahrain agreement is the most recent, signed in 2001. BITs can be considered the lesser cousins of other investment agreements discussed throughout this book. BITs are the two-country version of the Multilateral Agreement on Investment. The investment rules of the North American Free Trade Agreement are also based on BITs, as is Bremer Order #39 in Iraq. As mentioned earlier, negotiations over similar investment rules at the WTO contributed significantly to the breakdown of talks in Seattle and Cancun and were therefore removed altogether from negotiations in Hong Kong. Most likely to avoid drawing undue attention to the investment provisions in the MEFTA process, individual BITs have largely been bypassed in favor of subsuming the investment provisions directly into the individual Free Trade Agreements themselves.

MEFTA Step Five: Free Trade Agreements

The Free Trade Agreements (FTAs) are at the heart of the MEFTA process, turning pledges into legal commitments. Rich Mills of the U.S. Trade Representatives office told me that the FTAs represent "heavy economic, political, and foreign policy significance to the United States." The United States will enter into these agreements, Mills added, only with governments that are willing to "meet our needs."

The United States has signed FTAs with five Middle Eastern countries: Israel, Jordan, Morocco, Bahrain, and Oman. The last three were signed after the 2003 invasion of Iraq. Negotiations with the United Arab Emirates are under way and near completion. The first post-invasion agreement was signed with Morocco. As one of the friendliest nations to U.S. interests in the region, it provided "low-hanging fruit" with which to set the tone for subsequent negotiations. The Morocco FTA became law using Fast Track in July 2004. FTAs were signed with Bahrain and Oman shortly thereafter, in September 2004 and October 2005, respectively. The Bahrain FTA became law, again using Fast Track, in December 2005, while Oman's agreement awaits congressional approval.

In a piece entitled, "Ba-ba-ba Ba-ba-Bahrain," Dean Kleckner of the pro-agricultural industry website *AgWeb.com*, "The Homepage of Agriculture," expressed U.S. industries' interest in the U.S.–Bahrain FTA. The deal "is a piece of common sense," Kleckner explains, as the "biggest beneficiaries probably will be American manufacturers. Farmers will make gains as well because the agreement forces Bahrain to remove tariffs on 98 percent of our agricultural exports right away, and to phase out the rest over a ten-year period. It will also expand the financial-services sector and strengthen intellectual property rights."

The U.S.–Middle East Free Trade Coalition weighed in by sending a letter to every member of Congress, encouraging "rapid action" on the agreement. In a press release, the group stressed the agreement's importance to both U.S. economic and foreign policy interests. "It creates momentum behind our efforts to encourage market-oriented economic reforms and demonstrates the benefits of trade liberalization in the Middle East," explained William A. Reinsch, president of the National Foreign Trade Council. "It also supports our two nations' cooperation on the war on terrorism." The Coalition pointed out that Bahrain is the headquarters for the U.S. Navy's Fifth Fleet and "has been a strong supporter in the global war on terrorism and the military operations in Afghanistan and Iraq."[19]

The ever-present link between "free trade" and the War on Terror ensured passage of the Morocco and Bahrain FTAs, and will likely do the same with each subsequent MEFTA nation. Over the years, congressional opposition to corporate globalization agreements has intensified (although more slowly than that of the American public). The

2005 U.S.–Central American Free Trade Agreement, for example, passed by just one vote. However, Congress has been extremely reticent to debate, much less oppose, the administration's MEFTA process. The administration has therefore successfully tied together its War on Terror with its corporate globalization agenda, forcing elected officials in the United States and abroad to prove that they accept both or be revealed as foes on both counts.

OMAN'S FREE TRADE AGREEMENT. According to the U.S. Trade Representative's office, the U.S.–Oman FTA includes far-reaching investment, national treatment, agriculture, intellectual property rights, services, and market access provisions—many that go beyond WTO rules. The agreement eliminates all tariffs on all industrial and consumer products between the two nations and virtually all tariffs on agricultural products. It provides increased access to the U.S. textile and apparel industry in Oman, but only case-by-case access for Oman's products to the United States. Oman commits to "provide substantial market access across its entire services regime," which could mean everything from oil to water and from hospitals to schools. The U.S.–Middle East Free Trade Coalition's statement, urging congressional approval of the agreement, stressed these and other advantages. "Oman is a likely market for U.S. oil and gas equipment and services . . . water and environmental technology . . . power generation and transmission equipment and services, telecommunications equipment and services, financial services, franchising and U.S. poultry and beef."

The agreement has broad national treatment provisions, even surpassing those of the WTO, making it impossible for Oman to ensure local benefits from foreign companies by requiring that, for example, they hire local managers, professionals, or even specialty personnel, or that they use local products in their work. The elimination of the government's right to require local content in production will likely yield the same results in Oman as those experienced by Jordan—foreign companies shipping their own products to be assembled in Oman and then sold elsewhere, thereby shutting out Omani industry and creating jobs only for low-skilled assembly-line work. Government contracts cannot favor local over foreign companies for any reason. The agreement appears to incorporate the most far-reaching investment rules for foreign companies and investors, similar to those in Bremer Order #39.

Oman's FTA includes U.S. standards on intellectual property rights, copyrights, patents, trademarks, and the like. As discussed in earlier chapters, these provisions have been used by multinational corporations to reduce access to life-saving medications and to patent traditional medicines that are then priced out of reach of local populations. They have also been used to eliminate the ability of farmers to use patented seeds.

The agreement does include some provisions long sought by advocates of trade agreements that are "fair" rather than simply "free." For example, it has language requiring that each nation uphold its existing environmental and labor laws. However, this provision is very difficult to uphold. What if Oman were to eliminate a government safety regulation for factory workers one year after the signing of the U.S.–Oman FTA? To have the regulation reintroduced on the basis of this protection, one would have to prove that the *only reason* the government eliminated the safety regulation was because of the U.S.–Oman FTA—a very difficult thing to prove. The provision is further eroded by a statement on environmental laws, which the agreement says should be "married with provisions that promote voluntary, market-based mechanisms . . ." The agreement also includes long-sought changes to the dispute settlement provision, which open hearings to the public, allow friend of court submissions, and the release of government legal documents. These are, unfortunately, but a handful of the many fundamental changes necessary to make the dispute settlement mechanism transparent and democratic.

In the end, the Oman-FTA, like the Moroccan and Bahrain FTAs before it, meets criteria, as described to me by Jeffrey Donald of the U.S.–Middle East Free Trade Coalition, for a "perfect" trade agreement: They surpass even the WTO in areas such as services, investment, patents and trademarks, and government procurement and lock in a standard to which all other nations of the region will likely now have to accede. They provide new freedoms and expanded protections to multinational corporations, while offering workers, consumers, and the environment minimal protections that are difficult to defend. The FTAs set the stage for a radical form of economic imperialism that, once they are all melded together, will become the U.S.–Middle East Free Trade Area.

~

Like Mexico with NAFTA or China with the WTO, Egypt is now considered the next hot country on the horizon, promising a large cheap labor pool and a friendly government. What is missing is a free trade agreement to open the door. With the December 2005 Hong Kong ministerial meeting of the WTO ending in a stalemate, U.S. corporations and the Bush administration intend for the U.S.–Middle East Free Trade Area to be the agreement that opens up Egypt and the rest of the Middle East. With the War on Terror as the best selling point, they are on track to achieve their goal. Where multilateralism fails, Zoellick's competition strategy and the administration's unilateralism will continue to push the Bush Agenda forward.

The individual Middle East Free Trade Area agreements are paving the way for a radical, thoroughly U.S.-centric corporate globalization agenda for the Bush administration to carry from country to country in the Middle East and then well beyond. The president has forced into acquiescence the growing wave of criticism against these economic policies, both within the United States and abroad by linking them to the defeat of terrorism. It is economic imperialism in its truest form: Governments the world over are forced to adopt economic policies that benefit the growth and power of one nation with a threat of military action if they do not accede, all in the name of "world peace." The result for the people of the Middle East will likely be increased subservience to U.S. corporations rather than the "freedom" promised by Bush. The result for the people of the United States and its allies has already been revealed since the economic invasion of Iraq: intensified anti-U.S. sentiment, increased insecurity, and escalating acts of deadly violence.

EIGHT

THE FAILURE OF THE BUSH AGENDA

A WORLD AT GREATER RISK

> Our message is clear: you will not be safe until you withdraw from our land, stop stealing our oil and wealth and stop supporting the corrupt rulers.
>
> —*Ayman al-Zawahri, al-Qaeda, August 4, 2005*[1]

> We're not facing a set of grievances that can be soothed and addressed. . . . No act of ours invited the rage of the killers—and no concession, bribe, or act of appeasement would change or limit their plans for murder. . . . We will never back down, never give in, and never accept anything less than complete victory.
>
> —*President George W. Bush, October 6, 2005*[2]

U.S. MARINE CORPORAL SEAN O'NEILL

On March 17, 2003, Sean O'Neill and his Marine battalion were preparing to cross from Kuwait into Iraq in the first offensive of the Iraq war. From the top of a tank, Sean's regimental commander bellowed to his troops: "We are going to do this to make America safe. Saddam's going to kill Americans if we don't stop him!"

After crossing into Iraq, Sean's first order was to guard an oil facility. That was his first clue that this war was not only about defending America, and it was all downhill from there. "Believe me," Sean told me later. "Every Marine wanted to be the guy who found Hussein's WMDs." Once in Iraq, Sean personally searched chicken coops and pigpens looking for Iraq's weapons of mass destruction: "We looked everywhere. But none of us found any. Turns out there weren't any to be found." In the end, it was all "a cruel hoax. It was a sick joke played out on us. There were lots of sick jokes being played out on us."

Sean signed up for the military when he was just seventeen, still in high school and living with his parents in Freemont, California. He chose the Marines because "They are the baddest ass group." In Sean's own words, he suffered from "*Beau Geste* Syndrome," a reference to the 1939 movie in which Gary Cooper, Ray Milland, and Robert Preston redeem themselves by joining the French Foreign Legion and prove their loyalty, patriotic honor, and self-sacrifice.

After graduating high school in September 2000, Sean went straight to boot camp. Since then, he has served three tours of duty in the War on Terror. He was in Kuwait in 2002 preparing for the invasion of Iraq; he was in the first wave of soldiers in Iraq in March 2003; and his third deployment came on Valentine's Day 2004. Sean spent his twenty-first birthday in Tikrit with dysentery. His twenty-second birthday "went better except it was during the Fallujah offensive and I spent it laying siege to the village of Al Karma." After his third tour, Sean was awarded a Purple Heart.

Sean considered himself a Republican when he entered Iraq, but now he is not so sure. The reason, he said, was all of the lies, particularly those told by President Bush: "Everything that man said when I was over there pissed me off to no end because I was the one that had to live with the consequences."

A Danger to One

Sean was not alone. More than one million American soldiers have fought in Iraq and Afghanistan since 2001 and over 2,400 have died. America's dead are joined by more than 35,000 Iraqi and Afghani civilians, over 4,535 Iraqi and Afghani military and police, soldiers from Australia, Britain, Bulgaria, Canada, Denmark, El Salvador, Estonia, France, Germany, Hungary, Italy, Kazakhstan, Latvia, Netherlands, Norway, Poland,

Slovakia, Spain, Thailand, and Ukraine, and more than 390 private contractors from at least forty different countries who have all lost their lives.[3]

Of the many lies told by the members of the Bush administration to lure these men and women to war and to justify the death of the innocent, the one that may ultimately have the most far-reaching and devastating effect on us all is: "We are going to do this to make America safe." The only thing that the war effort has made definitively safer is U.S. corporate access to Iraq, Saudi Arabia, Morocco, Bahrain, and Oman, while the rest of the Middle East races to open its doors. U.S. oil company interests have been particularly well secured in this process. But beyond the administration's promise of making America safer, is the much grander and oft-repeated promise to make the whole world safer. It is the ultimate lie of *Pax Americana:* World peace will prevail as long as the United States can spread its military and economic dominance without impediment.

The truth is that we are less safe as a result of the Bush Agenda—whether "we" are Americans, Iraqis, Turks, Indians, or people from any other corner of the globe. As the Bush administration flexed its imperial muscle, the world may have initially shrunk back in terror, but it has increasingly fought back in anger. Terrorism against the United States and its allies did not originate during the presidency of George W. Bush, but his administration's policies of imperial expansion and corporate globalization have served both to intensify and increase it.

The number of serious acts of international terrorism has increased in every year of Bush's reign except 2001, which was the deadliest year on record due to September 11. The reason for this increase is not, as Bush has argued, because the terrorists "hate our freedoms," but because they hate our policies.

In *Imperial Hubris: Why the West Is Losing the War on Terror,* Michael Scheuer, a twenty-two year veteran of the CIA, who served as the agency's lead expert on al-Qaeda for four years, argues that under the Bush administration, "U.S. forces and policies are completing the radicalization of the Islamic world, something Osama bin Laden has been trying to do with substantial but incomplete success since the early 1990s. As a result, I think it fair to conclude that the United States of America remains bin Laden's only indispensable ally." Scheuer resigned his post in late 2004 when he decided that "being a good citizen was no longer compatible with being a good member of the CIA's Senior Intelligence Service."[4]

This "radicalization" in Iraq is painfully evident. The majority of Iraqis did not welcome the American invasion and certainly did not come out to greet the soldiers with flowers and kisses as Donald Rumsfeld all but promised. Many, however, were willing to give the Americans the benefit of the doubt once they were on Iraqi soil. This goodwill was quickly eroded by U.S. troops and Bush administration policies. Sean described the change among Iraqis from his first to his second tour: "When I went back in February [2004], everything had changed. The attitudes of the Iraqis were dramatically different. We were pissing people off a little bit more every day, and it all adds up." Sean added, "The war was over now. This stuff was supposed to stop. But we were still there and still doing the same shit. I could tell the Iraqis were asking, 'Why are they still here?' We were asking ourselves the very same thing."

Much had changed in that year, and Sean's observations were confirmed in polls taken of Iraqis. From January to May 2004, the number of Iraqis expressing confidence in the Coalition forces declined by nearly two-thirds, from 28 to 10 percent. By August, a full 80 percent said they wanted Coalition forces to leave immediately or directly after the January elections.[5] Instead, the number of foreign troops in Iraq increased, even though the country had supposedly been "liberated" long before. At the same time, the number of Iraqi resistance fighters quadrupled, increasing from 5,000 in November 2003 to 20,000 in September 2004, according to Pentagon estimates. The deputy commander of Coalition forces in Iraq, British Major General Andrew Graham, estimated a figure more than twice as large, between 40,000 and 50,000.[6]

A more important consideration, however, may be all that remained the same in Iraq. When Sean returned, unemployment in Iraq was stuck at 50 percent. Water, electricity, and sewage services were below prewar levels. U.S. corporations with multibillion-dollar contracts were everywhere, but the benefits of their work were virtually nowhere to be seen. Iraqis were overjoyed to have Hussein disposed but angry to have, in UN Special Advisor Lakhdar Brahimi's words, a "new dictator of Iraq" in Paul Bremer and his radical transformation of their laws. While Iraq's oil was flowing at prewar levels, Iraqis were waiting in long lines for expensive gasoline and wondering where the fruits of their oil were falling. Thus, more than half of Iraqis say their lives are either worse or have stayed the same since the fall of Hussein.[7]

This was then the second lie perpetrated against the Iraqis and the soldiers sent to implement the Bush Agenda—that the economic invasion would bring prosperity, security, and peace. The opposite is true. The economic campaign is viewed as a takeover of Iraq and its resources, generating hatred toward the occupiers, not allegiance. This is what Iraqis believe motivates the violent insurgency: Nearly 80 percent of Iraqis polled believe the insurgents attack Coalition forces because they think the "Coalition is trying to steal Iraq's wealth," while only 20 percent think the insurgents "want a return to Saddam and the Ba'ath Party," as the Bush administration regularly argues.[8] People across the Middle East look at Iraq and see a fate that may yet await them, whether through the U.S.–Middle East Free Trade Area or more militarily driven means.

And although President Bush argues that we are fighting the terrorists in Iraq so we do not have to fight them here (apparently assuming a shared disregard for the lives of Iraqis lost in the process), the fight has moved well beyond Iraq's borders. From Saudi Arabia to Indonesia, Morocco to Great Britain, Spain to Egypt, and Tunisia to Pakistan, acts of serious terrorism have spread, increased, and become more deadly the world over. There were more than *three and a half times* more serious incidents of international terrorism in 2004 than in 2003, resulting in nearly three times more deaths, according to the U.S. government's National Counterterrorism Center and the U.S. State Department. In 2004, there were 651 serious terrorist attacks, killing 1,907 people, compared to 175 attacks in 2003, which already marked a twenty-year high, killing 625 people. In 2002, there were 139 serious terrorist attacks compared to 123 in 2001. Of course, the 2001 death toll was the highest ever recorded, with 3,295 people killed.

"Significant" terrorist attacks are those that cause civilian casualties or fatalities or substantial damage to property. They *do not* include attacks on uniformed military personnel or attacks against people of the attackers' own nationality. Even with those caveats, the

2004 numbers represent a ninefold increase in the number of terrorist attacks in Iraq and a doubling of attacks in Afghanistan. Figures for 2005 are set to be record-breaking yet again. The Bush administration has succeeded in simply shifting the primary terrorist training ground from Afghanistan to Iraq, while increasing the factors that are known to motivate more people to adopt terrorist activities.

THEY HATE OUR POLICIES, NOT OUR FREEDOMS

President Bush has steadfastly maintained that the attacks against the United States and its allies are a result of those who "hate our freedoms." The president's words delivered in a 2002 speech in Wisconsin are typical: "My most important job, I think, is to make sure that I protect—do everything I can to protect the homeland and make sure that an enemy which hates freedom doesn't hit us again. They're out there. They can't stand the thought of America being a hospitable society to many cultures. They can't stand the thought of a society which worships an almighty God in different ways, a society which is tolerant to different approaches to religion. They really hate the thought of us being a society in which we are able to speak our mind, a society which values the individual worth of each person. And so long as we uphold those values, which we will do, they're going to try to strike us."

Three years later, almost to the day, the president conveyed the same message in Idaho: "Our enemies murder because they despise our freedom and our way of life. See, they're coming into Iraq because they fear the march of freedom."

The President's assertions, however, are directly contradicted by analysis performed by his own administration. The Defense Science Board, a federal advisory committee appointed by the U.S. Defense Department to report to then-Secretary Paul Wolfowitz, conducted a study of Muslim attitudes toward the United States in 2004.[9] The report flatly concludes, "Muslims do not 'hate our freedoms,' but rather, they hate our policies." The report, commissioned by the U.S. Defense Department to improve its "communication tools" with Muslims, paints a stark picture.

It finds that the War on Terror and, in particular, the invasions of Iraq and Afghanistan, have increased antipathy toward the United States and support for radical Islamists:

> In the war of ideas or the struggle for hearts and minds, American efforts have not only failed, they may also have achieved the opposite of what they intended. . . . American direct intervention in the Muslim world has paradoxically elevated the stature of, and support for, radical Islamists, while diminishing support for the United States to single digits in some Arab societies. . . . In the eyes of Muslims, the American occupation of Afghanistan and Iraq has not led to democracy there, but only more chaos and suffering. U.S actions appear in contrast to be motivated by ulterior motives, and deliberately controlled in order to best serve America national interests at the expense of truly Muslim self-determination.

The report also provides a serious warning against practicing the ideas of *Pax Americana:*

> There is no yearning-to-be-liberated-by-the-U.S. groundswell among Muslim societies. . . . Americans are convinced that the U.S. is a benevolent "superpower" that elevates values emphasizing freedom . . . deep down we assume that everyone should naturally support our policies. Yet the world of Islam—by overwhelming majorities at this time—sees things differently. Muslims see American policies as inimical to their values, American rhetoric about freedom and democracy as hypocritical and American actions as deeply threatening.

The report's conclusions are repeated in poll after poll, and study after study, of Muslims and people living in the Middle East. These findings also reveal a deep mistrust of U.S. intervention in their countries (particularly U.S. support for Israel and the occupation of Palestine), an extreme disregard for what they see as American arrogance, and anger at U.S. policies viewed as intent on enriching America at the expense of native homelands (including U.S. support for regimes considered brutal to their populations such as Saudi Arabia and Egypt).

It is worth returning for a moment to a CIA study quoted at the opening of this book. The study is conducted every five years and is used by the agency to predict the most serious threats to U.S. security fifteen years hence. In 2000, the report pointed the finger squarely at the policies of corporate globalization:

> The rising tide of the global economy will create many economic winners, but it will not lift all boats. . . . [It will] spawn conflicts at home and abroad, ensuring an even wider gap between regional winners and losers than exists today. . . . Regions, countries, and groups feeling left behind will face deepening economic stagnation, political instability, and cultural alienation. They will foster political, ethnic, ideological, and religious extremism, along with the violence that often accompanies it.[10]

President Bush has expanded the policies of corporate globalization through the barrel of a gun. Extreme benefits have accrued to his corporate allies, but the costs are being born out by us all.

The CIA's latest study was released in 2005. It predicts that the significant characteristics of the world in 2020 are likely to include "a more pervasive sense of insecurity, including terrorism." Virtually repeating its 2000 findings, the report predicts "the benefits of globalization won't be global. . . . And large pockets of poverty will persist even in 'winner' countries. . . . The gap between the 'haves' and 'have-nots' will widen. . . . Globalization will profoundly shake up the status quo—generating enormous economic, cultural, and consequently political convulsions." It concludes, "The key factors that spawned international terrorism show no signs of abating over the next fifteen years."[11]

Furthermore, those who perpetrate acts of terrorism have been painfully explicit in explaining the reasons for their attacks. In the examples offered here, the attackers cite retaliation to U.S. policies in the Middle East, particularly the wars in Iraq and Afghanistan, as the reason for their acts of deadly violence. They also provide regular warnings of whom, where, and why they plan to attack next. I do not point to the statements of those claiming responsibility for these horrific and deadly attacks to provide any sort of apology for their methods, but rather to offer insight into how they may be stopped in the future. Their statements flatly contradict the president's contentions that the terrorists oppose "our freedoms" and that their demands lack clarity. Simply put, the statements make brutally clear that the President's actions and policies are making America and its allies more hated and less safe.

Michael Scheuer observes, "Bin Laden has been precise in telling America the reasons he is waging war on us. None of the reasons have anything to do with our freedom, liberty, and democracy, but have everything to do with U.S. policies and actions in the Muslim world."

THIS IS "BLOWBACK"

Madrid, March 11, 2004

> I saw many things explode in the air, I don't know, it was horrible. People started to scream and run, some bumping into each other and as we ran there was another explosion. I saw people with blood pouring from them, people on the ground.
>
> —*Juani Fernandez, Atocha station, Madrid, March 11, 2004*[12]

It was rush hour in Madrid. At 6:30 A.M. on March 11, 2004, tens of thousands of people were passing through Atocha, one of Madrid's major stations in the center of the city. Hundreds were standing on the platforms waiting for their trains. The cars were crowded and filled with people just starting their commutes to work or school or ending a night of fun. Juani Fernandez was on his way to work in the Spanish government. Isabel Vega was on her way to school. Ken Seal was woken from his bed near the station. At 6:39 A.M., the train pulled into Atocha station. As the doors opened to let on new passengers, three bombs exploded in the third, fourth, and sixth cars. Almost simultaneously, four more bombs exploded in the first, fourth, and sixth cars of a second train, about 500 meters outside the station. The second train was late. Investigators believe the bombers intended to set the bombs off on both trains inside the station simultaneously to maximize their power. Student Isabel Vega said, "It has been chaotic, horrific. You could hear people screaming, on the platform there were dead and injured, people were running covered in blood."

At 6:41 A.M., two bombs went off in the fourth and fifth cars of a third train as it passed through El Pozo station, back down the line from Atocha. At 6:42 A.M., a fourth train was passing through Santa Eugenia station when a bomb exploded in its fourth carriage. The bombers were not on the trains. At Alcala de Henares at 6:00 A.M., they had put rucksacks containing about 10 kg (22 lbs) of explosives each into the cars. The bombs were then detonated by mobile phone as the assailants sat in a van outside of the station. Emergency worker Oscar Romero said the devastation was the worst he had seen in his job: "There were people destroyed, blown up, without legs. There were two cars in pieces with bodies underneath." More than 2,000 people were injured and 191 died.

The Abu Hafs al-Masri Brigades claimed responsibility for the attacks. The group takes its name from Mohammad Atef, Osama bin Laden's second in command and apparent successor before he was killed by U.S. forces in Kabul in November 2001. His alias was "Abu Hafs," while he was known as "al-Masri"—"the Egyptian" in Arabic. The Abu Hafs al-Masri Brigades were founded shortly after his death and are considered an al-Qaeda organization. In a video made by the alleged perpetrators and found shortly after the attacks, the Brigades warn Spain to remove its troops from Iraq and Afghanistan immediately: "Should you not do this within the space of a week, starting today, we will continue our jihad until martyrdom." They also claim responsibility for the attacks in this video: "You know that you are not safe, and you know that Bush and his administration will bring only destruction. We will kill you anywhere and in any manner."[13] According to the BBC, the alleged bombing mastermind, Serhane ben Abdelmajid Fakhet, known as "the Tunisian," is said to be the man in the video.

The Brigades also released a written statement, again explaining that the attacks were in response to Spain's support of the war in Iraq. The group warned of more attacks in any country supporting the Iraq war, specifically listing the United States, Japan, Italy, Britain, Australia, and Saudi Arabia.[14] In fact, the Brigades had made a similar warning four months prior to the Madrid attacks.

Istanbul, November 15 and 20, 2003, and August 10, 2004

In November 2003, the Abu Hafs al-Masri Brigades claimed responsibility for two sets of deadly attacks in Istanbul in partnership with a local suspected terrorist organization, the Islamic Great Eastern Raiders' Front, also known as IBDA-C. In the first attack on November 15, two suicide bombers driving trucks filled with some nine hundred pounds of explosives pulled up alongside two synagogues, exploding the trucks within minutes. Twenty-five people were killed and at least three hundred were injured.

Five days later, using the same method of attack, the group struck the British Consulate and the London-based HSBC bank in Istanbul. Along with U.S.-based JP Morgan Chase, HSBC is one of the first western banks to operate in Iraq after the invasion. Twenty-seven people were killed, including the British consul-general and his personal assistant, and approximately 450 were injured. The bombings coincided with a visit by President Bush to London.

In a statement, the Brigades said the attacks were "to let Britain and the Brits know that their alliance with America will only bring them economic ruin and death to their sons."[15] According to the *UK Telegraph*, the statement also "gave warning of more attacks around the world," including car bombs in the United States and attacks on Japan. Jewish synagogues were targets, according to the statement, in opposition to Israel's occupation of Palestine.

Five months after the Madrid bombings, the Brigades struck Istanbul again, this time sending a new form of distress around the world by successfully targeting a liquefied petroleum gas plant. On the same day, August 10, 2004, they also attacked two small tourist hotels, killing two and injuring eleven people.

In a statement, the group said it had "carried out the first of a series of operations that will be launched in the face of European states . . . and the upcoming attacks will be more violent. . . . The bitterness that Muslims are tasting in Iraq and Palestine will be tasted by everyone living in Europe, in Istanbul, Rome and the rest of the countries that are following the policy of the United States." The group also warned, "There will not be safety [in Europe] unless your governments withdraw from the land of Iraq. . . . There will be no safety unless you reject the American policy based on slaughtering, killing and torturing prisoners."[16]

London, July 7, 2005

As threatened, terrorist attacks struck Britain's soil the following year. The London attacks bore a striking similarity to those in Madrid. On July 7, 2005, at 8:50 A.M. during the morning rush hour, three bombs simultaneously exploded in three separate cars of London's subway—the Piccadilly line traveling south from King's Cross station to Russell Square, the Circle line traveling east from Liverpool Street station to Aldgate, and the Circle line train traveling west from Edgware Road station to Paddington. A fourth bomb exploded one hour later on a double-decker bus at Tavistock Square and Upper Woburn place. There were fewer causalities than in Madrid—fifty-six including the attackers, who were suicide bombers.

A videotape of Mohammad Sidique Khan, one of the four suicide bombers, aired shortly after the attacks. In the transcript, Khan provides a deadly warning to his audience: "Your democratically elected governments continuously perpetuate atrocities against my people all over the world. And your support of them makes you directly responsible, just as I am directly responsible for protecting and avenging my Muslim brothers and sisters. Until we feel security, you will be our targets. And until you stop the bombing, gassing, imprisonment and torture of my people we will not stop this fight."

One month later, Ayman al-Zawahri, considered Osama bin Laden's chief deputy, released his own video. He would later claim full responsibility for the attacks on behalf of al-Qaeda. According to *Al-Jazeera,* al-Zawahri said the London attacks were the result of Blair's decision to invade Iraq: "Blair has brought you destruction to the heart of London, and he will bring more destruction, God willing." Al-Zawahri also warns the United States of more military casualties in Iraq if it does not withdraw its forces: "What you have seen, O Americans, in New York and Washington and the losses you are having in Afghanistan and Iraq, in spite of all the media blackout, are only the losses of the initial clashes. If you continue the same policy of aggression against Muslims, God willing, you will see the horror that will make you forget what you had seen in Vietnam." He continues, "There is no way out of Iraq without immediate withdrawal, and any delay on this means only more dead, more losses. If you don't leave today, certainly you will leave tomorrow, and after tens of thousands of dead, and double that figure in disabled and wounded." And finally, as quoted earlier, al-Zawahri concludes, "Our message is clear: you will not be safe until you withdraw from our land, stop stealing our oil and wealth and stop supporting the corrupt rulers."

Closer to Home

Sean O'Neill and Hamid Hayat are both twenty-three years old. They live just fifty miles apart in Northern California, Sean in Freemont and Hamid in Lodi. Each young man has become embroiled in Bush's War on Terror.

Hamid and his father Umer, both U.S. citizens, were arrested in Lodi by the FBI in June 2005. Lodi is a small agricultural town on the road between San Francisco and Sacramento. Before the arrest of the Hayats, it was probably best known as home of the "best dirt motorcycle racing track in the state." Hamid Hayat is charged with providing material support to terrorists by undergoing training at an al-Qaeda terrorist training camp in Pakistan and reentering the United States to await instructions on waging jihad on Americans. He is also charged with lying to the FBI to conceal his activities. Umer Hayat is charged with lying to the FBI to conceal his firsthand knowledge of terrorist training camps in his native Pakistan and his son's alleged training at one of them.[17]

Hamid reportedly signed a sworn affidavit, saying he attended al-Qaeda training camps in Pakistan in 2003 and 2004, where he received training in weapons, explosives, hand-to-hand combat, and other paramilitary exercises. During his weapons training, photos of various high-ranking U.S. political figures including President Bush were reportedly pasted onto the targets as he learned how to "kill Americans."[18] The arrests made public an ongoing FBI investigation into a potential terror cell in Lodi, which locals describe as "terrorizing" the South Asian community there numbering in the several thousands, many of whom have called Lodi home for three generations.[19] Both Hayats have pled not guilty to all charges and have been in jail awaiting trial for eight months. In January 2006, a Sacramento federal judge set February 14 as the trial date for the Hayats.

Statistics on FBI investigations into terrorism in the United States are not public, nor are statistics on attempted and/or thwarted terrorist attacks. However, in October 2005, the White House released two lists of "ten serious al-Qaeda terrorist plots since September 11—including three al-Qaeda plots to attack inside the United States" and five attempts by al-Qaeda to "case targets in the United States or infiltrate operatives into our country."[20]

Hamid Hayat, his father, and the suspected terrorist cell in Lodi were not on the White House lists, nor was another alleged terror cell run by an inmate at California State Prison-Sacramento. In August 2005, the FBI charged Kevin James with founding a "radical Islamic organization known as Jam'iyyat Ul-Islam Is-Saheeh" and plotting terrorist attacks on "U.S. military facilities, Israeli government facilities and Jewish synagogues in the Los Angeles area," according to the U.S. Department of Justice. James and three other young men, ages twenty-one to twenty-five, were charged with, among other things, "conspiracy to levy war against the United States government through terrorism."[21] Each pled not guilty and awaits trial.

The three plots against the United States listed by the White House are the "West and East Coast Airliner Plots," involving attempts to attack targets on each coast using hijacked commercial airplanes in 2002 and 2003, respectively, and the Jose Padilla Plot. In May 2002, the United States reportedly disrupted a plot that involved blowing up apartment buildings in the United States. One of the alleged plotters, Jose Padilla, also purportedly discussed the possibility of using a "dirty bomb" in the United States.

The five "casing and infiltrations," provided by the White House are:

1. *The U.S. government and Tourist Sites Tasking:* In 2003 and 2004, an individual was tasked by al-Qaeda to case important U.S. government and tourist targets within the United States.
2. *The Gas Station Tasking:* In 2003, an individual was tasked to collect targeting information on U.S. gas stations and their support mechanisms on behalf of a senior al-Qaeda planner.

3. *Iyman Faris and the Brooklyn Bridge:* In 2003, and in conjunction with a partner nation, the U.S. government arrested and prosecuted Iyman Faris, who was exploring the destruction of the Brooklyn Bridge in New York. Faris ultimately pleaded guilty to providing material support to al-Qaeda and is now in a federal correctional institution.
4. *Tasking:* In 2001, al-Qaeda sent an individual to facilitate post-September 11 attacks in the United States. U.S. law enforcement authorities arrested the individual.
5. *Tasking:* In 2003, an individual was tasked by an al-Qaeda leader to conduct reconnaissance on populated areas in the United States.

As described, just one of these cases has resulted in a guilty ruling. There is no way to corroborate or contradict these reports. Whether the accusations are true or false, they certainly do not support the contention that the Bush Agenda has made America safer or that we are successfully fighting terrorists in Iraq "so that we don't have to fight them here." Furthermore, there are already clear losers in the United States, including those who may be wrongly accused of these crimes, their families, and their communities.

Then there are the approximately 750 people who have been held at the U.S. military facility in Guantanamo Bay, Cuba. Those classified as "illegal combatants," rather than prisoners of war, have been denied their rights under the Geneva Conventions, including the ability to meet with attorneys.

Since September 11, approximately 1,200 people in the United States—mostly Arab, South Asian, and Muslim men—have been detained and about half have been deported, mainly on minor immigration violations. Although not citizens themselves, many are married to U.S. citizens, have children who are citizens, are long-term residents, business owners, and taxpayers. Authorities have refused to disclose the identities and locations of those detained. Families are still trying to obtain information about those who remain in detention, as well as many who have been deported. To date, only a few of them have been charged with criminal activity tied to the government's antiterrorism offensive.

In the United States, anti-Arab sentiment has become far more pervasive than most people choose to admit, sometimes resulting in violent attacks and murders, and appearing in every form of discrimination, including a fourfold increase in reported cases of employment discrimination among Arab Americans.[22]

In San Francisco, there are announcements posted throughout the public transit system, warning riders to keep their eyes on their neighbors and report all "unusual activity," a phrase that covers most everything that happens in San Francisco. I have seen similar signs across the nation, all invoking some form of "citizen enforcement"—a kind word for tools that seem deeply prone to inviting the worst forms of vigilantism.

As the Bush Agenda continues to spread across the Middle East and the world, I wonder, then, if our fate is to become a country whose citizens spend every day fearful of riding a bus or sitting in a restaurant, suspicious of all strangers, particularly those with different skin colors, nationalities, or faiths. Are we to be a country that is hated because of its illegal invasions and occupations, economic brutality, greed, hypocrisy, violence, and disregard for the lives of others in pursuit of its own security? Whose young are forced to fight in wars to defend these policies, ushering thousands to an early death? Are we already that country?

The deadly terrorist acts just described and the hundreds of others like them around the world are horrific examples of "blowback." They are unthinkably horrific acts that are taken in retaliation to the policies of the U.S. government and the governments that support those policies. If the perpetrators are true to their words, the attacks will not stop until the policies that drive them stop.

Al-Qaeda experts agree that the United States lost its best opportunity to capture Bin Laden and handicap, if not destroy, al-Qaeda in the immediate aftermath of the September 11 tragedies. At that moment, world-wide support was fully united behind the United States. Unfortunately, the Bush administration squandered this opportunity largely because its sights had already turned to Iraq, draining first intellectual energy and then military and financial commitments, away from the hunt for al-Qaeda. As a result, al-Qaeda operatives were dispersed to Iraq and elsewhere around the world and increasing numbers of people turned to extremist beliefs and actions as a result of increasingly imperial American activities. The United States must look critically at all of its policies in the Middle East and around the world to determine which are necessary, just, and must be maintained, and which serve only the limited interests of the few and the powerful at the expense of the many.

~

The next chapter identifies alternative policies to the Bush Agenda, not in conciliation to the terrorists, but in consideration of the vast harms caused by these policies both at home and abroad in their immediate implementation and in the blowback they incur. In this book, I have described the devastating impacts of corporate globalization policy from South Korea to India, Zambia to Russia, Argentina to South Africa, Nigeria to Ecuador, Mexico to China, and Iraq to the United States. The Bush administration has contributed to the world its own imperial and militarily aggressive variant of corporate globalization—yielding the most violent response to U.S. policies in more than thirty years. The winners of the Bush Agenda are few, the losers many, and the costs great. But there are also a vast number of alternatives to the Bush Agenda. In addition to offering a new agenda, the concluding chapter describes the many peaceful ways in which people are already working to bring this new agenda about.

NINE

A BETTER AGENDA IS POSSIBLE

> There will always be cruelty, always be violence, always be destruction. . . . We cannot eliminate all devastation for all time, but we can reduce it, outlaw it, undermine its sources and foundations: these are victories.
>
> —*Rebecca Solnit,* Hope in the Dark[1]

Throughout his presidency, George W. Bush has guaranteed that we will live in a safer, more prosperous, freer, and more peaceful world if the United States remains at war and if countries throughout the world change their laws and adopt economic policies that benefit America's largest multinational corporations. The Bush Agenda has proven to have the opposite effect: increased deadly acts of terrorism, economic insecurity, reduced freedoms, and more war.

The Bush Agenda is a failure for all but its drafters—including executives of the largest multinational corporations. It is a danger for the rest of the world. In order to reduce terrorism both at home and abroad, and for Americans to live in a country that is not a source of fear and hatred but rather a welcome member of the global community, the Bush Agenda and its policies must be abandoned. The president's record-low approval ratings imply that three-fourths of Americans are likely to agree with this conclusion.

The Bush Agenda will not simply disappear, however, with the end of the Bush presidency in 2008, nor can we afford to wait that long for a new direction. The Agenda predates Bush and its adherents certainly hope it will outlast him. It is the work of some of the country's most durable politicians, including Dick Cheney, Zalmay Khalilzad, Eric Edelman, Robert Zoellick, Paul Wolfowitz, Donald Rumsfeld, Douglas Feith, Richard Perle, and Condoleezza Rice. It is also supported by current and former executives at the world's most powerful corporations, including Bechtel, Chevron, Halliburton, and Lockheed Martin. None of these power brokers are eager to give up the wealth and influence that the Bush Agenda accords them. The Agenda itself, not simply the president, must therefore be replaced.

To replace the Bush Agenda, we must address each of its key pillars individually—war, imperialism, and corporate globalization. The most urgent first step is ending the occupation of Iraq and replacing U.S. and Coalition military forces, in the immediate term, with a multinational peacekeeping force. In order for more nations of the world to be willing to partner with the United States, the ever-present threat of U.S. economic or military strikes must be removed. This means rejecting the Bush Agenda's unilateralism and use of prevention as a justification for war. Likewise, the economic invasion of Iraq must be brought to an end by canceling all U.S. corporate reconstruction contracts in Iraq and replacing them with an Iraq Reconstruction Trust Fund that directs resources to Iraqi companies and workers. The expansion of the economic invasion to the rest of the Middle East must be resisted—first by canceling the negotiation of the U.S.–Middle East Free Trade Area and then by supporting the development of meaningful alternatives to corporate globalization policy in the Middle East, around the world, and in the United States.

It is far easier to replace corporate globalization policy than its advocates would have us believe. Corporate globalization is a set of policies designed to reduce the ability of local communities and governments to set the rules by which foreign companies operate in their areas. The alternatives, therefore, are tools that allow local communities and governments to set the terms by which companies (both foreign and local) operate in their midst, in order to ensure that the companies serve the public interest. Far from being a new idea, this was in fact the guiding principal of international trade, investment, and corporate law until Reagan and Thatcher introduced corporate globalization policies that reached far beyond (and even contradicted) those envisioned by Harry White and John Keynes at Bretton Woods. The tools for local economic development are everywhere; they simply need to be chosen over those that are pushed by the Bush Agenda.

This is not to argue that international trade and investment should not occur, only that, when they do, there is a balance in interests served. In addition, both trade and investment should be used as tools of local, national, and global sustainability, as opposed to ends in themselves that serve only the interests of the few who profit directly from their use. Applying government and local control may reduce the likelihood that foreign companies and investors will come to a community, but it increases the likelihood that local companies and workers will benefit from the presence of foreign companies that do enter. The question is: *On which side does a community want to err?* And even more important: *Who gets to decide, the community or the foreign company?* Opposition to the rules of corporate globalization often rest on this issue; a decision between competing interests is posed, but only one set of factors is given weight. There is no balance, no debate, and in effect (if not in fact) just one set of actors decides the outcome. The policies simultaneously enrich corporations and increase their political influence, virtually erasing the democratic process and resulting in the corporate-government hybrid epitomized by the Bush administration.

This chapter offers examples of specific policies that can be introduced in order to unravel existing Bush Agenda polices and replace them with meaningful and achievable alternatives. The examples present a variety of methods for achieving these policies: Some methods require changes to be implemented from within the Bush administration;

others involve the work of local elected officials and members of Congress; still others demand the attention of people using their roles as workers, consumers, students, teachers, veterans, and concerned citizens to implement changes without the aid of government officials. All of these methods, however, involve steps that anyone reading this book can take right now to begin ushering in a better agenda.

IRAQ: END THE OCCUPATION—AND PAY FOR THE RECONSTRUCTION

In order for a real democratic debate to emerge in Iraq to determine what sort of economic, social, and political laws should be in force, the influence of the Bush Agenda must be reduced as much as possible. This can be achieved through the following steps: (1) end the military and economic occupation; (2) cancel the U.S. reconstruction contracts in Iraq and turn the work over to Iraqis; and (3) take reconstruction funds out of American hands. Any of these steps will reduce the overall influence of the Bush administration in Iraq.

Additional steps can also be taken to strengthen the power of local communities and local governments in Iraq and to reduce the control of foreign multinational corporations, particularly oil corporations, over the economy. While offering specific alternatives to the current Bush Agenda policies in force in Iraq, I present examples of alternative policies—in theory and in practice—that can replace the Bush Agenda in the United States and around the world.

Bring the Troops Home

More than 50 percent of Americans and over 70 percent of Iraqis support an immediate withdrawal of the troops from Iraq.[2] Without an end to the military occupation, there will be no positive change in Iraq. This means bringing the troops home now—all of the troops. Ending the occupation would bring an end to an illegal and immoral invasion, launched for reasons that were founded on deception, denial, and lies. The withdrawal of U.S. and Coalition troops would remove the one factor that is unifying and radicalizing the resistance and its supporters. It would allow Iraqis to participate in the rebuilding and governing of their nation without also being branded as collaborators of a hated occupation. It would likely lead to the removal from power of those officials appointed to serve the interests of the occupier, allowing space for new representatives to emerge. It would also free the tens of billions of dollars currently being spent supporting the U.S. occupation to fund the rebuilding of Iraq.

General George Casey, the most senior commander of Coalition forces in Iraq, told members of Congress in September 2005, "The perception of occupation in Iraq is a major driving force behind the insurgency."[3]

In November 2005, Congressman John Murtha of Pennsylvania shocked both political parties when he introduced a resolution calling for the immediate withdrawal of U.S. troops from Iraq. Murtha, a Democrat, is a Vietnam combat veteran and a retired Marine Corps colonel with thirty-seven years of military service followed with more than thirty years of congressional service. Considered a staunch military hawk, Murtha was such a strong supporter of the first Gulf War that Vice President Cheney said of him, "One of my strongest allies in Congress when I was secretary of defense was Jack Murtha, a Democrat who was chairman of the Defense Appropriations Subcommittee."[4] Murtha was, at most, a reluctant supporter of the second invasion of Iraq. After hearing from individual soldiers and generals about the realities on the ground of the Iraq War, Murtha concluded that a full U.S. withdrawal was immediately necessary.

Murtha explained his position in a speech in support of his resolution: "It is evident that continued military action in Iraq is not in the best interest of the United States of America, the Iraqi people or the Persian Gulf Region. . . . Our troops have become the primary target of the insurgency. They are united against U.S. forces and we have become a catalyst for violence. U.S. troops are the common enemy of the Sunnis, Saddamists, and foreign jihadists. I believe with a U.S. troop redeployment [out of Iraq], the Iraqi security forces will be incentivized to take control. . . . I believe we need to turn Iraq over to the Iraqis."

Murtha's resolution has yet to be voted upon, however, it has already ignited a firestorm of vocal congressional opposition to the war.

Murtha's arguments are echoed in a report by Phyllis Bennis and Erik Leaver of the Washington DC based think tank, the Institute for Policy Studies, "Without an outside enemy occupying the country, it is also more likely that the kind of secular nationalism long dominant in Iraq would again prevail as the most influential (though certainly not sole) political force in the emerging Iraqi polity, as opposed to the virulent Islamist tendencies currently on the rise among Iraqis facing the desperation of occupation, repression and growing impoverishment."[5] Ending the occupation is the first necessary step toward reducing such extremist elements in Iraq.

The presence of U.S. and Coalition troops also taints the actions of all of those who try to rebuild Iraq (regardless of their nationality) as collaborators of a hated occupation—be they aide workers, government officials, construction workers, judges, interpreters, police officers, political candidates, or antiwar activists. The rebuilding of Iraq is therefore effectively being put on hold while the occupation continues.

We must be realistic, however. The U.S. government and its corporations, supported a brutal dictator in Iraq, waged two wars against its people, and led over a decade of deadly economic sanctions. Iraq is in a desperate position for which there is no simple solution. Ending the occupation is however the necessary and vital first step that must now be taken to move the country in the right direction.

Introduce an International Peacekeeping and/or Security Force

In December 2005, the Brookings Institute estimated the total Iraqi security force, including police, national guard, and armed forces, at 223,700 people. The number of those considered "operational" is not available. The removal of U.S. authority over these troops, the end of the occupation, and the consequent likely reduction in insurgent attacks against security forces should lead to a significant increase in the number of Iraqis willing to join their national forces. All of these factors will make the domestic security forces more representative of the

Iraqi people. The end of the occupation will also increase the amount of money that the United States can provide for the training, equipment, pay, and benefits needed by Iraq's domestic security forces.

This transition will take some time. Therefore, a temporary international peacekeeping and/or security force will be necessary. The only international body with the authority, respect, and capacity to provide such a force is the United Nations. To provide regional accountability and legitimacy, the international force should be supported by forces accountable to the Arab League and/or the Organization of the Islamic Conference. As Bennis and Leaver point out, the use of regional forces will "reduce regional tensions and encourage neighboring countries to provide support throughout Iraq's reconstruction process."[6]

Bring the Military Contractors Home

The end of the military occupation will be the first step toward ending the economic invasion. The end of the military occupation will mean all U.S. contractors in Iraq that are providing support services to the military will automatically leave as their contracts become irrelevant. This includes much of Halliburton's contract, now worth over $11 billion, for construction, housing, meals, water, transportation, and infrastructure support for the troops; the $628 million contract to International American Products of Irmo, South Carolina for U.S. base camp construction and services; the $259 million contract to Advanced Systems Development of Arlington, Virginia, for information technology troop support; and dozens of others like these.

It will also mean the end of the second largest category of private military spending in Iraq: the collection and demolition of Iraqi weapons. Corporal Sean O'Neill told me that he and every Marine he knew spent a great deal of their time personally collecting and exploding weapons. Meanwhile, private companies were awarded contracts worth some $10 billion for this task, much of which has already been accomplished.

The end of these contracts will lead to savings of tens of billions of dollars, including money that remains unaccounted for or has been wasted and misused. For example, the U.S. General Services Administration's Federal Procurement Data System reveals that the U.S. Army has at least $1 billion worth of work that it contracted in Iraq in 2004 but does not know with whom. Then there are the expenditures that the U.S. Army can account for, such as the nearly $20,000 spent at the Home Depot, apparently for supplies in Iraq, over $11,000 spent on shredders, a $40,000 copier (plus supplies), and $3,040 paid to WECSYS of Robbinsdale, Minnesota, for the provision of twenty folding tables and thirty folding chairs in Iraq.

My personal favorite, however, is the nearly $150,000 American taxpayers spent so that Freeman's Funny Farm of Chehalis, Washington, could provide "technical assistance" to the U.S. Army in Iraq. I was intrigued to find out what exactly Freeman's Funny Farm did for this money but was stymied in all of my efforts. While this is the funniest example of expenditures that I came across, it is certainly not alone in the category of waste, fraud, and abuse of U.S. taxpayer and Iraqi money. The end of the military occupation will terminate all such contracts.

Bring the Construction Contractors Home

The reconstruction contracts are, hands down, the most expensive contracts in Iraq. The seven largest U.S. contracts in Iraq, excluding Halliburton's, are for reconstruction. The most recent report from the U.S. special inspector general for Iraq reconstruction reveals that the United States has appropriated nearly $30 billion in U.S. taxpayer funds for reconstruction projects in Iraq. Not a cent of this money is required to stay in Iraq as an investment in Iraq's long-term economic growth. These contracts are also the most unnecessary because they pay American firms bloated rates for work that Iraqis can and should be allowed to do themselves.

Nearly all of the U.S. appropriated dollars—more than 93 percent—have already been committed to specific companies. As of October 14, 2005, 2,784 projects had been started, 1,887 had been completed, and 897 are ongoing and are not expected to be completed until 2007.[7] Hundreds of additional projects—particularly those in the water and electricity sectors, were planned and even contracted for, but never got under way and have since been abandoned by the Bush administration.

All remaining undedicated funds should be made available to Iraqi companies and workers through the Iraq reconstruction fund proposed later. Existing contracts for unfinished work should be canceled and the money turned over to this fund. In addition, those U.S. contractors who have been paid for work that does not satisfy their contractual obligations should be fined and required to reimburse the U.S. taxpayer, and these funds should be added to the reconstruction dollars made available through the fund.

U.S. reconstruction has largely failed because the contractors have operated with a goal of potential future profits in mind, instead of the people of Iraq and their immediate needs, and because of the hatred and violence targeted at anyone and anything considered to be related to the occupation. Rebuilding in the aftermath of war cannot be equated with awarding a blank check to the winner as a spoil of war. Iraq's public and private sectors have the ability to rebuild on their own. What they need is money, equipment, and in some cases, training.

Tiri and Transparency International are two of the world's leading anticorruption organizations and experts on postwar reconstruction. They have found—through investigations in post-conflict countries such as Iraq, Lebanon, Timor-Leste, Afghanistan, Bosnia, and Eritrea—that the costs associated with bringing public infrastructure up to a workable standard after a conflict are on average *90 percent* less when local expertise and labor is used in place of industrial country standards and foreign contractors.[8]

The withdrawal of U.S. troops, occupation forces, and foreign contractors could also free many of the Americans who stay in Iraq as employees of Iraqi companies, aide workers, interpreters, reporters, and the like, from the need of private security companies. Thus, the contracts worth millions of dollars for private security companies such as Blackwater, Aegis, DynCorp, Custer Battles, CACI, Titan, and others would also be canceled and their forces (estimated at 30,000) sent home. Those companies that do remain would at least no longer be paid by U.S. taxpayers.

Bring the Economic Transformers Home

The economic invasion cannot end until those companies that have been brought to Iraq with the specific mandate of transforming the country's economic and political structures are sent home. All of the contracts designed to transform Iraq's economy, its schools and its media, and to control its political development, must be canceled. The transformation of an occupied country's laws is illegal under the Geneva Conventions and the U.S. Army's Rules of Land Warfare. Therefore, so too are these contracts. They take basic, vital decision making, democracy, and freedom away from Iraqis. There is no reason why American companies should be performing these tasks. After decades of despotic rule, Iraq may well be interested in seeking advice and support from the international community on restructuring its economy and its government. The invading nation, however, is hardly the appropriate country to take the lead in such an undertaking.

Bearing Point's contract to transform Iraq's economy from a state- to a market-controlled model, as well as Louis Berger Group's contract to provide similar work, must be canceled. In testimony to the Senate Democratic Policy Committee, former CPA adviser Franklin Willis described the dismay aroused by Bearing Point among even some CPA staff: "Part way through my time in Iraq, we were instructed to prepare problem areas we faced for evaluation by a company called Bearing Point, operating I think under a USAID contract. Their presence was utterly mystifying to us. . . . Their suggestions in my experience largely related to what we were already doing, or involved proposals so futuristic as to be impracticable. We wondered by what process and with what intent had Bearing Point been selected, and what costs for U.S. taxpayers were involved."

The intent of Bearing Point's contract was the implementation of Bush Agenda policies. I imagine these are the proposals that Willis considered "futuristic" in the face of the overflowing sewage, electricity blackouts, and increasing acts of deadly street violence across the country. Bearing Point's economic prescriptions are untimely, inappropriate, expensive, and illegal.

The Research Triangle Institute's $466 million contract to control local elected governments, their candidates, and structures must also be canceled. As Pratap Chatterjee, author of *Iraq, Inc.: A Profitable Occupation,* concluded, "What the RTI officials and the occupation authorities seemed to completely disregard was the fact that the use of the military to impose a council, in the face of local opposition, was not that different from the system imposed by Saddam Hussein, except that the dictator was more effective in suppressing dissent."

Then there is Creative Associates International, Inc. of Washington, DC, with its $273 million contract to rewrite Iraq's textbooks, reform its curriculum, and train its teachers. The United States should certainly pay for the rebuilding of schools and probably even for new textbooks and other necessities, but this does not give the United States the authority to write Iraq's history or dictate the lessons that will be taught to its children in the future. Iraqis deserve the right and honor of designing the education system for their children.

Iraqis also deserve the right to run their own free media. Immediately following the invasion, Coalition forces seized control of Iraq's state-run media network and established an Iraqi Media Network in its place. Science Applications International Corporation (SAIC) of California was awarded a contract to rebuild the infrastructure, run the newly refurbished newspapers and television outlets (including writing content), and train Iraq's journalists on how to work in a "free" media. SAIC was an odd choice because it had no media experience, a fact that became immediately obvious. Don North, a former SAIC employee in Iraq explained that the company was "an irrelevant mouthpiece for Coalition Provisional Authority propaganda, managed news and mediocre programs." When I mentioned to an Iraqi friend that I was writing about SAIC, he said, "Just spell the name backwards, without the 'S,' and you'll know what Iraqis think of them." These views were widely shared and SAIC failed to attract as many viewers as it did critics.

In January 2004, SAIC's contract was passed to Harris Corp. of Florida, another defense contractor lacking media experience. Just a few months after Harris took over, in April 25, 2004, the editor of the Iraqi Media Network, Ismail Zaher, and twenty staff members quit because of editorial interference from Harris. Iraqi Media Network is still considered the voice of the occupation, while Iraqis have thrived in a media environment no longer controlled by Saddam Hussein, with literally hundreds of independent newspapers springing up across the country. Harris's contract is simply inappropriate and unnecessary.

With these and all other "economic transformers" and U.S. contractors gone, there will be no more need for U.S. appointed auditors and inspector generals to remain imbedded in every Iraqi ministry.

Establish an Iraq Reconstruction Fund and International Oversight Board

An Iraq Reconstruction Fund, administered jointly by the newly elected permanent Iraqi government and the international community, is necessary for rebuilding Iraq. The Bush administration has squandered any right it may have originally possessed to dictate the use of U.S. taxpayer money in the reconstruction of Iraq. The Bush administration cannot account for $8.8 billion of the $18.4 billion Development Fund for Iraq put under its control after the invasion. Audits of U.S. taxpayer funds appropriated for the reconstruction have found contract files "unavailable, incomplete, inconsistent and unreliable."[9] In October 2005, the U.S. special inspector general for Iraq reconstruction was investigating fifty-four cases involving allegations of theft, false claims, bribery, and fraud of the U.S. taxpayer funds. However, the Geneva Conventions (not to mention basic morality) dictate that the United States, as well as all of the Coalition members that participated in the invasion, must pay for Iraq's reconstruction. Given the unique history of the U.S. government and its corporations with the former dictator of Iraq, the damage the U.S. military caused in the first Gulf War, and the devastation of the years of sanctions, the United States owes much more.

An international oversight board for a portion of Iraq's reconstruction already exists; however, it too has failed to adequately account for Iraq's money. The International Advisory and Monitoring Board of the Development Fund for Iraq (IAMB) was established under UN Security Council Resolution 1483 authorizing the occupation of Iraq. The IAMB had oversight over the Coalition Provisional Authority's expenditure of the Development Fund for Iraq (DFI). The DFI is comprised of money from Iraq's oil export sales, funds that remained from the UN Oil-for-Food Program, and Iraqi funds and financial assets frozen abroad. Thus, the missing $8.8 billion from the DFI is as much the fault of the IAMB as of the Bush administration. The IAMB's failure is due to the makeup of its board and the lack of domestic Iraqi input and accountability.

The IAMB was originally composed of representatives of the World Bank, IMF, the Arab Fund for Economic and Social Development, and the United Nations. The World Bank and IMF are simply too beholden to the Bush administration to provide unbiased oversight. This problem only worsened with the March 2005 appointment of Paul Wolfowitz as World Bank president. After the CPA was disbanded, the DFI was put into the control of the Iraqi interim government under the continued oversight of the IAMB. Other than adding a seat to the IAMB's board for an Iraqi government official, the IAMB, which had its authority extended to at least December 2006 by UN Security Council Resolution 1637, has done little else to insure that a repeat of its $8.8 billion failure with the CPA does not occur.

Had the process for establishing the new government of Iraq been less controlled by the occupation, I would have argued that the reconstruction funds be turned immediately over to the newly elected Iraqi government. Instead, I recommend an Iraq Reconstruction Fund administered by a newly established International Oversight Board to replace the IAMB. In making this recommendation, I draw heavily on the work of Tiri and Transparency International and their extensive analyses of postwar reconstruction efforts and government corruption.[10] The following discussion is offered as a starting point rather than a comprehensive proposal.

The International Reconstruction Fund would be comprised of all remaining unspent monies appropriated by U.S. taxpayers for the reconstruction, all funds made available when existing U.S. reconstruction contracts are canceled and U.S. contractors who have failed to meet their contractual obligations refund money paid to them, any future U.S. appropriated funds, funds from all Coalition countries that participated in the invasion and occupation, and donations from other countries. The DFI is Iraq's money and therefore must remain under Iraqi government control. However, the newly proposed International Oversight Board could take over the oversight auditing authority currently held by the IAMB.

The largest group of International Oversight Board members would be representatives of the Iraqi government and Iraqi civil society organizations, although they would not necessarily comprise the majority. The remaining members would be drawn from the appropriate UN development agencies, international nongovernmental organizations with a history of working in Iraq and/or particular expertise in post-conflict reconstruction, and a representative from the U.S. government. Representatives would be chosen for their knowledge of auditing, transparency, postwar reconstruction, and Iraq. Their task would be to assure full transparency and disclosure and the proper management and accountability of resources.

The main function of the Oversight Board would be to ensure that absolute preference is given to Iraqi organizations, companies, and workers, and that, wherever possible, companies are chosen from the localities within which they will work. Contracts to non-Iraqis would be treated like green cards—only going to those performing work that could not otherwise be done by Iraqis. The foreign investment rules implemented under Bremer Order #39 would therefore have to be waived, or, if the Iraqis so-choose, eliminated.

Local communities and nongovernmental organizations in Iraq have long histories of working closely with international and regional aid organizations. It has been personally devastating for many Iraqis to watch the slow and steady withdrawal of all of these groups from their nation. Agencies such as CARE, the World Health Organization, the International Red Cross, Christian Aid, and dozens of others with the experience, knowledge, will, and trust of local Iraqis can be enlisted to participate in the reconstruction. Many of these organizations chose not to do so originally because of the U.S. Pentagon's dominance over the reconstruction and they did not want to be viewed as allies of the U.S. military. With the reconstruction funds removed from U.S. control and the withdrawal of the U.S. military, these groups may be encouraged to return.

The Board would also be tasked with ensuring, as much as possible, the devolution of reconstruction decision making. The findings of Tiri and Transparency International clearly demonstrate the crucial importance of building local governance and devolving the reconstruction process instead of centralizing it. In this way, local communities are strengthened, reconstruction dollars reach a much wider segment of the population, local needs are far more likely to be directly met, and a broader segment of the population is involved in the reconstruction process and can therefore hold the government accountable if problems arise.

In addition, there is no shortage of strong civil society groups in Iraq who could be called on to compose their own board, which would provide independent monitoring of the funds in addition to the International Oversight Board.

Tiri and Transparency International also point out that not all of the reconstruction can or should be conducted at once. Immediate necessities should be met, but full reconstruction is a longer process. The perceived need to spend quickly and spend now breeds a wild-west approach to reconstruction, which fuels the acceptance of waste, abuse, and corruption. The groups point to "reconstruction trust funds," which have been advocated for some time, based on minimum ten- to fifteen-year commitments that enable postwar countries to plan and develop longer-term strategies of sustainable reconstruction. This option could be offered such that more money would be made available to the fund if the Iraqis were willing to receive the money over a delayed but specified period.

Finally, turning the U.S. reconstruction money over to the IOB and ending the U.S. military occupation would eliminate the need to maintain the largest U.S. embassy in the world in Iraq. Reducing the size and budget of the U.S. embassy would provide additional money for reconstruction while reducing the overall influence of the Bush administration in Iraq.

TRANSITION FROM THE OIL ECONOMY

Of all administrations in American history, the Bush administration may be the most beholden and interconnected to the energy sector. As mentioned earlier, in the 2000 election cycle, the oil and gas industry donated over thirteen times more money to the Bush/Cheney campaign than to its challenger. In 2004, the industry gave more than nine times more to Bush/Cheney.[11] The Bush administration itself represents the first time in history that the president, vice president, and secretary of state are all former energy company officials. The industry has subsequently been cuddled by the administration through subsidies, tax breaks, and deregulation. In just the most glaring example, the U.S. Energy Policy Act of 2005 granted $14 billion in new tax breaks and subsidies to oil and gas companies over the next ten years.[12] Also in 2005, oil and gas companies experienced the highest profits in corporate history, with the top three U.S. oil companies alone (ExxonMobil, Chevron, and ConocoPhillips) earning more than $64 billion between them. The Bush administration even waged a war for oil in Iraq.

Through decades of writing, policies, and wars, the leading members of the administration clearly demonstrated their shared desire to pursue greater U.S. control over Iraq's oil. The corporate and government officials on the Cheney Energy Task Force mapped out their specific areas of interest in Iraq and then the Bush administration waged a war to acquire them. The administration then set the conditions ensuring U.S. corporate access to Iraq's oil through the Bremer Orders, Iraq's new petroleum law, and the country's constitution.

The ultimate goal of this section is to outline a path away from the oil economy, which the people of Iraq may eventually welcome. For example, a friend from Iraq's oil-rich Kirkuk region recently told me that every day she wakes up and prays that the oil beneath her home will dry up forever so that she and her family, friends, and neighbors can one day live in peace. Her feelings are supported by the facts. Countries with oil are more than forty times more likely to be involved in civil wars. They also have higher rates of poverty, indebtedness, corruption, and totalitarian governments. Juan Pablo Perez Alfonzo, the cofounder of OPEC, once said, oil is the "devil's excrement."[13] The death and devastation brought by the oil industry to Nigeria and Ecuador discussed earlier provides ample justification for such a view.

An Alternative Path for Iraq

In the near term, however, countries such as Iraq, which are 95 percent dependent on oil wealth, will need to continue to export their oil. To date, few large oil contracts have been signed in Iraq and security remains too unstable for many foreign oil companies to get safely to work. Therefore, if the Iraqi public so chooses, there is time to rewrite the new Petroleum Law, the Bremer Orders, and even the Constitution. There are also many alternative policies for managing the country's oil wealth from which to choose.

U.S. oil companies and the Bush administration argue that Iraq needs foreign investment to harness its oil wealth and that Production Sharing Agreements (PSAs) are the best way to attract this investment. However, it is not at all obvious that Iraq needs foreign investors. Iraq's oil investment requirement is estimated at around $3 billion per year. This is well within the range of current budgetary allocations: The 2005 Iraqi oil investment budget was $3 billion (out of a total Iraqi budget of around $30 billion).[14] It is also highly likely that when the occupation ends and U.S. energy companies such as Halliburton depart, the attacks on Iraq's oil infrastructure will be significantly reduced and exports will increase while costs go down. Iraq's oil is relatively cheap to extract, the country is rich with the skill and expertise necessary to run its systems, and there is a hungry world waiting for its oil.

If Iraq does decide to seek foreign investment, however, PSAs are just one of several contractual options. PSAs favor private companies at the expense of the exporting governments and therefore just 12 percent of world oil reserves are subject to PSAs, compared to 67 percent developed solely or primarily by national oil companies.[15] PSAs are most often used in countries with small reserves or reserves that are expensive to extract—two conditions that could not be farther from Iraq's situation.

In a recent report, Greg Muttitt of the London-based Platform oil research organization lists three alternative options, which have recently been chosen by Iraq's neighbors—Iran, Kuwait, and Saudi Arabia: risk service contracts, buyback contracts, and development and production contracts. Muttitt explains that with each of these arrangements, oil remains the property of the state and the foreign company is paid as the state's contractor. All three give operatorship of the field to a foreign company, but with much more limited rights, and in the case of buybacks and development production contracts, for a much more limited period of time than PSAs. The foreign company does not have the opportunity to make excessive profits, as it is paid either a fixed fee or a fixed rate of return.[16]

People in the United States, including elected officials, can take the many steps outlined here to reduce and even eliminate the Bush administration's influence over the Iraqi economy, its laws, and its government. In this way, the Iraqi public will be freed to make its own decisions regarding its oil wealth. If Iraqis choose foreign investment, or if foreign investment happens against their wishes, Americans and all residents of nations with oil companies operating in Iraq can exert influence over those companies to insure they operate in both legal and socially and environmentally responsible ways and hold the companies accountable when they do not. We can also support demands of Iraqi people wrongly affected by the operations of our companies by holding the companies to account at home through consumer and political campaigns and acts of non-violent protest described in this chapter and throughout the book.

An Alternative Path for All

We need light, heat, electricity, and mobility—not oil. Hybrid cars, biofuels, and solar and wind power are not only viable technologies, they are already in use across the United States and the world. An aggressive ethanol subsidy by the Brazilian government has already replaced one quarter of that country's gasoline, and eliminated oil imports worth more than fifty times the investment by the Brazilian government. In 2003, seventeen times more biodiesel was produced in Europe than in the United States.[17]

One out of every seven barrels of oil in the world is consumed on America's highways alone. Transportation accounts for more than two-thirds of all oil consumption in the United States. Yet the American automobile industry is producing cars with worse fuel efficiency today than during the Reagan administration. The average fuel economy of 2005 vehicles was 21 miles per gallon (mpg), compared to 22.1 mpg in 1988. Using all available technologies, no new innovations are necessary for us to achieve a fuel economy of 50 mpg or more and greatly reduce our dependence on fossil fuels. In fact, improving fuel economy standards for passenger vehicles from 27.5 mpg to 40 mpg, and for light trucks (including SUVs and vans) from 22.2 mpg to just 27.5 mpg by 2015 would reduce U.S. gasoline consumption by one-third.

Fuel economy standards can be changed in two ways: through government regulation and by the automobile industry itself. Three organizations in San Francisco—Rainforest Action Network, Global Exchange, and the Ruckus Society—are working with consumers across the country in a campaign to put pressure on the U.S. auto industry to increase its fuel efficiency and reduce its use of petroleum. "Jumpstart Ford" focuses on the Ford Motor Company because, according to campaign materials, among all major U.S. automakers, Ford has the worst fleetwide fuel efficiency and the highest average vehicle greenhouse gas emissions. Consumers are demanding that Ford manufacture a fleet of cars, trucks, and even SUVs that average fifty miles per gallon by 2010 and achieve zero emissions of greenhouse gases by 2020—targets that are achievable today using existing technology. The campaign has already had victories. For example, in August 2004, Jumpstart Ford successfully convinced Ford to halt the destruction of the zero-gasoline, zero-emission Th!nk electric vehicles. In January 2005, following an eight-day "car sit" vigil, Ford reversed its unpopular decision to repossess and destroy its last zero-gasoline, zero-emission electric Ranger pickup trucks.

The government is also being called upon to increase its taxation, regulation, and oversight of the oil industry. For example, several proposals for a windfall profits tax have been introduced at both the state and national levels. The tax is a quick and easy policy to implement while U.S. oil companies are taking in the highest profits in history and as a direct result of a war for oil. Quite simply, each company's average profits over a five or ten year period are assessed and every dollar earned in profit over that average amount is taxed in addition to current tax obligations. The proceeds of the tax can then be specifically directed toward programs such as supporting non-fossil-fuel alternatives, public transit, and subsidizing home heating for the low income.

Since 1999, the U.S. oil sector has experienced a wave of megamergers. The industry is now over-concentrated such that an increasingly small number of giant oil corporations control an ever-larger percentage of global supply (the five largest oil companies produce 14 percent of the world's oil. Combined, they produce 10 million barrels per day, more than Saudi Arabia exports in a day); refining capacity (just five U.S. oil companies control more than 56 percent of domestic oil refinery capacity); and sales (U.S. oil companies are vertically integrated, with the same company controlling oil exploration, production, refining, marketing, and sales). They have used this power to inflate their profits (in 1999, U.S. oil refiners made 22.8 cents for every gallon of gasoline refined from crude oil. In 2004, they earned 40.8 cents, and in 2005, 99 cents) at the expense of consumers in the United States and around the world. In response, consumer advocacy groups such as Public Citizen are calling on Congress to re-evaluate the legality of these mergers, investigate anticompetitive practices in the industry, and aggressively enforce U.S. antitrust laws.

Washington, DC-based Oil Change International has launched a campaign called "Separation of Oil and State," modeled after the successful campaigns of the 1990s that led many politicians to reject campaign contributions from the tobacco industry. People across the country are calling on their elected officials at the local, state, and national levels to renounce oil company campaign contributions and support renewable energy incentives. The campaign will, at a minimum, raise awareness about the role of oil companies in the U.S. government and challenge their inordinate and destructive influence, while simultaneously increasing public knowledge about the many alternatives that exist to the oil economy. Given the umbilical connection between the Bush administration and the oil and gas industry, this campaign offers a profound opportunity to fundamentally destabilize the Bush Agenda.

ALTERNATIVE TOOLS OF LOCAL ECONOMIC DEVELOPMENT: REIGNING IN CORPORATE POWER

Corporate globalization policy is a central pillar of the Bush Agenda. Policies that reduce undue corporate influence over people, communities, and governments are therefore important replacements to the Bush Agenda. The corporate histories of Bechtel, Chevron, Halliburton, and Lockheed Martin reveal that corporations require more, not less, government oversight to ensure that they benefit the communities within which they operate. Iraq, for its part, has resources that foreigners want. Thus, Iraq—not the United States or its corporations—should be in a position to set the terms by which transactions for these resources take place. The Bremer Orders ensure that the benefits of Iraq's resources flow to foreign companies and a few local elites maintained in positions of power to secure foreign interests. Other policies can be chosen that direct resources inward, nourish industries beyond the oil sector, and direct the rules of foreign investment to ensure well-disbursed local benefits.

For example, tariffs and quotas in the manufacturing sector would give Iraq's businesses time to rebuild with the new resources made available to them through the reconstruction and the removal of sanctions. Colonial New England chose similar policies to bar cheaper British textiles, by imposing high tariffs in order to foster local industries. This helped to create a fiercely strong U.S. textile industry, on which much of the industrial revolution in the United States was based. In the same way, in the 1980's, countries such as Taiwan, Hong Kong, South Korea, and Singapore protected their local industries through tariffs, import quotas, limited trade rules, and capital controls. These devices helped to support nascent domestic industries, some of which then became global export producers, while others focused on serving the domestic economy. These countries were then dubbed the "Asian Tigers" because of their explosive growth. It was only after they were forced to reverse course and open their markets under IMF rules—particularly the elimination of capital controls—that the "Asian Tigers" became the "Asian Financial Crisis." As former World Bank senior vice president, Joseph Stiglitz once argued, "All the IMF did was make East Asia's recessions deeper, longer and harder."[18]

Policies can also be adopted to ensure that businesses, whether foreign or domestic, benefit the communities within which they operate. These include, for example, laws that require a company—whether foreign or national—to invest a certain portion of its profits locally, to hire a certain number of local employees, or to leave a certain percentage of profits in the local economy for a set period of time, even if the company itself leaves. Such policies could also be designed another way, offering higher tax breaks for long-term local investment and creating tax penalties for rapid withdrawal.

Communities can also design laws to keep out companies that continually act illegally. For example, in 1998, Pennsylvania's Wayne Township passed a city ordinance stating that any corporation with three or more regulatory violations over seven years is forbidden to establish operations in its jurisdiction.[19] Thus, laws can be passed that bar companies with records of using predatory business practices, overcharging for services, using environmentally destructive practices, having unfair labor practices, engaging in human rights abuses, or the like, from operating in towns, cities, states, or even countries.

Similarly, dozens of city zoning ordinances have been passed across the United States banning chains and "big box" stores. These localities are choosing laws that favor locally owned smaller businesses that benefit the local economy and can be held directly accountable for their treatment of workers, the environment, and the community. For example, in November 2005, voters in the Charlevoix Township of Michigan approved an ordinance that gives local government the authority to limit the size of big-box stores to 90,000 square feet, while the Township's neighbor, the City of Charlevoix, had passed an ordinance banning individual retail stores larger than 45,000 square feet. The Township's measure was initiated after Wal-Mart made an ultimately unsuccessful bid to open a 155,800 square foot superstore on a wooded 24-acre plot of land in the Township.

There is a strong and growing movement in the United States to use corporate charters as a means of holding companies accountable for illegal or otherwise unacceptable business practices, whether in the United States or abroad. In the United States, a corporation comes into being only when the government grants it a corporate charter. Without a charter, the corporation does not exist as a legal entity and therefore cannot own property, borrow money, sign contracts, hire and fire, or accumulate assets or debts. A state legislature can withdraw a corporation's charter at will. Historically, corporate charters were used in the United States by citizens to keep corporations on a short leash, spelling out the rules they had to follow and holding their owners liable for harm or injuries caused. People across the United States are now taking a renewed look at corporate charters as a point of influence over corporate activity.

Another policy option is to make investors liable for harms done by the companies in which they invest. This would make investing decisions a more serious affair and would greatly change financial calculations made by corporations when deciding what actions to take to protect people and the environment. Investors would be compelled to evaluate the environmental, labor, and human rights record of a corporation before becoming shareholders. Similarly, the CEO and management would give such concerns a higher priority if they were held liable for the company's illegal actions.

People around the world have also taken action as consumers. By making purchasing decisions based on how goods are produced, under what conditions, at what cost, and for whose benefit, consumers are demanding that corporations serve the public interest. A particularly powerful alliance is created when workers seeking change within the company they work for team up with consumers of the company's products.

A Victory: Coalition of Immokalee Workers

> To a significant extent, Taco Bell's tremendous global revenues are based on cheap ingredients for the food they sell, including cheap tomatoes picked by farmworkers in Florida paid sub-poverty wages. Well, we as farmworkers are tired of subsidizing Taco Bell's profits with our poverty. We are calling for this boycott today as a first step toward winning back what is rightfully ours—a fair wage and respect for the hard and dangerous work we do.
>
> —*Lucas Benitez, Coalition of Immokalee Workers, April 1, 2001*[20]

This is a victory marked by an agreement over just one penny. The Coalition of Immokalee Workers (CIW) is an organization of migrant farmworkers in Immokalee, Florida. Immokalee is farm country, located on Southern Florida's west coast, just over 100 miles from Miami. According to the U.S. Census Bureau, fewer than 20,000 people call Immokalee home, 70 percent of whom identify as Hispanic. Most of Immokalee's farmworkers migrate from Mexico, Guatemala, and Haiti for the nine-month harvest season. Most are young men in their early twenties who do not speak English and send their earnings home to support their families. They have no health benefits, sick leave, paid vacations, or pension plans. They do not have the right to form unions. Some have even been the victims of slavery: forced servitude by employers through threats and the actual use of violence. None of this has kept them from organizing and winning.

In one case, the Coalition helped convict three Florida-based agricultural employers in federal court on slavery, extortion, and weapons charges. They were sentenced to a total of almost thirty-five years in prison and the forfeiture of $3 million in assets. The men, who employed over 700 farmworkers, threatened to kill workers who tried to leave and pistol-whipped and assaulted at gunpoint passenger van service drivers who gave rides to farmworkers leaving the area. The case was brought to trial by federal authorities from the U.S. Department of Justice after two years of investigation by the CIW.

On March 8, 2005, the farmworkers won their first pay raise in twenty-five years, the result of a ten-year-long struggle, including a four-year national boycott of Taco Bell and its parent company, Yum! Brands, Inc. Based in Louisville, Kentucky, Yum! Brands is the world's largest restaurant company with nearly 34,000 restaurants in over 100 countries. It owns, among other restaurants, Taco Bell, KFC, Pizza Hut, A&W Restaurants, and Long John Silver's. Yum! Brands CEO and President David C. Novak is one of the country's most well-paid chief executives, receiving over $8 million in total compensation in 2004 alone. Despite its size relative to the corporation, the Coalition won every demand it made of Taco Bell and Yum!.

Taco Bell agreed to pay one penny more per pound of tomatoes from Florida tomato growers with a contractual agreement that this penny be passed on to the workers. In other words, Taco Bell will only buy tomatoes from companies that agree to put each additional cent toward their workers' wages. A penny per pound almost doubles the wages of the farmworkers. Before the agreement, they received 40 cents per 32-pound bucket of tomatoes. Now, the piece rate is around 72 cents per bucket, representing a tremendous increase in income, but not their only victory.

The agreement also ensures that workers' rights are protected in the fields where Taco Bell's tomatoes are picked. If there is a complaint about a human rights violation from any worker in a company that supplies Taco Bell, then Taco Bell and the CIW will investigate the case together; if evidence is found that a violation occurred, Taco Bell will stop buying tomatoes from that company. After ten years of refusing these terms, Taco Bell and Yum! Brands have pledged zero tolerance for slavery and other workplace abuses prohibited by law.

How was this victory won? The company's initial position was that it purchased tomatoes from independent suppliers whose mistreatment of workers was "unrelated to Taco Bell."[21] It eventually changed its position. One powerful agent of change was the repeated "Taco Bell Truth Tours" in which the workers and their allies marched on foot from Florida to

Taco Bell's headquarters in Irvine, California. The march stopped at Taco Bell restaurants along the way, where marchers were joined by local community members, students, church leaders, restaurant workers, elected officials, local and national unions, and others—all demonstrating their belief that Taco Bell should meet CIW's demands.

Consumers proved to be a powerful ally. The boycott was successful because so many people realized that they could exercise their own power over Taco Bell simply by choosing whether to eat there based on the company's practices. The next step was letting the company know about their decision. Students at 300 colleges and universities and more than fifty high schools participated in the boycott. They shut down or blocked the chain's restaurants on twenty-two campuses and formed their own network, the Student Farmworker Alliance, making clear the unity between the company's target market and the people who supply their food.

CIW member Gerardo Reyes Chavez explained in an interview with David Solnit, editor of *Globalize Liberation,* "By focusing our struggle on the demand that multinational corporations pay attention to human rights we are changing the way they do business and in some small way changing their mentalities, the responsibility of taking care of the people who work on the farms that produce their products. For the first time ever we were able to open a door that was before thought to be impassable. And we've opened it not just for farmworkers but also for all of the workers that make up these corporations."

Just as such organizing efforts were successfully used to change fundamental business practices of the world's largest restaurant, they can also be used to change the behavior of the world's largest energy services corporation (Halliburton), engineering and construction firm (Bechtel), arms manufacturer (Lockheed Martin), or oil company (ExxonMobil or its lesser cousin, Chevron).

A Victory: The SEMAPA Water Company

There is no reason why Iraq or any other country should be forced to privatize its services. Advocates of privatization argue that it is a more cost efficient means of running a service, particularly when a government has insufficient resources. However, privatizations of former government services, especially when forced on a country rather than chosen by its government, have proven time and again to be more costly, as private companies increase fees and reap the rewards as profits rather than as money to be reinvested in the public service. In Iraq's case, the United States and the Coalition are required to pay for the reconstruction of Iraq's basic services. This money can and should go to Iraqis who have the knowledge, expertise, and desire to rebuild and run the water, electricity, sewage, and other public systems themselves. Once the rebuilding is complete, Iraq might choose to privatize its services. Alternatively, it might prefer to follow the model of the United States, in which more than 85 percent of all water systems are publicly run. Among the other alternatives from which Iraq may choose, my favorite is that of the SEMAPA water company in Cochabamba, Bolivia.

Bechtel privatized the water systems of Cochabamba, Bolivia, in 1999 and then immediately increased the price of water by as much as 300 percent, without expanding services. In response, the people of Cochabamba forced the Bolivian government to cancel Bechtel's contract. In 2000, the Cochabambans established an alternative water system—a government-community-worker hybrid—which has become a model for water systems the world over. The company, SEMAPA (*Servicio Municipal de Agua Portable y Alcantarillado*), is run by a rotating board of seven directors: three from the community (who are publicly and democratically elected), two from the mayor's office, one from the professional schools, and one from the workers' union. The meetings take place every Tuesday in different neighborhoods to assess needs, prices, and overall functioning of the system. Extra meetings are called as needed. The wealthier citizens subsidize those with lower incomes such that the company has stabilized prices while successfully expanding services to the city's poorest neighborhoods, many of which had never received water before.

SEMAPA has received economic and technical assistance from groups all over the world. While SEMAPA faces financial and other difficulties, it continues to provide water more universally, fairly, and reliably than the systems that proceeded it. The secret, according to those who run the company, is that all the people who use the water can choose to become directly involved in the decisions on how the water is allocated, priced, distributed, and maintained. This model of "direct democracy" has encouraged the city residents to support the company both financially and with their time.

"FREE TRADE" IS NOT NECESSARY (NOR DOES IT EXIST): REJECT THE MIDDLE EAST FREE TRADE AREA— REINVIGORATE THE UNITED NATIONS

The Bush Agenda's imperial ambitions are spread not only through war, but also through economic agreements such as the U.S.–Middle East Free Trade Area (MEFTA). With a stronger, more practical focus on local economic development and a reigning in of multinational corporate power, the current level of international trade and investment activity—as well as the need to regulate it—will likely be significantly reduced. However, international trade and investment will and should continue. As long as it does, international trade and investment rules will be necessary. Agreements such as the MEFTA, which are designed to increase the access of foreign multinational corporations into the Middle East by reducing the ability of national and local governments to control their behavior, are not necessary. Such agreements only allow one side—the one favoring increased corporate freedom above all else—to participate in the debate over economic policy. The MEFTA, for one, is not a democratic process and it should be rejected.

Similarly, the World Bank, IMF, and WTO are not democratic institutions. They have also failed miserably at their agendas. They are creating more poverty, inequality, and instability than they are preventing and should therefore be decommissioned. This is not to say that the baby should be thrown out with the bath water.

Forty years ago, developing countries argued for rules to direct the terms of trade between nations within the context of democracy, health, labor rights, equality, stability, and poverty alleviation. They successfully established international bodies within the United Nations to address these issues, such as the UN Centre for Trade and Development, the World Health Organization, the International Labor Organization, the Food and Agriculture Organization, and the UN Development Program. The United Nations, particularly its Security Council, needs reform. It needs to be "de-corporatized" with more financial resources, greater public attention, greater transparency, more democracy, and more influence. Yet it remains the institution with the broadest mandate and, despite its considerable flaws, is more open and democratic than any of the Bretton Woods organizations. In practice, it has given much greater weight to human, social, and environmental priorities. Where international trade and investment rules must be written, a reformed United Nations is the place to do it.

In the past ten years, I have been fortunate to work with some of the greatest minds writing alternatives to corporate globalization policy. I spent five of those years working at the International Forum on Globalization, where I participated in the writing of a book dedicated exclusively to this topic, *Alternatives to Economic Globalization: A Better World is Possible, Second Edition.* Many ideas briefly mentioned here are greatly expanded on in this book.

CANCEL THE DEBT—ALL THE DEBT

Iraq needs money to rebuild. It cannot be wholly dependent on the United States and the international community for this money. While the Bush administration certainly wants to control the rebuilding effort and ensure U.S. corporations benefit from the process, it does not want to shoulder all of the financial burden. Thus, President Bush became an early and outspoken advocate of debt cancellation for Iraq. Iraq faced a crushing $121 billion debt burden at the start of the occupation. Most of Iraq's debt is owed to Russia, Saudi Arabia, and Kuwait, while approximately one-third was owed to Japan, France, Germany, the United States, and the United Kingdom. For the first time, President Bush adopted the language of campaigners against international poor country debt when he argued that Iraq's debt endangered its "long-term prospects for political health and economic prosperity," and that the world must not allow the financial obligations to "unjustly burden a struggling nation at its moment of hope and promise."[22] In November 2004, Bush persuaded Japan, France, Germany, and the United Kingdom to join the United States in canceling 80 percent of the debt owed to them by Iraq, but they included a catch: 30 percent was canceled immediately; the next 30 percent was tied to Iraq's implementation of an IMF economic program; and the final 20 percent will only be granted after the IMF certifies the success of the program.

A sure way to bring billions of dollars to Iraq—and toward its reconstruction—is to cancel its entire debt burden outright. Iraq's debt was incurred during the rule of Hussein. Those who had to suffer under his rule should not now be forced to pay his debts or to bear IMF conditions in order to have those debts removed.

The cancellation of Iraq's debt should open the door to all nations suffering from similar debt burdens while reducing the overall influence of the World Bank and IMF around the world.

A Victory: Debt Cancellation

> The issue of debt provides one clear instance in which a network of international activists has affected governmental decision-making and in doing so has opened real possibilities for human development.
>
> —*Mark Engler, "A Movement Looks Forward," May 19, 2005*

A recent victory in international debt cancellation provides tools for the further dismantling of the Bush Agenda. The World Bank and IMF continue to exercise enormous control over nations the world over, due in significant part to the crippling levels of debt that nations have owed the institutions for decades. Indebted countries give out far more in debt repayments then they take in from new aid, while new loans add to their overall debt burden. More money is spent paying off their debts than on their local economies; in turn, the economies continue to shrink, adding to their indebtedness. For example, sub-Saharan Africa alone pays almost $15 billion in debt service to wealthy nations and financial institutions every year, while Africa as a whole spends four times as much on debt repayment as it does on healthcare.[23]

Under their debt burdens, countries are desperate for cash and must cede to the requirements forced on them by multinational corporations, international trade and investment agreements and institutions, and foreign governments.

During the mid-1980s, developing countries began to demand that the World Bank, IMF, and wealthy country governments cancel their debts. Through dedicated organizing, this message reached people in lender nations. Groups in debtor and lender nations then joined forces and became part of a growing anti-corporate globalization movement that lobbied elected officials, formed international organizations such as 50 Years Is Enough and Jubilee 2000, conducted analyses and released reports on the impacts of debt on poor countries, held public teach-ins and press conferences, and protested at the meetings of the World Bank and the IMF. One columnist commented, "If you make a campaign out of it or use extreme language . . . the very people you want to influence, the ministers and officials of the rich democracies, stop listening to you."[24] How wrong he was.

Each piece of education and activism helped to raise public awareness and give legitimacy to the cause, as well as to galvanize world leaders and celebrities such as Pope John Paul II and U2's Bono. The debt cancellation movement won a partial victory in 1996 when the IMF instituted the Heavily Indebted Poor Countries Initiative (HIPC) for the poorest of the poor nations. While welcome, this program tied debt relief to the very same IMF programs that caused much of the debt crisis in the first place. A larger victory occurred when, in 1999, President Clinton agreed to cancel 100 percent of bilateral debts—debts owed to the U.S. government by other country governments. Clinton was followed by the governments of the United Kingdom, France, Germany, Japan, and others. However, the substantially larger debts that countries owed to the World Bank, IMF, and commercial banks remained virtually untouched.

In February 2005, members of the Group of Eight industrialized nations (G8) announced their intention to provide "as much as 100% multilateral debt relief for the 42 HIPC nations." "Multilateral debt" is debt owed to the World Bank, IMF, and other international financial institutions. While the agreement does not address commercial debt or the debt of the hundreds of severely poor nations that do not qualify as HIPC countries, a tremendous victory has already been won that can be both expanded on and emulated. More important for our purposes here, the agreement and the discussion that brought it about have provided the most significant admission by these governments that the economic policies of the World Bank and IMF have failed to better the lives of people around the world and have in fact done just the opposite. The discredit brought to these policies, many of which are central to the Bush Agenda's "free trade and free markets," and the ongoing campaigns seeking their removal can be used in the process of dismantling the Bush Agenda itself.

ENDING THE WAR

I have learned one very important lesson in my years of work on public policy and in social movements: Change is slow, but it does happen everywhere all of the time. As the examples offered throughout this book make clear, real change comes about when hundreds of small steps made by some and enormous steps made by others all come together—sometimes over decades, sometimes over minutes—to achieve the conditions necessary for change.

In 2002, hundreds of people gathered in their local communities in small antiwar groups. From those meetings, thousands met in Porto Alegre, Brazil, and agreed to call for a global day of protest against the war in Iraq. They went home and tens of thousands of people the world over began holding teach-ins on Iraq and the war, calling friends, reading books, arranging for permits, meeting with other groups, debating with their parents, and holding fundraisers. On February 15, 2003, they took to the streets on every corner of the globe, including Antarctica, to demonstrate their personal commitment to and belief in change.

For some, it was their first act of protest; for others it was one step among many. I marched in San Francisco with 150,000 other people. My youngest sister, still in high school, marched in Boulder, Colorado. Another sister participated in her first major protest in New York City.

My father, my oldest sister, and her young son marched together in Los Angeles. We were six of fifteen million people demonstrating to each other and to the hundreds of millions of people watching on television that we were not alone in opposing the war—and we were not afraid to let George Bush, Dick Cheney, Donald Rumsfeld, Paul Wolfowitz, and the rest of the Bush administration know it.

Many people have since formed new networks or joined existing groups like United For Peace and Justice, Not in Our Name, Direct Action to Stop the War, Code Pink, and thousands of others across the country and the world. They've banded together with neighbors and other members of their communities in weekly protests. They've taken their protests to the doors of the corporations driving the war. They've inspired others to write letters to their representatives in Congress, to the editors of their local newspapers, or to the White House. Some have raised the issue at the dinner table for the first time. Some have held antiwar rock concerts or written books.

In groups such as Iraq Veterans Against the War, Iraqi war veterans have organized in opposition to the war far more quickly and in greater numbers than veterans did during Vietnam. Community networks have emerged to stand in support of soldiers who refuse to fight, to challenge recruitment in their high schools and colleges, and to support the soldiers when they return home. The families of soldiers have also joined the antiwar movement in unprecedented numbers, founding groups such as Military Families Speak Out and Gold Star Families for Peace. Cindy Sheehan spoke out with and on behalf of the thousands of others who have lost family members in the war. In August 2005, with the support and participation of activists from across the country and the world, she established a camp in her son Casey's honor in front of President Bush's Crawford, Texas, ranch. She waited to no avail for the president to tell her why her son died in Iraq. All of this activity has contributed to the growing worldwide opposition to the war and the urgent demand for alternatives.

Polls in the United States show that the majority of Americans now believe that the United States should never have entered the war in the first place and want an immediate withdrawal of troops. Members of

Congress, even members of the president's own party, have demanded a timetable for withdrawal. Other members have demanded an immediate withdrawal. Thousands of people around the world have reported on and protested the war profiteering of Halliburton, Bechtel, Chevron, Lockheed, and others. The public is now demanding an accounting. Dozens of congressional hearings have taken place in response. The contracts and the corporations' support for the war are under a greater level of scrutiny than at any other time since planning for the war began. A global network, including Iraqis and Americans, has formed to advocate for alternatives to war and the Bush Agenda. As a result, the Bush Agenda has not yet forced its way into Iran or Syria as the president has threatened. A real debate has emerged.

In response to Rebecca Solnit's statement, which opens this chapter, we might argue that "There will always be kindness, always be justice, always be renewal. They cannot eliminate all benevolence for all time, but they can reduce it, outlaw it, undermine its sources and foundations: these are our challenges. We can meet them. We require hope to do so."

~

I close with a story told by Neil Gaiman in his 1990s series *The Sandman.* It dates as far back as Homer's *Iliad,* if not earlier, and is a literary version of "rock-paper-scissors."[25] A contest between two players, Choronzon and the Sandman, is taking place. It is a test of creativity and will. The winner is the person who outlives the other by choosing the unbeatable element.

Choronzon begins by imagining himself as a wolf. He is bested by Sandman who, as a hunter on horseback, shoots the wolf. Choronzon chooses a horse-stinging fly that throws the hunter from his horse to his death. Sandman becomes a spider who consumes the fly, only to be eaten by Choronzon as a snake. Sandman becomes an ox and crushes the snake. Choronzon chooses anthrax, destroying the ox and all life. He is bested by Sandman as a world that nurtures the return of life. Choronzon becomes a supernova that destroys the world. Sandman is a universe embracing all life.

For the first time in the contest, Choronzon pauses, unsure what to choose.

His response: "I am anti-life. The beast of judgment. I am the dark at the end of everything. The end of universes, gods, worlds, of everything."

Sandman's reply comes quickly.

"I am hope."

And the game is won.

AFTERWORD

WHAT A DIFFERENCE A YEAR MAKES

At the time I finished writing *The Bush Agenda* in January 2006, Republicans controlled all three branches of the United States federal government. Approximately 50 percent of Americans still supported the president's decision to go to war against Iraq and his handling of the war on terror.[1] Few outside of the political left publicly described the Iraq war as a war for corporate—particularly oil company—gain. The U.S.–Middle East Free Trade Area was on a clear trajectory toward completion. The Bush Agenda seemed impenetrable as a sense of powerlessness took hold of much of the nation. And the war in Iraq raged on.

With the passage of just one year, President Bush and his agenda stood virtually isolated. Democrats had taken control of both the House and Senate. Sixty-four percent of Americans considered the Iraq war a mistake (a higher percentage than voiced the same misgivings during the Vietnam war) and did not believe that it had made them safer from terrorism.[2] The administration's corporate agenda in Iraq, including the oil agenda, had become mainstream discourse. The president's free trade agenda was sinking. Only a handful of the original Bush agenda adherents remained in the White House. A feeling of empowerment swept across the nation, leading to the largest antiwar protests in Washington, DC's history in January 2007, and certainly the largest since the Iraq war began. And the war in Iraq raged on.

In January 2007, ABC news was blunt in its assessment of the president's record low approval ratings: "The root of Bush's problems can be summed up in three words: Iraq, Iraq and Iraq."[3] Not only did the American public oppose the war, so did U.S. soldiers and Iraqi citizens. In February 2006, 72 percent of U.S. soldiers stationed in Iraq said that the United States should leave Iraq within a year.[4] Sixty-five percent of Iraqis said that they wanted an immediate withdrawal of U.S. troops in November.[5] In fact, it seemed the only people who supported the continuation of the war were sitting inside the Bush administration or in the headquarters of their corporate cohorts. The question on everyone's mind was, "Why?" If we now knew that the reasons Bush gave for going to war were false, what were the real reasons for the war—and, more important, for refusing to end it? The answers offered here in *The Bush Agenda* have grown only more evident and pertinent with the passage of time.

In chapter 9, I wrote that to replace the Bush Agenda, we must address each of its key pillars individually—war, imperialism, and corporate globalization. Now I offer an assessment of how we have done thus far, where the pillars stand, and what should come next.

In Iraq, the Bush administration's corporate globalization agenda succeeded in giving U.S. corporations free reign but failed to create a free market haven. U.S. corporations are therefore cutting and running from Iraq and taking their billions with them. They're leaving the nation in many ways worse off than when they found it, as they turn toward the rest of the Middle East.

The Bush administration and its corporate allies are closing in on Iraq's oil—the big prize for which they've been fighting. And Iran, with the world's third largest oil reserves, is emerging as the next war on America's horizon. Bush Agenda adherents may yet achieve their imperial ambitions for the region, or, at least, find their wallets well stuffed.

CORPORATE GLOBALIZATION FAILS IRAQ

Upon his return from Iraq on October 5, 2006, Senator John Warner, Republican of Virginia and a veteran of both World War II and Korea, remarked, "There is progress being made in certain areas, but you just find that so many communities don't even have drinking water. . . . It seems to me that the situation is simply drifting sideways."[6]

Senator Warner's comments are, if anything, an understatement. Across Iraq, the provision of basic services is insufficient and in most cases worse than before the war. The lack of services and the knowledge that U.S. corporations have received tens of billions of dollars for this failure helped fuel Iraqi opposition to the occupation and significantly contributed to overall instability in Iraq. The Bush Agenda, the Bremer Orders, and U.S. corporations are fully to blame.

The Bush administration did not enter Iraq with plans to rebuild, but with plans to remake Iraq into a U.S. corporate free-for-all. And it succeeded—to a point. The Bremer Orders enabled U.S. corporations to receive preference over Iraqi companies for reconstruction contracts. The U.S. companies did not have to partner with Iraqi companies, hire Iraqi workers, use Iraqi products, or reinvest any of their profits into Iraq's economy. They did not answer to the Iraqi government, and they were not bound by Iraqi laws. As a result, U.S. companies took over all of the reconstruction in Iraq and ran with it.

U.S. corporations had their own agendas upon entering Iraq. Many hoped to privatize the systems they were rebuilding, to maximize their profits under their cost-plus contracts, and to make themselves attractive to other nations in the Middle East, as they looked toward the U.S.-MEFTA. Many others simply hoped to cash in on the reconstruction bonanza.

The focus of these companies, therefore, was not on immediate nuts-and-bolts rebuilding. Rather, many took their time doing assessments and attempting to build state-of-the-art facilities. The delay in rebuilding, combined with the passing over of hundreds of Iraqi companies and millions of Iraqi workers who were more than capable of taking over the reconstruction, turned the Iraqis against the U.S. corporations. The companies became targets of the violent insurgency, their ability to do their jobs became increasingly difficult, and a vicious cycle was born. As the insurgency grew, billions of dollars originally budgeted for reconstruction were diverted to the training of Iraqi soldiers, and the cycle of violence and deteriorating reconstruction went on unabated.

Public pressure in the United States to expose the corporate abuses perpetrated in Iraq led Congress to establish the independent Special Inspector General for Iraq Reconstruction (SIGIR) in October 2004 . Continued public pressure on SIGIR has helped it to emerge as the most meaningful source of oversight of the U.S. reconstruction effort. As such, it very nearly fell prey to an attempt by some Republican members of Congress to shut it down well before its work was complete. Undeterred, SIGIR has opened 256 investigations into criminal fraud by U.S. contractors in Iraq, four of which have resulted in convictions. The findings of SIGIR's January 2007 report to Congress, discussed below, detail the full extent of the failure of U.S. reconstruction in Iraq.

Electricity

While nearly $3 billion has been spent on rebuilding the electricity sector, less electricity is being generated today than prior to the invasion. And, while Bechtel and many other U.S. corporations are leaving Iraq, only 65 percent of the electricity projects they were hired to perform are complete. Many of those projects considered "complete" by U.S. contracting agencies have been identified by SIGIR as poorly built, badly run, and mismanaged. "Complete" electricity projects, for example, mean little when contractors have failed to build transmission and distribution lines to connect new generators to homes, businesses, and hospitals.

Nationally, Iraqis have an average of just eleven hours of electricity a day, and in Baghdad, the heart of the nation's instability, Iraqis have an average of four to eight hours of electricity per day. Before the war, Baghdad averaged twenty-four hours of electricity per day.

Water and Sewage

The U.S. government originally allocated $4.33 billion to the reconstruction of water and sewage services in Iraq. But when more money was needed to train Iraqi soldiers, more than $2 billion was taken from this sector, cutting the overall allocation by nearly 50 percent and reducing the total number of planned projects. Today, $1.4 billion has been spent and approximately 80 percent of the water and sewage projects are complete. This rebuilding is undermined greatly, however, by the lack of electricity. Without electricity, clean water cannot be pumped into homes and the sewage cannot be pumped out.

Health Care

Put simply, the health-care sector was never a high priority for the U.S. reconstruction. While Iraqis suffer from daily acts of staggering violence, little has been done to ensure that they have services to meet their health-care needs. Only 4 percent of the total reconstruction budget was allocated to health care—about $800 million—and to this day, less than 70 percent of that money has been spent. Only 38 percent of planned health-care projects are complete, including the construction of just eight of 150 planned Primary Health Care Centers (PHCs) and twelve of twenty hospitals scheduled for rebuilding.

This sector has been particularly marred by the actions of U.S. corporations. Fortunately, SIGIR has taken action. The Parsons Corporation, the second largest recipient of reconstruction dollars after Halliburton, with $5.3 billion in contracts, held the original contract for the 150 PHCs. The contract was cancelled when a SIGIR investigation revealed that after more than two years of work and $186 million spent, only six centers had been built, only two of which were serving patients. SIGIR cancelled Parsons' contract and turned it over to the Army Corps of Engineers, who then gave the contract to an Iraqi company.

The Bechtel Corporation was dropped from a $50 million contract to build a children's hospital in Basra after it went $90 million over budget and a year and a half behind schedule. SIGIR found that the project was not only mismanaged but that both Bechtel and the U.S. Agency for International Development (USAID) lied about its status in reports to Congress. In September 2005, the project was already ten months behind schedule. This delay alone added several million dollars to its estimated cost. In 2006, the onsite representative from the Army Corps of Engineers reported problems with construction and further delays. However, according to SIGIR, "neither USAID nor Bechtel reported any problems with the contract throughout this period [July to September 2005]."[7] While Bechtel increasingly came clean in its reporting to USAID, USAID continually misled Congress by reporting that the project was on budget and on time.

Ultimately, SIGIR found that USAID, the State Department, and the U.S. Embassy in Iraq had all failed in their accounting and managerial systems. In fact, USAID still does not know exactly how much money it has disbursed for the project. SIGIR recommended that Bechtel's contract be canceled and given to the Army Corps of Engineers, which is likely to turn the project directly over to Iraqis. SIGIR estimates that this transfer will save approximately $90 million "exclusively from the reduction in contractor overhead."

For all of SIGIR's critical work, it is understaffed, underfunded, limited in its scope, and does not always name the corporations whose projects it has assessed. Bechtel's $50 million in Basra is just the tip of the iceberg. It represents less than 2 percent of the company's $2.85 billion award for Iraq reconstruction, while Bechtel's total award represents just 8 percent of all U.S. reconstruction dollars. However, to date, Basra Hospital is the only Bechtel project to receive SIGIR's investigative attention by name. (A full audit of Bechtel's projects is reportedly under way but behind schedule for completion.) In fact, of the 12,395 projects completed and ongoing by U.S. contractors in Iraq, SIGIR has fully assessed only eighty.

Fixing the Failed Reconstruction

The solution I proposed more than a year ago to address the reconstruction debacle makes even greater sense today: both the military and corporate occupations of Iraq must be brought to an immediate end, and Iraqi companies and Iraqi workers given the full job of reconstruction. In fact, SIGIR has found that when Iraqi companies receive contracts (rather than subcontracts from U.S. companies), their work is faster, less expensive, and less prone to insurgent attack. Three steps are necessary to turn reconstruction over to Iraqis: (1) Existing U.S. contracts must be cancelled; (2) U.S. companies must return all misspent and unused funds; (3) These funds, plus a good deal more money, must be turned over to Iraqi companies and Iraqi workers. I still propose an International Reconstruction Fund as the best means to accomplish this transfer of funds.

Do not doubt that this can happen. In July 2006, the activists who had been working for years to expose Halliburton's abuses in Iraq achieved an important victory: the U.S. Army cancelled Halliburton's largest government contract for worldwide logistical support to U.S. troops (LOGCAP). Halliburton continued its current Iraq contract (under dozens of government investigations), but LOGCAP will now be broken into smaller parts and competitively bid out to other companies. By the end of 2006, Halliburton's shares had dropped by more than 17 percent and its profits were down by 4 percent.[8] The Democrats who now chair the key oversight and investigative committees in Congress began holding hearings on the contractor debacle in Iraq within the first month of taking power in January 2007.

Overall, corporate globalization policy in Iraq has had the anticipated impact: the economy as a whole is languishing, with high inflation, low growth, and unemployment rates estimated as high as 70 percent in some areas and 30 to 50 percent for the nation. The IMF and World Bank have stepped in to enforce structural adjustment programs in Iraq that mirror much of the Bush administration's corporate globalization agenda, including Iraq's admission to the WTO. Bush Agenda adherents are not likely to give up their pursuit of establishing a corporate free trade haven in Iraq while at the same time helping U.S. corporations extradite themselves without consequence. However, their focus in Iraq is concentrating increasingly on one ultimate goal: winning Iraq's oil prize.

OWNING IRAQ'S OIL

To the adherents of the Bush Agenda, Iraq is truly an oil bonanza in waiting. It is the nation with at least the second largest oil reserves in the world and quite possibly the largest. The oil is right below the surface and bursting at the seams. It is cheap to produce, yet highly valuable to sell. Of Iraq's eighty known oil fields, only seventeen have even begun to be developed. Gaining control of that oil serves several interests. It serves the interests of oil and energy services corporations that both support and comprise the Bush administration, as well as the administration's goals of imperial power and global dominance: whoever controls the oil can deny it to those who do not and dictate the terms on which they receive it. A friendly government in Iraq granting access to its oil provides a support mechanism for the administration's regional interests, including the protection of Israel. And Iraq's oil offers literally trillions of dollars in raw profit.

It should come as little surprise that the Bush administration has spent more than four years trying to gain control of Iraq's oil. The Bremer Orders laid the groundwork for a corporate-friendly haven in Iraq. At the same time, the Bush administration and its oil company cohorts have worked toward the passage of a new oil law for Iraq that would turn its nationalized oil system over to private foreign corporate control. On January 18, 2007, the administration's plans came one step closer to fruition when an Iraqi negotiating committee of national and regional leaders approved a new hydrocarbon law. One month later, the law passed Iraq's cabinet and then moved to the Parliament. As I write, the Parliament is preparing to take up consideration of the law.

The law would represent an unqualified victory for U.S. oil companies. It would transform Iraq from a nationalized oil system all-but-closed to U.S. oil companies, into a commercial industry, all-but-privatized and open to U.S. corporate control.

The Iraq National Oil Company would only have exclusive control of Iraq's seventeen developed fields, leaving two-thirds of Iraq's known fields and all of its as-of-yet undiscovered fields open to foreign control. As under the Bremer Orders, U.S. (and all foreign) companies would not have to invest their earnings in the Iraqi economy, partner with Iraqi companies, hire Iraqi workers, or share new technologies. They could even ride out Iraq's current instability by signing contracts now, while the Iraqi government is at its weakest, and then wait at least two years before even setting foot in Iraq, leaving its oil under the ground when it is most needed to service Iraq's economic development.

The foreign companies will also be offered some of the most corporate-friendly contract terms in the world. The draft law proposes that Iraq use Production Sharing Agreements (PSAs)—the oil industry's preferred model—which grant long-term contracts (twenty to thirty-five years in the case of Iraq), and greater control, ownership, and profits to the companies than other models. The law grants foreign oil companies "national treatment," which means that the Iraqi government cannot give preference to Iraqi oil companies (whether public or privately owned) over foreign-owned companies when it chooses with whom to sign contracts. This provision alone will severely cripple the government's ability to ensure that Iraqis gain as much economic benefit as possible from their oil.

In order to determine "the best model for its future contracts with international oil companies" the Iraqi government has arranged for fact-finding teams from the Iraqi Oil Ministry to visit the United States, Britain, and Norway.

"Why," you may ask, "are the Iraqis turning north for answers rather than, say, next door?" Next door they would find that Kuwait, Iran, and Saudi Arabia all maintain nationalized oil systems and have outlawed foreign control over oil development. None use PSAs but rather hire foreign oil companies as contractors to provide specific services as needed, for a limited duration, without giving the foreign company any direct interest in the oil produced. In fact, a full 90 percent of all the world's oil resources are held under nationalized systems. The United States is the only nation in the world to maintain a virtually fully privatized system.

Instead of looking for answers from its neighbors in the Middle East, Iraq's fact-finding tour is headed straight to those nations whose governments and corporations are putting the most pressure on Iraq to pass the law. It appears, therefore, that the tour was simply another opportunity for these governments and corporations to apply direct pressure on the Iraqis while the law is still being considered.

Oil corporations and the Bush administration want to set back the clock on global ownership of oil. As stated in chapter 5, for much of modern history the world's oil reserves have been owned by a handful of private oil companies. This all began to change in the mid 1960s. Oil-rich nations came together, many as a part of the Organization of Petroleum Exporting Countries (OPEC), and took ownership of their oil by nationalizing their industries. The oil companies simply want to get "their" oil back, and, in the case of Iraq (and possibly Iran), they have found an American president willing to go to war to see this agenda fulfilled.

Most Iraqis remain in the dark about the new oil law. Iraq's oil workers had to travel to Jordan to learn details of the law from the London-based research organization Platform. As a result of the briefing, Iraq's five trade union federations, representing hundreds of thousands of workers, released a public statement rejecting "the handing of control over oil to foreign companies, whose aim is to make big profits at the expense of the Iraqi people, and to rob the national wealth, according to long-term, unfair contracts, that undermine the sovereignty of the state and the dignity of the Iraqi people."[9] They demanded a delay in consideration of any law until all Iraqis could be included in the discussion.

Access to direct information for Iraq's parliamentarians is not much better. In December 2006, I received a phone call from my colleague, Raed Jarrar, an Iraqi-American who directs the Iraq desk at Global Exchange, a human rights advocacy group. A member of the Iraqi parliament had phoned Jarrar earlier that day. The parliamentarian said that "the U.S. government" had told him that if Iraq did not pass the proposed oil law, "the IMF would cut off Iraq's loans and cripple the Iraqi economy." He hoped Jarrar could tell him whether or not this statement was true. Unsure of the answer, Jarrar called me. I told him, "No and no." While the IMF had set out passage of a new hydrocarbons law in its (publicly available) loan conditions for Iraq, this did not mean that the parliament had to pass this particular draft of the law—a draft that was just about the most draconian version possible. Moreover, even if the IMF did cut off Iraq's promised future loans, I continued, the IMF's $705 million (not even half the price of Bechtel's reconstruction contracts in Iraq) was a pittance in comparison to the trillions of dollars the Iraqis would give up from their oil profits by signing the law. Jarrar thanked me, hung up, and called the parliamentarian, who then shared this new information with his colleagues.

The Bush administration and U.S. oil companies have been increasing public pressure on the Iraqis to pass the law. In July 2006, U.S. Energy Secretary Samuel Bodman announced at a press conference in Baghdad that oil executives had told him that their companies would not enter Iraq without passage of the new oil law.[10] A few months later, *Petroleum Economist* magazine reported that U.S. oil companies put passage of the oil law before security concerns as the deciding factor over their entry into Iraq.[11] The big push, however, came with the results of the November midterm elections and the American public's strong statement against the war. House Minority Leader John Boehner (R-OH) commented that he believed his party was tossed out of power largely because of the "people's views of what was happening in Iraq."[12] Time, it seemed, was running out on the administration's oil clock.

In late November, the press began reporting on the president's plan to increase troops in Iraq by some twenty thousand soldiers. Then, on December 6, the highly anticipated report of the bipartisan Iraq Study Group was released. With longtime Bush family favorite James A. Baker III as cochair and Lawrence Eagleburger as a group participant (both of whom I discuss in depth in chapter 5), I was not at all surprised to see that page 1, chapter 1 of the report laid out Iraq's importance to its region, the United States, and the world with this reminder: "It has the world's second-largest known oil reserves."

The report specifically (and publicly) called on the Bush administration to "assist Iraqi leaders to reorganize the national oil industry as a commercial enterprise" and to "encourage investment in Iraq's oil sector by the international community and by international energy companies." Recommendation number 63 also calls on the U.S. government to "provide technical assistance to the Iraqi government to prepare a draft oil law."

This last step was already under way. The Bush administration had hired Bearing Point more than a year earlier to advise the Iraqi Oil Ministry on drafting and passing a new national oil law. The rest of the *Iraq Study Group Report* was ignored, but the oil recommendations were in lock step with the Bush Agenda. The president followed the release of the report with his first public demand of the Iraqi government to pass the oil law. Shortly thereafter, U.S. Ambassador to Iraq Zalmay Khalilzad and General George W. Casey Jr., the senior American commander in Iraq, made the same public demand. That same month, the al-Maliki government in Iraq announced that it would be working to complete the law by year's end. It was unsuccessful.

On January 10, in a nationally televised address, President Bush announced that he was increasing the number of troops on the ground in Iraq by 21,500. The White House also released a fact sheet identifying the president's benchmarks for the Iraqi government, including the enactment of a "hydrocarbons law to promote investment, national unity, and reconciliation." One week later, the law made it through its first official government hurdle in Iraq when it passed out of the drafting committee.

If the law passes, oil corporations will then sign contracts with the Iraqi government. The corporations that appear first in line are ExxonMobil, Chevron, ConocoPhillips, Marathon, Shell, and BP. All of these companies will then need to get to work, but they will require a certain level of security to do so. What better security force is there than 150,000 American soldiers? This oil timeline now dictates the conclusion of the war—at least from the perspective of the Bush administration. It holds our soldiers and the Iraqi people hostage to the Bush Agenda.

Iran may be next. As described in chapter 5, for most of its history until the Iranian revolution in 1979, Iran's oil has belonged to U.S. and British oil companies. The oil companies want the oil back. The administration would also like to see "regime change" reach Iran. It also wants to provide further protection for Israel. President Bush sent two Navy carrier battle groups to the Persian Gulf in October 2006 and January 2007. Also in January, the president announced deployment of Patriot Air Defense Systems to the region "to reassure our friends and allies" while U.S. forces raided an Iranian government office in Iraq. Bush also appointed a naval aviator, Admiral William Fallon, as the new commander of U.S. Central Command in the Middle East. In response, the *New York Times* reported, "Admiral Fallon's appointment comes amid a series of indications that the Bush administration is increasingly focused on putting pressure on Iran and, perhaps, veering toward open confrontation."[13]

For four years, the peace movement has kept the Bush administration out of Iran by demonstrating that the United States simply has had no stomach for a second war in the gulf. With the administration's time running out, however, its stance against Iran is hardening and talk of war has increased.

THE DISAPPEARING FREE TRADE AGENDA

The result of the 2006 midterm elections reflected more than the American public's opposition to the war; it was also a firm stand against bare-knuckled corporate globalization. As Daniel Griswold of the conservative Cato Institute commented, "The election means that the president's trade agenda has come to a screeching halt."[14] The *Wall Street Journal* said it all with this November 11 headline—"Slow Track: Democratic Gains Raise Roadblocks to Free-Trade Push." How is it that the election that ushered into power the party that brought us NAFTA and the WTO was apparently heralding the demise of free trade? The global justice movement is "what happened."

Decades of organizing, educating, lobbying, and protesting described in *The Bush Agenda* worked. Not only have the Democrats in Congress increasingly taken up the banner of "fair" versus "free" trade, but also many of the newly elected members of Congress in 2006 can easily be characterized as global justice advocates themselves. For this reason they are increasingly referred to as a new breed of "economic populists."

Public Citizen, a research and advocacy organization, reported that in eighteen races for the House of Representatives in the midterm elections, "fair traders" replaced "free traders" and not a single "free trader" beat a "fair trade" candidate.[15] In every Senate seat that changed hands, fair traders beat free traders.

The Bush administration succeeded in getting the U.S.-Oman Free Trade Agreement through the House and Senate and signed into law in September 2006. But it was a narrow victory, with 205 members of the House voting against the bill. With the new Congress now in place, the fate of the U.S. FTA with the United Arab Emirates is far less secure. In January 2007, the undersecretary of the UAE's Ministry of Finance and Industry told reporters that with Democrats now controlling Congress, the odds of working out a new free trade agreement with the United States were less likely. That has not stopped Bechtel from benefiting from the mere negotiation of the FTA. The company was hired last year by the Abu Dhabi Ports Co. of the UAE to manage the construction of a major new industrial zone at the country's Khalifa Port. Bechtel's success aside, since passage of the Oman FTA, further negotiations on the U.S.-Middle East Free Trade Area have all but stalled.

A core tenet of the Bush Agenda, the linking of free trade and freedom as two interdependent attributes, is failing. The president argued that those who stood in opposition to free trade were also in opposition to freedom, and anyone in opposition to freedom was automatically considered a terrorist. Fortunately, the peace and global justice movements have largely repelled this particular strain of the president's propaganda.

In fact, across the globe, peoples' movements for global justice have swept in elected officials who represent their views. These elected officials have brought their resistance to free trade and the Bush Agenda into the institutions of corporate globalization. Peoples' movements across Central and South America have completely rewritten the political map with the election of radical global justice leaders in Bolivia, Argentina, Venezuela, Brazil, Ecuador, and Nicaragua. As Lori Wallach and Deborah James of Washington, DC's Public Citizen write, "Even Costa Rica, Peru, and Mexico, traditionally neoliberal strongholds, have experienced presidential elections almost entirely dominated by debate over trade liberalization."[16]

Remember Bolivia's water warriors from chapter 3? One of them, Evo Morales, is now the President of Bolivia. He has joined with Venezuela, Cuba, Nicaragua, and Ecuador not only to oppose the Bush free trade agenda, but also to launch an alternative trade association called the Bolivarian Alternative for the Americas (ALBA in Spanish), based upon cooperation, solidarity, and complementarity. ALBA includes provisions in which member nations collaborate in industrial and infrastructure development projects and the greatest benefits are provided to the smallest economies.

Five years and three failed ministerial meetings later, the World Trade Organization negotiations launched in Doha are all but over, with nations unable to agree on any of the key negotiating topics. Most believe the end of the Doha Round is sounding the death knell for the WTO as an institution. One reason for its demise is the increasing number of developing countries whose leaders are now opposed to corporate globalization.

Meanwhile, the IMF and World Bank are being increasingly sidelined with countries refusing to repay loans, take new ones, or make their contributions to the institutions. The presence of Paul Wolfowitz has done nothing to bolster the power of the World Bank and may, in fact, have backfired due to increasing global opposition to the Bush administration, its policies—particularly the war in Iraq, and Wolfowitz's deep connection to all three.

HOW WE GOT HERE—MOVING FORWARD

Every one of the 15 million people who marched on Saturday, February 15, 2003, woke up on March 19, 2003, devastated to learn that, in spite of their efforts, the Bush administration invaded Iraq. Many stayed home and turned their backs on activism. But many others did not. They continued organizing and raised their voices against the administration and its war. Some joined grassroots campaigns already under way with peace groups like Code Pink to demand a "pink slip" for Donald Rumsfeld. Some worked to defeat Bush and Cheney in the 2004 presidential election.

When, on November 4, they learned that their efforts had failed, again many went home and turned their backs on activism. But still many others did not. They kept protesting, lobbying, writing, reading, arguing, refusing to fight in the war, resigning from a government that supported the war, and learning. In 2004, more groups joined the campaign to demand that Rumsfeld be fired. In May, people across the country joined the Win Without War Coalition, moveon.org, and a network of religious and human rights organizations in a "Civic Action Campaign to Fire Rumsfeld and Change Course." Even when this call was joined by military publications, U.S. generals, and members of Congress, the president refused to relent. Many people went home defeated.

Many others turned their attention to the 2006 midterm elections. On November 4, they emerged victorious when the nation voted against the president's party, his war, and his agenda. On November 8, Donald Rumsfeld was handed his pink slip.

In fact, the failure of the Bush Agenda is evident in the long list of its key architects and advocates who are no longer a part of the administration: Paul Wolfowitz, Donald Rumsfeld, Scooter Libby, Richard Perle, Douglas Feith, Robert Zoellick, and John Bolton. Francis Fukuyama, a father of neoconservatism, a founding member of PNAC, and an early Bush administration adviser, was ultimately so appalled at the administration that he wrote *America at the Crossroads: Democracy, Power, and the Neoconservative Legacy*, denouncing both the administration and the very term "neoconservative" so he would never again be associated with either.

The midterm elections made it clear that the American public had listened to the protestors, to the evidence exposing the administration's false claims for launching an illegal war, to the soldiers returning from battle and those who refused to fight, to the families who lost loved ones, to the Iraqis suffering the costs of war, to the Bush administration's own findings that the war is increasing the risk of terrorism against the United States, and to the news that day in and day out revealed the war's failure.

In the history of social change, three years, eight months, and eighteen days is a very short time for an activist "cause" to have a major political "effect." The election did more than usher in a Democratic majority in both the House and the Senate: it renewed a feeling of empowerment across America. People who firmly believed that their actions could not possibly make a dent in the Bush corporate machine suddenly proved themselves wrong. The oil companies, the defense contractors, the engineering giants, and other powerful actors all put their money behind the Republicans, but the Republicans lost anyway. A sense of helplessness has been replaced by one of excitement and expectation. Millions of people across the nation are now asking, "What can we do next?"

To be sure, the Bush administration has not given up on its agenda of corporate globalization, imperialism, and war. But political realities are forcing it to hone its ambitions. Controlling oil tops the list now and doing so will require the continuation of one war and quite possibly the start of another. Nor will the adherents of the Bush Agenda stop their pursuit simply because President Bush is disempowered or leaving office; after all, many are still "winning." According to their respective annual reports, in 2006, Chevron's $17 billion in profits were 22 percent higher than in 2005; Lockheed Martin's shares jumped an astounding 43 percent in value; and Bechtel's profits, while not yet posted, are following the same trajectory. For its part, ExxonMobil, for the third straight year, broke its own record and earned the highest profits of any corporation in world history—this time with nearly $40 billion in pure profit. The corporations are not going to stop trying to use this unparalleled wealth to purchase officials to do their bidding and policies to serve their interests.

We must therefore congratulate ourselves for all we have done to take down and replace the Bush Agenda and acknowledge that our efforts are working. Then we must give ourselves a collective shove to get up to do more. As I told congresswomen Maxine Waters and Lynn Woolsey as I participated in a congressional briefing on the war in Iraq in January, the next time we bring half a million people to Washington, DC, and surround the United States capitol, we're going to sit down, bring our sleeping bags, and stay until the war in Iraq *stops* raging on.

NOTES

Chapter One: The Bush Agenda

1. *CBS News, New York Times Poll,* September 9–13, 2005.
2. Constanza Vieira, "Talks Between US, Andean Countries in Final Stretch," *InterPress New Service,* September 21, 2005.
3. *Global Trends 2015: A Dialogue about the Future with Nongovernment Experts,* approved for publication by the National Foreign Intelligence Board under the authority of the Director of Central Intelligence, NIC 2000-02, December 2000.
4. The Center for Responsive Politics, www.opensecrets.org.
5. Tom Petruno, "Taking Aim at Oil's Riches," *Los Angeles Times,* October 26, 2005; "ExxonMobil Profits Exceed $25 billion," *BBC News—World Edition,* January 31, 2005; and "ExxonMobil Sees Record Profit for U.S. Company," *Associated Press,* January 30, 2006.
6. Office of Trade and Industry Information, Manufacturing and Services, International Trade Administration, U.S. Department of Commerce, data collected December 19, 2005.
7. Luis Hernández Navarro, "Mr. Lee Kyung Hae," *La Jornada* (Mexico), September 23, 2003.
8. Anuradha Mittal, "Food First Daily Report from the WTO in Cancun #4," *Institute for Food and Agriculture Policy,* September 11, 2003.
9. Anuradha Mittal, "Food Sovereignty: A New Farm Economy to Challenge Economic Globalization," *Oakland Institute,* July 26, 2004.
10. *Third World Network,* Cancun News Update, September 14, 2003.
11. Robert Zoellick, "Countering Terror with Trade," *Washington Post,* September 20, 2001.
12. Quoted in E. J. Dionne, "Trade and Terror," *Washington Post,* October 2, 2001.
13. President George W. Bush, address before a joint session of the Congress on the State of the Union, January 29, 2002.

Chapter Two: Ambitions of Empire

1. "Rebuilding America's Defenses: Strategy, Forces, and Resources for a New Century," a report of the Project for the New American Century, September 2000.
2. Arthur M. Schlesinger Jr., *War and the American Presidency* (New York: W.W. Norton, 2004), p. 65.

3. Robert D. Kaplan, "Supremacy by Stealth: Ten Rules for Managing the World," *Atlantic Monthly,* July/August, 2003.

4. Edward Gibbon, *The Decline and Fall of the Roman Empire,* abridged (New York: Dell, 1963), p. 66.

5. Ibid., p. 33.

6. Steven Kreis, "Augustus Caesar and the *Pax Romana,*" The History Guide, Lectures on Ancient and Medieval European History, www.historyguide.org/ancient/lecture12b.html, May 13, 2004.

7. See note 4, p. 53.

8. John A. Garraty and Peter Gay, ed., *The Columbia History of the World* (New York: Harper & Row, 1972), p. 213.

9. Christopher Madison, "Haig's Planning Chief Finds Rewards, Risks in Helping Keep State Straight," *National Journal,* April 10, 1982.

10. Michael Goldsmith, "Military Cooperation Talks Went Well," *Associated Press,* April 27, 1982.

11. James Mann, *Rise of the Vulcans: The History of Bush's War Cabinet* (New York: Penguin, 2004), p. 172.

12. David Armstrong, "Dick Cheney's Song of America: Drafting a Plan for Global Dominance," *Harper's Magazine,* October 2002.

13. Kevin Phillips, *American Dynasty: Aristocracy, Fortune, and the Politics of Deceit in the House of Bush* (New York: Viking Books, 2004), p. 157.

14. Barton Gellman, interviewed on "Frontline," *PBS,* January 29, 2003.

15. Barton Gellman, "Keeping the U.S. First," *Washington Post,* March 11, 1992.

16. See note 12.

17. Zalmay Khalilzad, "Afghanistan: Time to Reengage," *Washington Post,* October 7, 1996.

18. Joe Stephens and David Ottaway, "Afghan Roots Keep Adviser Firmly in the Inner Circle," *Washington Post,* November 23, 2001.

19. "Rebuilding America's Defenses," a report of the Project for the New American Century, September 2000.

20. "Letter to the Honorable William J. Clinton" from eighteen members of the Project for the New American Century, January 26, 1998.

21. See note 13, p. 140.

22. George Lardner Jr. and Lois Romano, "Bush Name Helps Fuel Oil Dealings," *Washington Post,* July 30, 1999.

23. Craig Unger, *House of Bush House of Saud* (New York: Scribner, 2004), p. 120.

24. See note 13, p. 113.

25. Charles Lewis, *The Buying of the President 2004* (New York: Perennial, 2004), p. 4.

26. Bob Woodward, *Plan of Attack* (New York: Simon & Schuster, 2004), p. 203.

27. Carl von Clausewitz, *On War* (New York: Penguin, 1968), pp. 102, 123.

28. Joseph Kahn, "Globalization Proves Disappointing," *New York Times,* March 21, 2002.

Chapter Three: A Model for Failure: Corporate Globalization

1. "Globalizing Poverty," *Ecologist Report,* September 2000.
2. R. Bruce Craig, *Treasonable Doubt: The Harry Dexter White Spy Case* (Lawrence, KS: University of Kansas Press, 2004), p. 146.
3. Ibid., p. 140.
4. James Boughton, "Harry Dexter White and the International Monetary Fund," *Finance & Development: A Quarterly Magazine of the IMF,* September 1998, vol. 35, no. 3.
5. Daniel Drache, "The Short but Significant Life of the International Trade Organization," *Robarts Centre for Canadian Studies,* November 2000.
6. Walden Bello, "Building an Iran Cage," in *Views from the South: The Effects of Globalization and the WTO on Third World Countries,* Sarah Anderson, ed., Institute for Food and Development Policy, 2000.
7. Quoted in Daniel Yergin, *The Prize: The Epic Quest for Oil, Money, and Power* (New York: Simon & Schuster, 1991), p. 662.
8. Ibid., p. 625.
9. Ibid., p. 685.
10. Jim Vallette and Stephen Kretzmann, "The Energy Tug of War: The Winner and Losers of World Bank Fossil Fuel Finance," Sustainable Energy and Economy Network/Institute for Policy Studies, April 2004.
11. Ibid.
12. William T. Onorato, "Legislative Frameworks Used to Foster Petroleum Development," *World Bank,* February 1995.
13. William Easterly, "The Failure of Development," *Financial Times,* July 4, 2001.
14. Mattis Lundberg and Lyn Squire, "The Simultaneous Evolution of Growth and Inequality," *World Bank,* 1999.
15. Lishala Situmbeko and Jack Jones Zulu, "Zambia: Condemned to Debt," *World Development Movement,* April 2004.
16. Ibid.
17. Ibid.
18. D. Logie, "Health in Zambia and the UN AIDS Conference in Lusaka: Report on a Meeting Held on 09/02/00," University of Glasgow, Section of General Practice and Primary Care, 2000; quoted in Lishala Situmbeko and Jack Jones Zulu, "Zambia: Condemned to Debt," *World Development Movement,* April 2004.
19. "Women Standing Up to Adjustment in Africa," a Report of the *African Women's Economic Policy Network,* July 1996.
20. Robin Broad and John Cavanagh in *Alternatives to Economic Globalization: A Better World Is Possible,* John Cavanagh and Jerry Mander, eds. (San Francisco, CA: Berrett-Koehler, 2004), p. 58.
21. "Russian Oil Reserves Three Times Higher," *Moscow News,* April 30, 2004.
22. David M. Kotz, "Russia and the Crisis of Neoliberalism," *Z Magazine,* January 1999.

23. Joseph Stiglitz, "What I Learned at the World Economic Crisis," *New Republic,* April 6, 2000.
24. Ibid.
25. Mark Weisbrot, testimony before the General Oversight and Investigations Subcommittee, House of Representatives, Committee on Banking and Financial Services, September 10, 1998.
26. See note 23.
27. See note 22.
28. See note 20, p. 59.
29. Patricio McCabe, "Argentina's New Forms of Resistance," in *Globalize Liberation: How to Uproot the System and Build a Better World,* David Solnit, ed. (San Francisco, CA: City Lights Books, 2004), p. 344.
30. "Is Wal-Mart Good for America?" *Frontline,* November 16, 2004.
31. Evelyn Iritani, Nancy Cleeland, and Tyler Marshall, "Scouring the Globe to Give Shoppers an $8.63 Polo Shirt," *Los Angeles Times,* November 24, 2003.
32. *Wal-Mart Fact,* provided by Wal-Mart, www.walmartfacts.com.
33. Jeff Faux, "NAFTA at 10," *Nation,* February 2, 2004.
34. Public Citizen, *NAFTA at Ten Series,* January, 2004.
35. David Leonhardt, "Two-Tier Marketing," *Business Week,* March 17, 1997.
36. "Dumping Without Border," briefing paper, *Oxfam International,* August 2003.
37. Tim Weiner, "Wal-Mart Invades, and Mexico Gladly Surrenders," *New York Times,* December 6, 2003.
38. See note 30.
39. See note 31.
40. Patrick Bond, "ANC Privatizations Fail to Deliver in South Africa," *CorpWatch,* August 18, 2004.
41. Andrew Nowicki, "What Went Wrong in 'the New South Africa,' " *Z Magazine,* October 22, 2003.
42. Patrick Bond, "From Racial to Class Apartheid: South Africa's Frustrating Decade of Freedom," *Monthly Review,* November 10, vol. 55.
43. Ibid.
44. Ibid.
45. Ibid.
46. "Great Leap into Stagnation Courtesy of World Bank," *Bloomberg News Service,* April 25, 2002.

Chapter Four: The Corporations: Bechtel, Chevron, Halliburton, and Lockheed Martin

1. John Gibson, "Sustainability," *The Leading Edge,* Society of Exploration Geophysicists, February 2004.

2. Sarah Anderson, John Cavanagh, et al., "Executive Excess 2004," Institute for Policy Studies and United for a Fair Economy, Washington, DC, August 31, 2004.
3. Edward Wolff, "Recent Trends in Wealth Ownership," *Jerome Levy Economics Institute,* Bard College, 2000.
4. Sarah Anderson, John Cavanagh, Scott Klinger, and Liz Stanton, "Executive Excess 2005," Institute for Policy Studies and United for a Fair Economy, Washington, DC, 2005.
5. Center on Budget and Policy Priorities, "Economic Recovery Failed to Benefit Much of the Population in 2004," August 30, 2005; and Holly Sklar, "Growing Gulf Between Rich and Rest of Us," *Knight Ridder/Tribune Information Services,* October 3, 2005.
6. Robert Wade, "Winner and Losers," *Economist,* April 28, 2001.
7. John B. Judis, "Below the Beltway," *American Prospect,* May 2003.
8. Company financial data and U.S. General Services Administration, Federal Procurement Data System, Fiscal Year 2004 Contracts Performed in Iraq.
9. Daniel Yergin, *The Prize: The Epic Quest for Oil, Money, and Power* (New York: Simon & Schuster, 1991), p. 147.
10. The Paleontological Research Institution of Ithica, NY, www.priweb.org/ed/pgws/history/signal_hill/signal_hill.html.
11. See note 9, p. 82.
12. Earth Rights International, "Bowoto v. ChevronTexaco," September 2, 2005.
13. Ibid.
14. Amy Goodman, *The Exception to the Rulers* (New York: Hyperion, 2004), p. 78.
15. See note 12.
16. Rick Jurgens, "Chevron Scrutinized for Role in Nigeria," *Contra Costa Times,* May 18, 2003.
17. Transcript of Oral Argument on Chevron's Motion to Dismiss Plaintiff's Motion for Leave to Amend Complaint, U.S. District Court, Northern District of California, Charles A. Legge, Judge, May 12, 2000.
18. D'arcy Doran, "Poverty Spurs Nigeria Oil Standoff," *Associated Press,* July 17, 2002.
19. "Oil Deal 'Off' Nigerian Women Say," *British Broadcasting Company,* July 16, 2002.
20. See note 18.
21. Lawsuit for Alleged Damages filed before the President of the Superior Court of Nueva Loja in Lago Agrio, Province of Sucumbios, on May 7, 2003, by forty-eight inhabitants of the Orellana and the Sucumbio Pronvince.
22. Results of several studies, including Anna-Karin Hurtig and Miguel San Sebastián, "Geographical Differences in Cancer Incidence in the Amazon Basin of Ecuador in Relation to Residence Near Oil Fields," *International Journal of Epidemiology,* 2002, vol. 31, pp. 1021–1027, 2002.
23. Bob Herbert, "Rain Forest Jekyll and Hyde?" *New York Times,* October 20, 2005.
24. Letter to Edward B. Scott, Chevron Vice President and Legal Council, from Sarah C. Aird, Amazon Watch Legal Counsel, October 25, 2005.
25. Jad Mouawad, "Big Oil's Burden of Too Much Cash," *New York Times,* February 12, 2005.

26. Dan Briody, *The Halliburton Agenda: The Politics of Oil and Money* (Hoboken, NJ: John Wiley & Sons, 2004), pp. 3–12.

27. Jane Mayer, "Contract Sport," *New Yorker,* February 16, 2004.

28. "Halliburton's Cheney Sees Worldwide Opportunities, Blasts Sanctions," *Petroleum Finance Week,* April 1, 1996.

29. "Nigeria Bans Halliburton from New Contracts on Safety Concerns," *Halliburton Watch,* September 20, 2004.

30. Simon Romero and Craig Smith, "Halliburton Severs Link with 2 Over Nigeria Inquiry," *New York Times,* June 19, 2004.

31. Russell Gold, "Halliburton Uncovers Talk of Bribes," *Wall Street Journal,* September 2, 2004.

32. Nigeria House of Representatives Petition Committee, *Interim Report: The Halliburton/TSKJ/LNG Investigation, Summary of Facts,* September 2004.

33. Anthony Barnett and Martin Bright, "Cheney in Firing Line over Nigerian Bribery Claim," *London Observer,* June 20, 2004.

34. Doug Ireland, "Tricky Dick," *Nation,* August 21, 2000.

35. James Risen, "Cheney's Path: From Gulf War to Mideast Oil; In Business, He Benefited from His Pentagon Days," *International Herald Tribune,* July 28, 2000.

36. Laton McCartney, *Friends in High Places: The Bechtel Story* (New York: Ballantine Books, 1988), p. 80.

37. Ibid., pp. 18–22.

38. Book referenced in "The Earth Wrecker," by Pratap Chaterjee, *San Francisco Bay Guardian,* May 31, 2000.

39. "Incompetence, Wheeling and Dealing, the Real Bechtel," *Multinational Monitor,* August 1989.

40. Clifford K. Beck, director, Office of Government Liaison, U.S. Nuclear Regulatory Commission, "Situation in Tarapur," memorandum, December 27, 1972.

41. See note 39.

42. Pratap Chaterjee, "Bechtel's Nuclear Nightmares," special to *CorpWatch,* May 1, 2003.

43. See note 36, p. 225.

44. See note 42.

45. Bill Bartleman, "Company Eager to Begin Cleanup," *Paducah Sun,* October 31, 1999.

46. See note 42.

47. Testimony, Pamela Gillis Watson, U.S. Senate Governmental Affairs Committee Hearings on Oak Ridge, Tennessee, and Portsmouth, Ohio, Department of Energy Gaseous Diffusion Plants, March 29, 2000.

48. E-mail from Dennis Hill, Bechtel Jacobs Co., LLC, Media Relations, to author, December 14, 2005.

49. Micah L. Sifry and Christopher Cerf, eds., *The Iraq War Reader* (New York: Touchstone, 2003), pp. 132, 139.

50. Tim Weiner, "A Vast Arms Buildup, Yet Not Enough for Wars," *New York Times,* October 1, 2004.
51. E-mail correspondence with Frida Berrigan of the Arms Trade Resource Center, June, 2005.
52. Amy Goodman, "The Price of Nuclear Weapons: The Case of Paducah, Kentucky," *Democracy Now!* interview, September 1, 1999.
53. Senator Grassley letter to Spencer Abrams, August 2002.
54. See note 52.
55. Elian Robbins, "Lockheed Martin Exposes Workers to Plutonium; NRDC Sues," *Amicus Journal,* Winter 2000.
56. Joby Warrick, "In Harm's Way, But in the Dark," *Washington Post,* August 8, 1999.
57. Ibid.
58. "Weapons of Mass Destruction Discovered at Lockheed Martin," *Direct Action to Stop the War Press Release,* April 22, 2003.

Chapter Five: "A Mutual Seduction": Turning Toward Iraq

1. Kenneth T. Derr, "Engagement—A Better Alternative," speech to the Commonwealth Club of California, San Francisco, CA, November 5, 1998.
2. "Halliburton Execs Want More Work in Iraq," *Oil Daily,* May 8, 2003.
3. Michael A. G. Bunter, "Early Concessions in Iraq and the Middle East," *Oil, Gas, & Energy,* vol. 1, no. 01.
4. Edward Hudson, "C-130's on the Market for the Airlines," *New York Times,* August 30, 1964.
5. Phebe Marr, *The Modern History of Iraq,* 2nd ed. (Boulder, CO: Westview Press, 2004), p. 102.
6. Dilip Hiro, *Iraq: In the Eye of the Storm* (New York: Nation Books, 2002), pp. 25–31.
7. David Morgan, "Ex-U.S. Official Says CIA Aided Baathists," *Reuters,* April 20, 2003.
8. Daniel Yergin, *The Prize: The Epic Quest for Oil, Money and Power* (New York: Simon & Schuster, 1991), p. 596.
9. Ibid., p. 625.
10. Michael C. Jensen, "Attacks on Oil Industry Grow Fiercer," *New York Times,* February 3, 1973.
11. "More Layoffs at Ford," *New York Times,* January 25, 1974, and Ibid.
12. See note 10.
13. William Robbins, "Oil Profits Up 46% on 6% Volume Rise," *New York Times,* January 22, 1973.
14. Richard Witkin, "Iran Offers To Fully Back C-5A Output Resumption," *New York Times,* December 2, 1974.
15. "Lockheed Experts Will Train Iranians Under U.S. Project," *New York Times,* November 26, 1976.

16. Robert Bryce, *Cronies: Oil, the Bushes and the Rise of Texas, America's Superstate* (New York: Perseus Publishing, 2004), pp. 120, 124.
17. See note 8, p. 672.
18. Ibid., p. 685; and Joel Jacobson, "Government and the Oil Industry: The Myth and the Reality," *New York Times,* January 11, 1981.
19. Guy Gugliotta, Charles R. Babcock, and Benjamin Weiser, "At War, Iraq Courted U.S. into Economic Embrace," *Washington Post,* September 16, 1990.
20. Dilip Hiro, *The Essential Middle East* (New York: Carroll & Graf Publishers, 2003), pp. 154–157.
21. Jim Vallette, Stephen Kretzmann, and Daphne Wysham, "Crude Vision: How Oil Interests Obscured U.S. Government Focus on Chemical Weapons Use by Saddam Hussein," *Sustainable Energy and Economy Network/Institute for Policy Studies,* March 24, 2003.
22. Joe Conason, "The Iraq Lobby: Kissinger, the Business Forum & Co.," in *The Gulf War Reader,* Micah L. Sifry and Christopher Cerf, eds. (New York: Random House, 1991).
23. See note 19.
24. Robert D. Hershey Jr., "Fledgling U.S.-Iraqi Trade Group Says It Feels Betrayed and Embarrassed," *New York Times,* August 20, 1990.
25. See note 22.
26. Ibid.
27. Leslie Gelb, "Kissinger Means Business," *New York Times,* April 20, 1986.
28. Ibid.
29. Don Oberdorfer, "Amid Praise for Eagleburger, Ties to Kissinger Clients Questioned," *Washington Post,* March 16, 1989.
30. "Angered by Questions on Consulting Firm, Journalists' Probing Outrages Kissinger," *Los Angeles Times,* March 23, 1989.
31. Alan Friedman, *Spider's Web: The Secret History of How the White House Illegally Armed Iraq* (New York: Bantam Books, 1993), p. 163.
32. See note 19.
33. See note 21.
34. See note 16, pp. 126–130.
35. Murray Waas, "What Washington Gave Saddam for Christmas," in *The Gulf War Reader,* Micah L. Sifry and Christopher Cerf, eds. (New York: Random House, 1991).
36. Bruce Jentleson, *With Friends Like These: Reagan, Bush, and Saddam, 1982–1990* (New York: W.W. Norton, 1994), p. 50.
37. Ibid., pp. 44–46; see note 31, pp. 60–127.
38. See note 35.
39. See note 36, p. 42.
40. See note 31, p. 141.
41. See note 36, p. 116 and note 31, p. 117.

42. See note 31, p. 171; and Robert Bryce, *Cronies,* pp. 85, 205.
43. Ibid., p. 133.
44. Ibid., p. 157.
45. James Mann, *Rise of the Vulcans: The History of Bush's War Cabinet* (New York: Penguin Books, 2004), p. 183.
46. See note 31, pp. 161, 144.
47. Bob Woodward, *The Commanders* (New York: Simon & Schuster, 1991), p. 211.
48. See note 31, p. 166.
49. R. W. Apple Jr., "Bush Invokes U.S. Values," *New York Times,* August 16, 1990.
50. See note 31, p. 173.
51. See note 16, pp. 161–162.
52. See note 45, p. 192.
53. Christian Parenti, *The Freedom: Shadows and Hallucinations in Occupied Iraq* (New York: New Press, 2004), p. 15.
54. See note 45, p. 190.
55. Paul Wolfowitz, "Victory Came Too Easily," *National Interest,* Spring 1994.
56. Colum Lynch, "Firm's Iraq Deals Greater than Cheney Has Said," *Washington Post,* June 23, 2001.
57. Carola Hoyos, "A Discreet Way of Doing Business with Iraq," *Financial Times,* November 3, 2000.
58. Henry Kissinger, "Our Shilly-Shally 'Strategy' on Saddam," *Washington Post,* March 23, 1998.
59. See note 47, p. 39.

Chapter Six: The Economic Invasion of Iraq

1. Anonymous, *Imperial Hubris: Why the West Is Losing the War on Terror* (Dulles, VA: Potomac Books, 2004), p. xvi.
2. *National Development Strategy 2005–2007,* Republic of Iraq, Iraqi Strategic Review Board, Ministry of Planning and Development, June 30, 2005.
3. David Leigh, "General Sacked by Bush Says He Wanted Early Elections," *Guardian* (United Kingdom), March 18, 2004.
4. Greg Palast, "Iraq for Sale," *BBC* television interview with Jay Garner, March 14, 2004.
5. President George W. Bush, remarks by the president from the USS *Abraham Lincoln,* May 1, 2003.
6. Lakhdar Brahimi, statement made at UN press conference, June 2, 2004.
7. Coalition Provisional Authority Regulation Number 1, "Coalition Provisional Authority," CPA/REG/16 May 2003/01, L. Paul Bremer, Administrator, Coalition Provisional Authority.

8. Barbara Slavin, "U.S. New Transition Chief 'Brings a Lot to the Table,'" *USA Today*, May 13, 2003.
9. Paul Bremer, "New Risks in International Business," *Viewpoint*, no. 2, 2001, Marsh and McLennan Companies.
10. Chris Foote, "Reviving the Iraqi Economy in the Aftermath of War," *Regional Review*, 2003, Q3, vol. 13, no. 3.
11. Dan Baum, "Nation Builders for Hire," *New York Times*, June 22, 2003.
12. Award/Contract, Bearing Point, Inc. and USAID/Iraq, "Technical Assistance for Economic Recovery, Reform, and Sustained Growth in Iraq," July 18, 2003; Darwin G. Johnson, senior vice president, Bearing Point, Inc., Anne Quinlan, contracting officer, USAID.
13. Laura Peterson, "Outsourcing Government," *Center for Public Integrity*, October 30, 2003.
14. "Stimulating Economic Recovery, Reform, and Sustained Growth in Iraq," Statement of Work, Bearing Point Inc., February 21, 2003.
15. The Coalition Provisional Authority, CPA Official Documents, www.iraqcoalition.org /regulations/index.html#Regulations.
16. John Barry and Evan Thomas, "The Unbuilding of Iraq," *Newsweek*, October 6, 2003.
17. Report on Development in Iraq, *UNICEF*, 2002.
18. United Nations Annex II of S/1999/356, 30 March 1999, report of the second panel established pursuant to the note by the president of the Security Council of 30 January 1999 (S/1999/100) concerning the current humanitarian situation in Iraq.
19. Coalition Provisional Authority Order Number 86, "Traffic Code," CPA/ORD/20 March 2004/86, L. Paul Bremer, Administrator, Coalition Provisional Authority.
20. The Interim Constitution of Iraq, 1990, translated into English by the International Constitutional Law Project, www.oefre.unibe.ch/law/icl/iz01000_.html.
21. Coalition Provisional Authority Order Number 1, "De-Ba'athification of Iraqi Society," CPA/ORD/16 May 2003/01, L. Paul Bremer, Administrator, Coalition Provisional Authority.
22. Coalition Provisional Authority Order Number 2, "Dissolution of Entities," CPA/ORD/23 May 2003/02, L. Paul Bremer, Administrator, Coalition Provisional Authority.
23. Peter Slevin, "Wrong Turn at Postwar Crossroads?" *Washington Post*, November 20, 2003.
24. "Iraqi Disarmament, Demobilization, and Reintegration Program," contract #DASW01-03-P-0366, Ronco Consulting Corporation and Defense Contracting Command-Washington, effective date, March 14, 2003.
25. Iraq Index, Brookings Institution, January 2006.
26. Coalition Provisional Authority Order Number 12, "Trade Liberalization Policy," CPA/ORD/7 June 2003/12, L. Paul Bremer, Administrator, Coalition Provisional Authority.

27. Coalition Provisional Authority Order Number 54, "Trade Liberalization Policy 2004," CPA/ORD/24 February 2004/54, L. Paul Bremer, Administrator, Coalition Provisional Authority.
28. "Report from Iraq: Working Conditions and Labor Rights Under the Occupation," *U.S. Labor Against the War,* October 2003.
29. Ariana Eunjung Cha, "Iraqis Face Tough Transition to Market-Based Agriculture," *Washington Post,* January 22, 2004.
30. Dahr Jamail, "Bechtel's Dry Run: Iraqis Suffer Water Crisis," *Public Citizen,* April 2004.
31. Coalition Provisional Authority Order Number 14, "Prohibited Media Activity," CPA/ORD/10 Jun 2003/14, L. Paul Bremer, Administrator, Coalition Provisional Authority.
32. Nimrod Raphael, "Understanding Muqtada al-Sadr," *Middle East Quarterly,* Fall 2004, vol. 11. no. 4.
33. Coalition Provisional Authority Order Number 17 (Revised), "Status of the Coalition Provisional Authority, MNF—Iraq, Certain Missions and Personnel in Iraq," CPA/ORD/27 June 2004/17, L. Paul Bremer, Administrator, Coalition Provisional Authority.
34. Kamal Ahmend, "Iraqis Lose Right to Sue Troops over War Crimes," *Observer,* May 23, 2004.
35. Coalition Provisional Authority Order Number 37, "Tax Strategy for 2003," CPA/ORD/19 September 2003/37, L. Paul Bremer, Administrator, Coalition Provisional Authority.
36. Coalition Provisional Authority Order Number 49, "Tax Strategy of 2004," CPA/ORD/19 February 2004/49, L. Paul Bremer, Administrator, Coalition Provisional Authority.
37. Dana Milbank and Walter Pincus, "U.S. Administrator Imposes Flat Tax System on Iraq," *Washington Post,* November 2, 2003.
38. Coalition Provisional Authority Order Number 40, "Bank Law," CPA/ORD/19 September 2003/40, L. Paul Bremer, Administrator, Coalition Provisional Authority.
39. Coalition Provisional Authority Order Number 94, "Bank Law of 2004," CPA/ORD/6 June 2004/94, L. Paul Bremer, Administrator, Coalition Provisional Authority.
40. Jessica Woodroffe, "GATS: A Disservice to the Poor," *World Development Movement,* January 2002.
41. Coalition Provisional Authority Order Number 62, "Disqualification from Public Office," CPA/ORD/26 Feb 2004/62, L. Paul Bremer, Administrator, Coalition Provisional Authority.
42. Coalition Provisional Authority Order Number 65, "Iraqi Communications and Media Commission," CPA/ORD/20 March 2004/65, L. Paul Bremer, Administrator, Coalition Provisional Authority.
43. Coalition Provisional Authority Order Number 57, "Iraqi Inspectors General," CPA/ORD/5 February 2004/57, L. Paul Bremer, Administrator, Coalition Provisional Authority.

44. Phebe Marr, *The Modern History of Iraq*, 2nd ed. (Boulder, CO: Westview Press, 2004), p. 27.
45. Coalition Provisional Authority Order Number 77, "Board of Supreme Audit," CPA/ORD/18 April 2004/77, L. Paul Bremer, Administrator, Coalition Provisional Authority.
46. Coalition Provisional Authority Order Number 80, "Amendment to the Trademarks and Descriptions Law No 21 of 1957," CPA/ORD/26 April 2004/80, L. Paul Bremer, Administrator, Coalition Provisional Authority.
47. Coalition Provisional Authority Order Number 81, "Patent, Industrial Design, Undisclosed Information, Integrated Circuits and Plant Variety Law," CPA/ORD/26 April 2004/81, L. Paul Bremer, Administrator, Coalition Provisional Authority.
48. Coalition Provisional Authority Order Number 83, "Amendment to the Copyright Law," CPA/ORD/20 May 2004/83, L. Paul Bremer, Administrator, Coalition Provisional Authority.
49. Coalition Provisional Authority Order Number 97, "Political Parties and Entities Law," CPA/ORD/7 June 2004/97, L. Paul Bremer, Administrator, Coalition Provisional Authority.
50. Coalition Provisional Authority Order Number 100, "Transition of Laws, regulations, Orders, and Directives Issued by the Coalition Provisional Authority," CPA/ORD/28 June 2004/100, L. Paul Bremer, Administrator, Coalition Provisional Authority.
51. Coalition Provisional Authority Order Number 39, "Foreign Investment," CPA/ORD/19 September 2003/39, L. Paul Bremer, Administrator, Coalition Provisional Authority.
52. David Bacon, "Um Qasr—From National Pride to War Booty," *CorpWatch*, December 15, 2003.
53. See note 9.
54. Clayton Hirst, "Iraqis Investigate Halliburton over Allegations of Bribery," *Independent*, April 25, 2004.
55. Daud Salman, "Iraq Draws Up Plan to Privatize State-Owned Firms: Foreign Investors Solicited for Cash," *Beirut Daily Star*, May 17, 2005.
56. *National Development Strategy 2005–2007*, Republic of Iraq, Iraqi Strategic Review Board, Ministry of Planning and Development Cooperation, June 30, 2005.
57. Opening remarks by Her Excellency Jowan Masum, Iraqi minister of communications, June 28, 2005, available on the Iraq Development Program website, http://www.iraqdevelopmentprogram.org.
58. "Overview of Commercial Law in Pre-War Iraq" (draft), prepared by the Office of the Chief Counsel for International Commerce in the Office of General Counsel at the U.S. Department of Commerce, September 12, 2003.
59. Sheila McNulty, "Working in Iraq Boosts Income at US Group," *Financial Times*, April 29, 2004.
60. The Center for Responsive Politics, opensecrets.org, online database.
61. Riverbend, *Baghdad Burning: Girl Blog from Iraq* (New York: Feminist Press, 2005), p. 36.

62. James Glanz and Erik Eckholm, "Reality Intrudes on Promises in Rebuilding of Iraq," *New York Times,* June 30, 2004.
63. See note 28.
64. Russell Gold, "The Temps of War: Blue-Collar Workers Ship Out for Iraq," *Wall Street Journal,* February 25, 2003.
65. See note 25.
66. Dan Baum, "Nation Builders for Hire," *New York Times,* June 22, 2003.
67. Dahr Jamail, "Her Name Is Ahlam Abt Al-Hassan," *Dahr Jamail's Iraq Dispatches,* May 19, 2005.
68. International Republic Institute poll, February 27–March 5, 2005.
69. See note 25.
70. Jason Vest, "Fables of the Reconstruction," *Village Voice,* April 20, 2004.
71. Pratap Chatterjee, *Iraq, Inc.: A Profitable Occupation* (New York: Seven Stories Press, 2004), p. 68.
72. "Solicitation, Offer, and Award," contract no. EEE-C-00-03-00018-00, Negotiated (RFP), USAID and Bechtel National, Inc., February 7, 2003.
73. "Iraq Infrastructure Reconstruction," contract no. EEE-C-00-03-00018-00, USAID and Bechtel National, Inc., April 17, 2003.
74. David Baker, "Bechtel Under Siege," *San Francisco Chronicle,* September 21, 2003.
75. Christian Parenti, "Fables of the Reconstruction," *Nation,* August 30/September 6, 2004.
76. See note 74.
77. James Glanz, "New Election Issues: Electricity and Water," *New York Times,* January 26, 2005.
78. John Hendren and Ashraf Khalil, "The Conflict in Iraq," *Los Angeles Times,* December 13, 2004.
79. See note 30.
80. Pratap Chaterjee and Herbert Docena, "Occupation Inc.," *Southern Exposure Magazine,* Winter 2003/2004.
81. Ariana Eunjung Cha, "Iraqi Experts Tossed with the Water," *Washington Post,* February 27, 2004.
82. Erik Eckholm, "Showcase: Basra Revival, But It's Harder than Expected," *New York Times,* January 19, 2005.
83. T. Christian Miller, "Millions Said Going to Waste in Iraq Utilities," *Los Angeles Times,* April 10, 2005.
84. Ibid.
85. See note 54.
86. Stephen Pelletiere, "A War Crime or an Act of War?" *New York Times,* January 31, 2003.
87. Quoted in "Of Water and Wars," *Frontline Magazine,* April 24–May 7, 1999, vol. 16, no. 9.
88. Shawn Tully, "Water, Water Everywhere," *Fortune Magazine,* May 15, 2000.
89. David Baker, "Bechtel in Iraq," *San Francisco Chronicle,* December 26, 2003.

90. U.S. General Accounting Office, "Rebuilding Iraq: Resource, Security, Governance, Essential Services, and Oversight Issues," *GOA-04-902R*, June 2004.

91. Audits performed by the Special Inspector General for Iraq Reconstruction, including "Audit Report: Administration of Iraq Relief and Reconstruction Fund Contract Files," report number 05-007, April 30, 2005.

92. *Iraq Weekly Status Reports*, U.S. State Department, July 2005–February 2006.

93. See note 25; and note 56.

94. Special Inspector General for Iraq Reconstruction, *January 2006 Quarterly Report and Semiannual Report to Congress*, January 6, 2006; and Jonathan Finer, "Report Measures Shortfall in Iraq Goals," *Washington Post*, January 27, 2006.

95. Rajiv Chandreasekaran and Walter Pincus, "U.S. Edicts Curb Power of Iraq's Leadership," *Washington Post*, June 27, 2004.

96. Briefing by Lakhdar Brahimi, Special Adviser to the Secretary General for Iraq, Baghdad, unofficial transcript, *UN News Service*, June 2, 2004.

97. Rajiv Chandrasekaran, "Former Exile Is Selected as Interim Iraqi Leader," *Washington Post*, May 20, 2004 and Dexter Filkins and Warren Hoge, "Iraqi with Close U.S. Ties Chosen to be Prime Minister," *New York Times*, May 28, 2004.

98. Annex to the Law for the Administration of Iraq in the Transitional Period, "On the basis of Article 2(1)B of the Law of Administration for the State of Iraq in the Transitional Period, the Governing Council decided in its session of 1/6/2004 to issue this Annex."

99. Patrick J. McDonnell and T. Christian Miller, "Election in Iraq," *Los Angeles Times*, February 14, 2005.

100. U.S. Department of Defense News Transcript of Donald Rumsfeld, July 27, 2005.

101. Nathan J. Brown, "The Final Draft of the Iraqi Constitution: Analysis and Commentary," Carnegie Endowment for International Peace, September, 2005.

102. Ibid.

103. Statement by Peter J. Robertson, vice chairman, ChevronTexaco Corporation, at the Middle East Petroleum and Gas Conference, Dubai, United Arab Emirates, September 8, 2003.

104. Gal Luft, "Iraq's Oil Sector One Year After Liberation," Brookings Institution, Saban Center Middle East Memo #4, June 17, 2004.

105. Section 2207 First Quarterly Report to Congress on Use of Iraq Relief and Reconstruction Funds, Office of Management and Budget, April 5, 2004.

106. *Iraq Weekly Status Report*, U.S. State Department; and Lawrence Kumins, "Iraq Oil: Reserves, Production, and Potential Revenues," Congressional Research Service Report for Congress, April 13, 2005.

107. *Monthly Energy Review*, Energy Information Administration, U.S. Department of Energy, November 2005.

108. Office of Trade and Industry Information, Manufacturing and Services, International Trade Administration, U.S. Department of Commerce, data collected December 19, 2005.

109. Verne Kopytoff, "Iraqi Oil Reaches California," *San Francisco Chronicle,* October 15, 2003.

110. American Enterprise Institute, transcripts, "A Conversation with Adel Abdul Mahdi," October, 2004; Congressional Research Service Report for Congress, "Iraq Oil: Reserves, Production, and Potential Revenues," April 13, 2005; and "Iraqi Oil & Gas: A Bonanza-in-Waiting," Special Report, Energy Intelligence Research, Spring 2003.

111. Greg Palast, "OPEC on the March," *Harper's Magazine,* April 2005.

112. Greg Muttitt, "Crude Designs," Platform, Global Policy Forum, Institute for Policy Studies and Oil Change International, October 2005.

113. "Iraqi Oil & Gas: A Bonanza-in-Waiting," Special Report, Energy Intelligence Research, Spring 2003.

114. See note 112.

115. "Iraqi Plan for Radical Oil Reform Runs into Controversy," *Iraq Oil Daily,* September 30, 2004.

116. See note 113.

117. Transcript, National Press Club Afternoon Newsmaker News Conference with Alan Larson, undersecretary of state for economic, business, and agricultural affairs and Adil Abd Al-Mahdi, Iraqi minister of finance, Washington, DC, December 22, 2004.

118. See note 112.

119. See note 56.

120. David Baker, "Seeking Iraq's Oil Prize," *San Francisco Chronicle,* January 26, 2005.

Chapter Seven: Exporting "Free Trade" in Place of "Freedom" to the Middle East

1. Jacob Jordan, "Some South Carolina Grads Forego Ceremony in Protest of Bush," *Savannah Morning News,* May 10, 2003.

2. Richard F. Grimmett, "U.S. Arms Sales: Agreements with and Deliveries to Major Clients, 1997–2004," Congressional Research Service Report for Congress, December 29, 2005.

3. The Office of U.S. Trade Representative, "USTR Zoellick Statement on World Trade Ministerial in Doha," September, 14, 2001.

4. Robert Zoellick, "Remember Seattle: Mixed Signals Are Bad for Trade," *Wall Street Journal,* October 5, 2004.

5. Gary Yerkey, "USTR Says Other Nations Must 'Compromise' or WTO Meeting in Doha Could End in Failure," *International Trade Daily,* October 31, 2001.

6. Al Kamen, "No-Show on Trade Burns Up Finance Panel," *Washington Post,* December 17, 2001.

7. Guy de Jonquières, "Brussels Resists Demands for Iraq WTO Seat," *Financial Times,* January 25, 2004.

8. Wendy Lubetkin, "Iraq Granted Observer Status at the World Trade Organization," *Embassy of the United States, Japan,* February 12, 2004.

9. "Causes of 9/11: U.S. Troops in Saudi Arabia?" Council on Foreign Relations, 2004.
10. See note 2.
11. "Saudi Business Delegation Introduces $623 Billion in Foreign Investment Opportunities in Saudi Arabia," press release issued by the Information Office of the Royal Embassy of Saudi Arabia in Washington, DC, May 13, 2005.
12. Ibid.
13. U.S. Trade Representative's Office fact sheet, "Accession of the Kingdom of Saudi Arabia to the World Trade Organization," September 9, 2005.
14. Phillip Kurata, "Egypt, U.S. Consider Opening Free Trade Talks in 2005," U.S. State Department, January 14, 2005.
15. Debra Glassman, "Bush Tries to Recycle Cold War-Era Policy," *Seattle Post-Intelligencer,* July 1, 2003.
16. "Egypt Hopes for Jobs Boost as It Signs Trade Deal with Israel and U.S." *Daily Star,* December 15, 2004.
17. "Changing Minds Interview," *Al-Ahram Weekly On-Line,* no. 741, May 5–11, 2005.
18. Robert Zoellick, "Global Trade and the Middle East," as prepared for delivery at the World Economic Forum, Dead Sea, Jordan, June 23, 2003.
19. "U.S.–Middle East Free Trade Coalition Urges Congress to Act Swiftly on U.S.–Bahrain FTA," U.S.–Middle East Free Trade Coalition press release, September 14, 2005.

Chapter Eight: The Failure of the Bush Agenda: A World at Greater Risk

1. "Zawahri Warns the Britons Against Blair's Policy," video, *Al-Jazeera,* August 4, 2005.
2. President George W. Bush, remarks at National Endowment for Democracy, Washington, DC, October 6, 2005.
3. Iraq Index and Afghanistan Index, Brookings Institution, January 2006.
4. Michael Scheuer, "Why I Resigned from the CIA," the *Los Angeles Times,* December 5, 2004.
5. Iraq Index, Brookings Institution, January 2006.
6. "A Failed 'Transition': The Mounting Costs of the Iraq War," a study by *Foreign Policy in Focus* and the *Institute for Policy Studies,* September 2004.
7. See note 5.
8. Ibid.
9. Report of the Defense Science Board Task Force on Strategic Communication, Department of Defense, Office of the Undersecretary of Defense for Acquisition, Technology, and Logistics, September 2004.
10. *Global Trends 2015: A Dialogue About the Future with Nongovernment Experts,* approved for publication by the National Foreign Intelligence Board under the authority of the director of Central Intelligence, NIC 2000-02, December 2000.
11. Report of the National Intelligence Council's 2020 Project, "Global Trends 2020: Mapping the Global Future," NIC 2004-13, December 2004.

12. "Commuters Describe Madrid Blast Chaos," *BBC News,* March 11, 2004.

13. "Threat Video in Spain Flat Rubble," *BBC News,* April 9, 2004.

14. "Extract: 'Al-Qaeda' Warns of More Attacks," letter reprinted by *BBC News,* March 18, 2004.

15. Toby Harnden and Amberin Zaman, "Synagogue Bombs Were Ours, Says al-Qaeda Group," *Telegraph,* November 17, 2003.

16. "Al-Qaeda Linked Group Claims Istanbul Attacks," *Agense France Press,* August 10, 2004.

17. Denny Walsh, "Judge Sets Trial Date in Lodi Case," *Sacramento Bee,* January 7, 2006.

18. "FBI: Al-Qaeda Plot Possibly Uncovered," CNN.com, June 9, 2005.

19. Veena Dubal and Sunaina Maira, "White Hunt in Lodi," *Alliance of South Asians Taking Action,* June 23, 2005.

20. "Plots, Casings, and Infiltrations Referenced in President Bush's Remarks on the War on Terror," White House fact sheet, October 6, 2005.

21. "Four Men Indicted on Terrorism Charges Related to Conspiracy to Attack Military Facilities, Other Targets," U.S. Department of Justice press release, August 31, 2005.

22. "Report on Hate Crimes and Discrimination Against Arab Americans. The post September 11 Backlash: September 11, 2001–October 11, 2002," American-Arab Anti-Discrimination Committee, 2003.

Chapter Nine: A Better Agenda Is Possible

1. Rebecca Solnit, *Hope in the Dark: Untold Histories, Wild Possibilities* (New York: Nation Books, 2004).

2. *New York Times/CBS News Poll,* September 9–13, 2005; and Iraq Index, Brookings Institution, January 2006.

3. Quoted in the Honorable John P. Murtha, "War in Iraq," speech delivered to Congress, November 17, 2005.

4. Quoted in Ken Rubin, "Long-Time War Hawk, Murtha Is an Angry Dove," NPR.org, November 18, 2005.

5. Phyllis Bennis and Erik Leaver, "Ending the U.S. War in Iraq," Institute for Policy Studies, January 12, 2005.

6. Ibid.

7. Special Inspector General for Iraq Reconstruction, Report to Congress, October 30, 2005.

8. Daniel Large, ed., "Corruption in Post-War Reconstruction," Tiri and the Lebanese chapter of Transparency International, January 2005.

9. Audits performed by the Special Inspector General for Iraq Reconstruction, including "Audit Report: Administration of Iraq Relief and Reconstruction Fund Contract Files," report number 05-007, April 30, 2005.

10. See note 8; and "Annual Global Corruption Report 2004," Transparency International, March 2005.

11. The Center for Responsive Politics, www.opensecrets.org, online database.

12. Justin Blum, "Energy Tax Breaks Total $14.5 Billion," *Washington Post,* July 28, 2005.

13. Stephen Kretzmann and Irfan Nooruddin, "Drilling into Debt: An Investigation into the Relationship Between Debt and Oil," Oil Change International, the Institute for Public Policy Research, and the Jubilee USA Network, June 2005.

14. "Iraqi Oil Ministry Gets Big Leap in Funds," *International Oil Daily,* October 26, 2004.

15. Greg Muttitt, "Crude Designs," Platform, Global Policy Forum, Institute for Policy Studies and Oil Change International, October 2005.

16. Ibid.

17. Stephen Kretzmann, "A Ten-Step Program to End Oil Addiction," Oil Change International, October, 2005.

18. Joseph Stiglitz, "What I Learned at the World Economic Crisis," *New Republic,* April 6, 2000.

19. Robin Broad and John Cavanagh, in *Alternatives to Economic Globalization: A Better World Is Possible,* John Cavanagh and Jerry Mander, eds. (San Francisco, CA: Berrett-Koehler, 2004), p. 286.

20. David Solnit, ed., *Globalize Liberation: How to Uproot the System and Build a Better World* (San Francisco, CA: City Lights Books, 2004), p. 347.

21. Wes Smith, "Group Champions Migrants," *Orlando Sentinel,* June 6, 2005.

22. Mark Engler, "A Movement Looks Forward," *Foreign Policy in Focus,* May 19, 2005.

23. "Africa: Debt and AIDS," *Africa Policy E-Journal,* African Action, June 2, 2002.

24. See note 11.

25. Neil Gaiman, "A Hope in Hell," in *The Sandman: Preludes & Nocturnes* (New York: DC Comics Vertigo, 1991).

Afterword: What a Difference a Year Makes

1. *USA Today*/CNN Gallup Poll, January 2006.

2. Gary Langer, "State of the Union: Unhappy with Bush," *ABC News.com,* January 22, 2007.

3. Gary Langer, *ABC News.com.*

4. Zogby International Poll, February 28, 2006.

5. Amit R. Paley, "Most Iraqis Favor Immediate U.S. Pullout, Polls Show," *Washington Post,* September 27, 2006.

6. Anne Plummer Flaherty, "Sen. Warner Casts Dismal View of Iraq," *The Associated Press,* October 5, 2006.

7. Special Inspector General for Iraq Reconstruction, October 2006 Report to Congress.

8. "Halliburton announces full year and fourth quarter results," Halliburton press release, January 26, 2007.

9. Statement issued by the Iraqi Labor Union Leadership at a Seminar held from 10 to 14 December 2006, in Amman, Jordan to discuss the draft Iraqi Oil Law.

10. "US' Bodman Says No US Oil Companies Willing to Enter Iraq Now," *Platts,* July 18, 2006.

11. "Waiting for the Green Light," *Petroleum Economist,* October 2006.

12. Paul Kane, "GOP Lawmakers Reflect on Losses," *Washington Post,* January 26, 2007.

13. John Kifner, "Gunboat Diplomacy," *New York Times,* January 14, 2007.

14. "Bush's Trade Agenda in Doubt after Democratic Win," *Agence France Press,* November 9, 2006.

15. Public Citizen, "Election 2006: No to Staying the Course on Trade," December 2006.

16. Lori Wallach and Deborah James, "Why the WTO Round Talks Have Collapsed," *Common Dreams,* April 14, 2006.

INDEX